DICTIONARY OF CONTEMPORARY CHINESE MILITARY HISTORY

DICTIONARY OF CONTEMPORARY CHINESE MILITARY HISTORY

LARRY M. WORTZEL

Robin Higham, Advisory Editor

Greenwood Press
Westport, Connecticut • London

Library of Congress Cataloging-in-Publication Data

Wortzel, Larry M.
Dictionary of contemporary Chinese military history / Larry M. Wortzel.
p. cm.
Includes bibliographical references and index.
ISBN 0–313–29337–6 (alk. paper)
1. China—History, Military—1644–1912—Dictionaries. 2. China—History, Military—1912–1949—Dictionaries. 3. China—History, Military—1949- —Dictionaries. I. Title.
DS754.15.W67 1999
355'.00951'031—dc21 99–10655

British Library Cataloguing in Publication Data is available.

Library of Congress Catalog Card Number: 99–10655
ISBN: 0–313–29337–6

First published in 1999

Greenwood Press, 88 Post Road West, Westport, CT 06881
An imprint of Greenwood Publishing Group, Inc.
www.greenwood.com

The paper used in this book complies with the Permanent Paper Standard issued by the National Information Standards Organization (Z39.48–1984).

10 9 8 7 6 5 4 3 2 1

CONTENTS

Maps follow page xviii.

PREFACE

This book was originally conceived as a project to cover only twentieth-century Chinese military history. However, as I began the research, it became apparent very early that, if one wanted to understand how the Chinese military (and its peculiar problems with the military such as the phenomenon of warlordism) evolved in the twentieth century, one had to grasp the seminal events of the nineteenth century. The military forces, strategic orientation, policies, foreign and domestic, and industrial base of both the Republic of China and the People's Republic of China were shaped by the period of military attack and domination of the late Qing dynasty by foreign powers. While recent issues may be of more interest to readers focused on current events and how China will interact with the world in the twenty-first century, the period between the **Opium War** (1839–1842) and the **Republican Revolution** (1911) is a seminal period in modern China's military history. (Throughout this dictionary, **boldface** type highlights cross-references to other entries.) During the period between the Opium War and the **Boxer Rebellion** (1990), Western powers and Japan engaged in commercial and diplomatic initiatives backed up by military force in order to expand trade with China and to protect their citizens from Qing dynasty responses. Japan's interaction with China is particularly interesting because, although it is an Asian nation, it industrialized before China and acted in concert with the Western powers against the Qing. The successful foreign incursions into China led to the progressive weakening of dynastic rule in China and the overthrow of the Qing. The Republican Revolution, the **warlord** period of the 1920s, and the rise of Communist forces with the formation of the Communist Party and the Red Army must be understood in the context of the Opium War and foreign domination of parts of China. This period between the mid-1800s and the Boxer Rebellion is still interpreted in the People's Republic of China as the beginning of a period of domination by foreign powers that began to end only with the recovery of Hong Kong in 1997 and will not close until the eventual reunification of Taiwan with the mainland. Recent issues might be of more interest to readers focused on current events. But, to repeat, if one wants to understand how China will interact with the world in the twenty-first century, one must grasp the seminal events that gripped China in the period between the Opium War and the Nationalist Revolution. The period between the late nineteenth century and the early twentieth century forms the underpinning of the security posture of the People's Republic of China and is strongly reflected in the white paper on *China's National Defense* published by the State Council Information Office in July 1998.

By the beginning of the nineteenth century, the Qing dynasty began to weaken under the weight of internal corruption and the strength of foreign pressures on China. As the Qing administration attempted to react to corruption and suppress the opium trade, it also acted against foreign enclaves in China. Through its own centrist lenses, after centuries of self-imposed isolation from the world order and the European system of states and diplomacy, the Qing rulers continued to view China as the traditional "Middle Kingdom" to which other states paid homage and tribute. This attitude is probably best represented by the response by Emperor Qianlong to a diplomatic mission by Lord MacCartney of England in 1793. MacCartney visited China representing King George III of England, seeking to establish trade missions at ports along the Chinese coast, including Ningbo and Tianjin. Because MacCartney brought along gifts from George III, the Chinese emperor interpreted the visit as fitting into China's own tribute system. Emperor Qianlong's response was to accept George III's "respectful spirit of submission" to the Qing dynasty and to tell the British king that China had all of the things it needed "in abundance" and therefore had no need to trade with England. The Western powers, however, were at the height of their own industrialization and expansion and by the mid-eighteenth century had the military wherewithal to react to the Qing rulers. With the attempt to end opium trade in China spearheaded by Lin Zexu in 1839, the Qing dynasty used various forms of military pressure and force to suppress foreign business encroachment in South China.

England, especially, having been victorious against Napoleon and the French on the European continent, would not tolerate mistreatment of its citizens and trading companies by Qing forces. British fleets and ground forces acted against China in the Opium War, demonstrating that the British monarchy was not going to kowtow to the Qing, attacking up the Pearl River into Guangzhou (Canton) and up the Chinese coast to impose England's will on the Qing. France, the United States, Germany, Japan, Russia, and other European powers soon took advantage of British successes at securing trading rights and concessionary enclaves, securing their own privileges through what are known today in China as the "unequal treaties." The Qing rulers, having isolated themselves from the system of diplomatic discourse in the West, first agreed to treaties imposed on them by military force and then ignored these treaties, refusing to honor some of the provisions. This led to the events of the **Arrow War** (or Second Opium

War, as it is sometimes called), further attacks by Western powers, and the division of parts of China into foreign spheres of influence.

While the Qing dynasty and a few of its administrators were dealing with the encroachment into China by the West, they simultaneously had to face domestic rebellion. The Muslims in the west, before the Opium War, had challenged the Qing rule, and after the Opium War the **Taiping Rebellion** and the **Nian Rebellion** occupied regional governors in South-Central China and North China, respectively. Ironically, even while they were undertaking punitive expeditions against the Qing dynasty, the French, British, and even the American governments supported the Qing fights against the Taiping and Nian rebels, since these rebellions interfered with trade. A few regional leaders, assisted by foreign mercenaries or military leaders sent by the French and British, managed the rebellions. These same Qing local governors, particularly **Li Hongzhang** and **Zeng Guofan**, saw the need to develop an indigenous arms industry and build China's own military strength and industrial base. The Qing dynasty efforts, termed the **self-strengthening movement** (*Zi Qiang Yundong*), proved to be unsuccessful, however, and China's military forces were shown to be hollow. Japan's military forces had no trouble on land or at sea defeating Chinese naval and ground forces during the **Sino–Japanese War** (1894–1895). China's defeat by the Japanese, reversed, in part, by international pressure against Japan, created only more internal unrest. More peasant uprisings destabilized the country, and the Boxer Rebellion broke out. Attacks by Boxers on foreigners were tacitly supported by the Qing dynasty, until the entire foreign legation area was under siege in Beijing. The lifting of the siege in August 1900 and the suppression of the Boxers were accomplished by the combined military forces of England, France, Germany, the United States, Japan, Russia, and Austria. The march from Tianjin to Beijing and the storming of the walls of the Forbidden City by the combined Western armies were the death knell for the Qing.

Moving forward into the twentieth century, China was still dominated by sectors of foreign encroachment and influence. As the Nationalist Revolution took place, military rulers with local armies and their own arsenals, the "warlords," continued to control vast portions of China. These warlord forces evolved from the local, personal armies organized in the late nineteenth century to fight the Taiping and Nian rebels. The division of China into personal, armed fiefdoms left little choice but to use military force to unite the country. **Chiang Kai-shek** tried to do this with the **Northern Expedition** (1926). Meanwhile, after the Russian Revolution, the Moscow-controlled Communist International (Comintern) began to work within the Nationalist government to establish a Leninist structure in the Guomindang (KMT). Concurrently, the Comintern recruited and sent influential young leftist Chinese to France and other parts of Europe for an education. These young leftists, through work-study programs, were introduced to the ideas of Marxism and revolution and formed the Chinese Communist Party. The history of the twentieth century in China, from 1921 forward, reflects the tension between the two competing parties, the Nationalist and the Communist, for control of the country. Japanese aggression, beginning in Manchuria in the 1930s and expanding throughout China and Asia, only postponed the showdown between the Nationalists and the Communists, which is still going on as this book goes to press.

As the twenty-first century approaches, the main forces that buffeted China for the last 150 years are only in the process of being resolved. Nationalist–Communist tensions over the mainland's control of Taiwan and the question of sovereignty over territories that China claims are only in the process of being solved. This tension is reflected in the sorts of military exercises we saw in the **Taiwan Strait** in 1996. The actions by the Chinese military in the South China Sea, where sovereignty is also an issue, are also a reflection on the legacy of the nineteenth century in China.

The reader should note a few recurring themes that are important for understanding the events covered in this volume. The first is how Sinocentric arrogance and a refusal to interact with the outside world on an equal basis weakened the Qing dynasty. For a brief historical overview of how the weight of China's background is reflected in modern society and culture, the reader should first take the time to review the synopsis of China's earlier history provided in the Introduction to this book. Second, because of its own internal problems, like the centrifical tendencies exacerbated by military strongmen, the warlords, who held sway over large portions of the population, China never developed a rational, national economic infrastructure. Third, thousands of years of dynastic rule, perhaps reinforced by the need to organize and control a large population, have created a strong tradition of authoritarian leadership by a single individual. This theme is extremely important, because in many ways the Nationalist–Communist tensions are as much a reflection of traditional dynastic battles for power and central control as they are a battle of ideologies. In reading the individual entries of the book, it is also important to keep in mind that, even within the Communist Party, from its formation in 1921 to today, political and military battles for dominance have characterized the period of Communist rule. Events like the **Futian Incident** in 1930, when **Mao Zedong** directed the massacre of units of the Red Army that opposed him, reflect the trend of infighting in the Communist Party, as do the political fights during the **Long March**. The banishment of **Peng Dehuai** and the excesses of the **Cultural Revolution** are other examples of the use of the military as an instrument of political power to dominate the party.

This book concentrates on the military campaigns, doctrines, and personalities of the Communist forces and the People's Republic of China. Before reading the encyclopedic entries of this book, however, a brief review of China's history is useful, including some mention of the more important classical military texts. The reader will find this in the Introduction. The book tracks the evolution of the Red Army through the war against Japan and the Chinese **civil war** by focusing on major events, people, and issues that have produced the historical legacy of the **People's Liberation Army**. To do this, I have relied on discussion with American and foreign scholars, but especially friends in the Chinese military, in order to learn about what their own basic military education emphasized. I have tried to capture the main events that constitute the way that the Chinese military defines and understands itself. In recent years, since the mid-1980s, a few excellent volumes of broad military history have been published by the military academies and military presses in China. I have relied heavily on these books for cataloging and defining the main events. Unfortunately, Communist and Nationalist historiography has a way of eliminating facts that are ideologically unpalatable. Complicating the problem of finding reliable and verifiable primary sources is the fact that many Communist archives, especially those of the military and the party, are not open either to foreign or to Chinese scholars. The available books, therefore, must be read critically and compared to other foreign sources on any given subject. Fortunately, there is also a solid body of Western literature on which to draw. Change is coming, though, and, increasingly, there are a number of critical histories dealing with contemporary events published in China.

It will quickly become apparent to the reader that there has always been a great deal of tension within the Communist military over the relative importance of ideological rectitude and purity, as opposed to military expertise and professionalism in the practice of arms. This tension focused early in the Chinese Communist Party's history, before and during the Long March, on the positions taken by Mao Zedong and **Zhang Guotao**. However, the sections of this book on the **Lushan Conference**, when Peng Dehuai was purged by Mao, and the Cultural Revolution, when Mao eliminated many of his adversaries from the 1920s through the 1940s, are particularly instructive. These internal struggles demonstrate the personalized nature of power in China and the tendency to form factions struggling for supremacy.

Here I want to acknowledge the help of my wife, who typed most of the manuscript for this book. As she typed in our apartment in Beijing and read some of the sections of the book, she commented to me one day that "these Communists were really treacherous bastards to the people, to each other, and to their enemies, especially Mao Zedong." I can't think of a more worthwhile lesson to be drawn from the contemporary military history of China.

TRANSLITERATION OR ROMANIZATION OF THE CHINESE LANGUAGE

Chinese is an ideographic language, without an alphabet. Over the years, several systems of romanization have been used by Western countries to transliterate Chinese ideographs and pronunciation into sounds close to Western languages. The most popular and widely used of these, the Wade–Giles system, is used in most major libraries in the United States with extensive Chinese-language collections and in the Library of Congress. However, it is not used by the People's Republic of China, which uses the Pinyin system, which perhaps more closely approximates the pronunciation of Chinese. Pinyin has some very strange combinations for English speakers. For instance, the Wade–Giles combination *Hs*, which is pronounced like the letter ''s'' with a lisp, becomes an ''*x*'' in Pinyin (so that the city of *Hsi-an* is romanized in Pinyin as *Xi'an*. Yet one will still find books that use the romanization *Sian* for the city. Also prevalent in the United States is the Yale transliteration system, which is easy for Americans to pronounce but is not used anywhere else in the world. To complicate transliteration, slightly different systems were used in Germany and France from the seventeenth century through the mid-twentieth century.

The changes in dynastic order in China also serve to complicate how the dynasties are named. For instance, the Qing dynasty (1644–1911) will be seen transliterated as Ch'ing in some Western literature. But because it was not a Han Chinese dynasty but one from north of the Great Wall in Manchuria, it is sometimes called the Manchu dynasty, after the minority group from Manchuria. Likewise, the Yuan dynasty (1271–1368) may be transliterated as Yuen. But since it was another ''foreign'' dynasty of non-Han people's from Mongolia, it is also called the Mongol dynasty in some books. This text will sometimes refer to the Qing as the Manchu and the Yuan as the Mongol.

This book uses primarily the Pinyin system of romanization. The capital of China, in Pinyin, is transliterated as *Beijing*. In Wade–Giles this would be *Pei-ching*, but the Germans and French who originally did romanization once transliterated it as *Peking*, which still appears in many Western maps and in book titles. If one were pronouncing the word in Mandarin, the standard language for China (*Putonghua*), it would sound like *Bay-jing*. The Pinyin system is used primarily, but other systems are used in this book. Because Taiwan and Nationalist publications use primarily the Wade–Giles system, for the names of the most commonly known leaders of Nationalist China and Taiwan, I have retained the Wade–Giles form. There are so many historical texts that use these names that to the non-Chinese speaker, it would be confusing to see the names in a different form. Thus, Chiang Kai-shek is transliterated in Wade–Giles, instead of the Pinyin *Jiang Jieshi*. The name would appear in Wade-Giles in most books published in English. I have also retained the Wade–Giles for the names of major battles conducted by Nationalist forces and for the Nationalist generals and political leaders.

Those of us in the West who have learned to speak and read Chinese are already confused. We have had to master several of the romanization systems in order to conduct research. For the non-Chinese speaker there will still be some points of confusion in the romanization process. Most people who grew up before or during the 1950s and 1960s will recognize the Nationalist Party as the *Kuomintang*, or the KMT, as it is often abbreviated. I have used the Pinyin term for the KMT, *Guomindang*, throughout the text of the book. However, one cannot control the titles of books already published or their content. Therefore, the reader will find many bibliographical references that use Wade–Giles, and even some of the less commonly used European romanizations (e.g., the *Fuzhou* Dockyard, in the diaries of Prosper Giquel, the Frenchman who helped establish it in the late nineteenth century, is romanized as *Foochow*). In the references and the bibliography I have retained the form of romanization in the published title of the book or article cited. Some examples: the *Ruijin Conference* (Pinyin romanization) will sometimes in references be spelled in its Wade–Giles form, *Juichin*; the *Dagu* forts (Pinyin) may appear in book titles and references as *Taku* (Wade–Giles); and the *Eyuwan* Soviet (Pinyin) may appear as *Oyuwan* (Wade–Giles). Consonants in Wade–Giles that are aspirated are romanized as (*ch', k', p', t'*) and are pronounced as they would be in English. Unaspirated consonants (*ch, k, p, t*), are pronounced as though they were (*j, j, b, d*). The Wade–Giles combinations *ts* and *tz* are pronounced as *dz* but appear in Pinyin as "*z*", while the combinations *ts'* and *tz'* are "*c*" and "*z*" in Pinyin.

These problems of romanization have been a source of frustration for scholars and students of Chinese for generations, and they will not go away. Here are a few examples:

Pinyin	**Wade-Giles**	**Alternative Forms**
Beijing	Pei-ching	Peiping
		Peking
Sun Zhongshan	Swun Chong-shan	Sun Yat-sen

Guomindang	Kuomintang	GMD, KMT
Fuzhou	Foo-chou	Foo-chow
Jiang Jieshi	Chiang Kai-shek	
Taibei	T'ai-pei	Taipei
Zhongguo	Chung-kuo	Chung Kuo
Chongqing	Ch'ung-king	Chungking
Dai Li	T'ai Li	T'ai Lee
Xi'an	Hsi-an	Sian
		Xian
Xizang	Hsi-tsang	Tibet
Qing	Ch'ing	Ching
Yan'an	Yenan	Yanan

ABBREVIATIONS

CCP	Chinese Communist Party
CMC	Central Military Commission
CPV	Chinese People's Volunteers
GAD	General Armaments Department
GLD	General Logistics Department
GPD	General Political Department
GSD	General Staff Department
KMT	Guomindang, or Nationalist Party
OSS	Office of Strategic Services of the U.S.
PLA	People's Liberation Army
PLAAF	People's Liberation Army Air Force
PLAN	People's Liberation Army Navy
PRC	People's Republic of China

ABBREVIATED BOOK TITLES IN ENTRIES
(See bibliography for full citation)

ZGJSSL	*Zhongguo Junshi Shilue*
ZGRMJFJDSD	*Zhongguo Renmin Jiefangjun Dashi Dian*
ZGDSJMCD	*Zhonggong Dangshi Jianming Cidian*
ZGRMJFJ	*Zhongguo Renmin Jiefangjun*
ZGRMJFJZS	*Zhongguo Renmin Jiefangjun Zhanshi*
ZGRMJFJZSJB	*Zhongguo Renmin Jiefangjun Zhanshi Jianbian*
ZGDBKQS	*Zhongguo Da Baike Quanshu*

China

RUSSIA
KAZAKSTAN
MONGOLIA
XINJIANG
GANSU
QINGHAI
XIZANG
SICHUAN
NEI MONGOL
HEILONGJIANG
JILIN
HENAN
HUBEI
HUNAN
YUNNAN
GUANGXI
NEPAL
BHUTAN
BANGLADESH
INDIA
BURMA
LAOS
VIETNAM
THAILAND
CAMBODIA
PHILIPPINES
KYRGYZSTAN
Lake Baikal
Lake Balkhash
Yellow Sea
East China Sea
South China Sea
Hainan Dao
Taiwan
Luzon
Barnaul
Bratsk
Irkutsk
Ulan Ude
Ulaanbaatar
Hailar
Harbin
Changchun
Shenyang
Almaty
Bishkek
Urumqi
Korla
Kashi
Hohhot
Baotou
Beijing
Tianjin
Dalian
P'yongyang
Seoul
Yinchuan
Shijiazhuang
Taiyuan
Jinan
Qingdao
Xining
Golmud
Lanzhou
Xi'an
Zhengzhou
Lianyungang
Nanjing
Hefei
Shanghai
Wuhan
Hangzhou
Chengdu
Chongqing
Changsha
Nanchang
Fuzhou
Taipei
Xiamen
Kao hsiung
Guiyang
Kunming
Guangzhou
Hong Kong
Nanning
Zhanjiang
Haikou
New Delhi
Agra
Kathmandu
Lhasa
Thimphu
Myitkyina
Mandalay
Hanoi
Vientiane
Rangoon
Hue
Bangkok
Manila
China
International boundary
Province-level boundary
National capital
Province-level capital
Railroad
Road
500 Kilometers
500 Miles

China: Administration

China-USSR Border: Eastern Sector

Manzhouli Zabaykal'sk Area

Amur-Ussuri Confluence Area

China-USSR Border: Western Sector

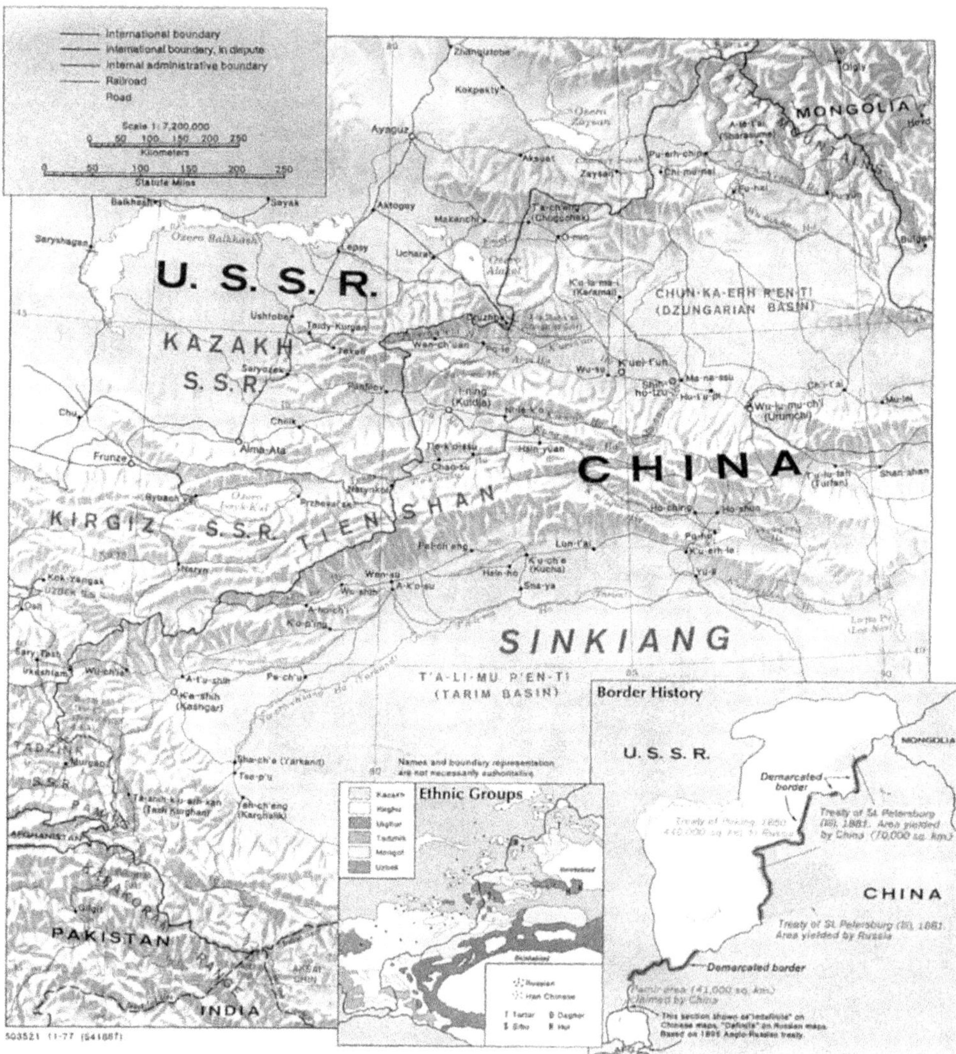

China-India Border

China-India Border: Eastern Sector

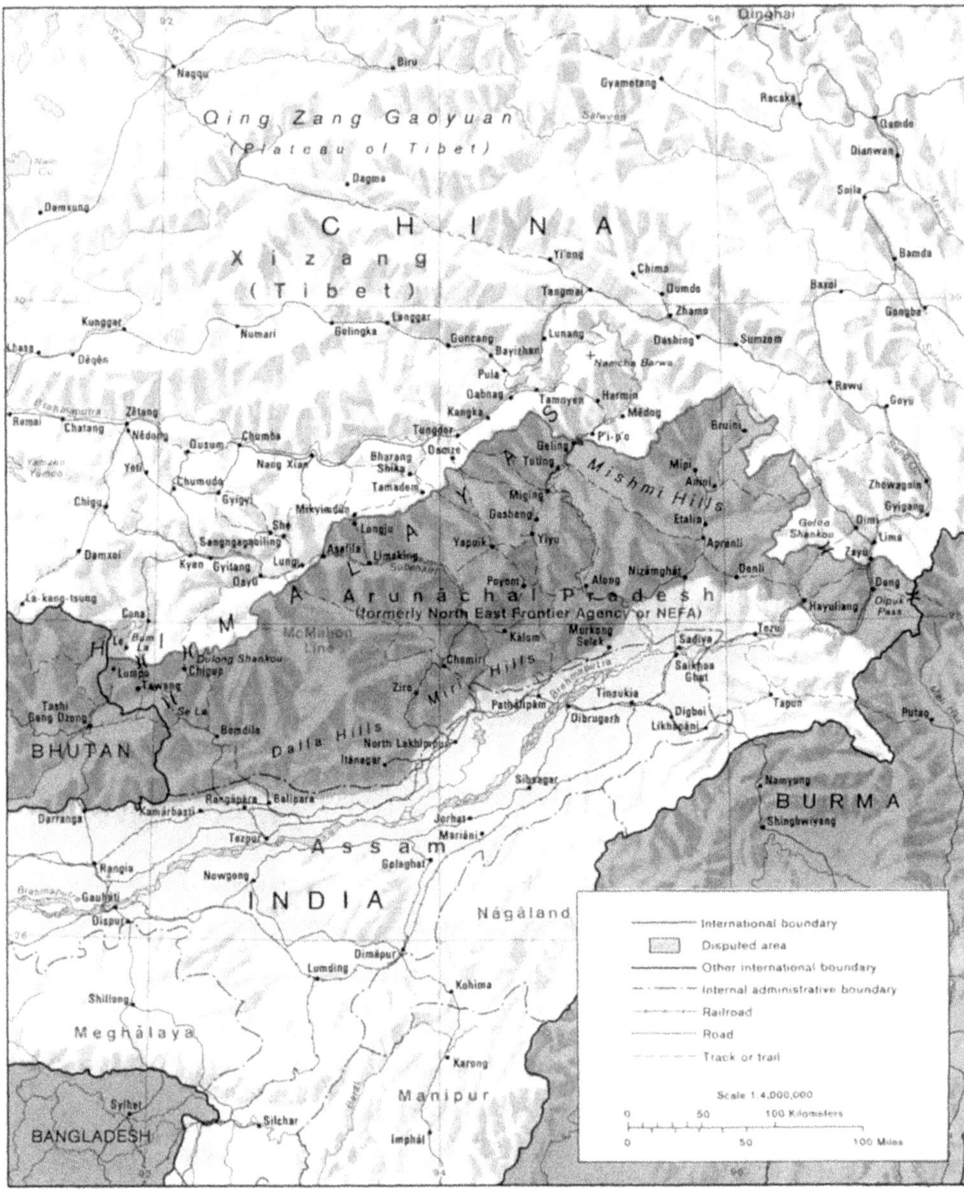

Eastern China

China-Vietnam Border

China: Industry

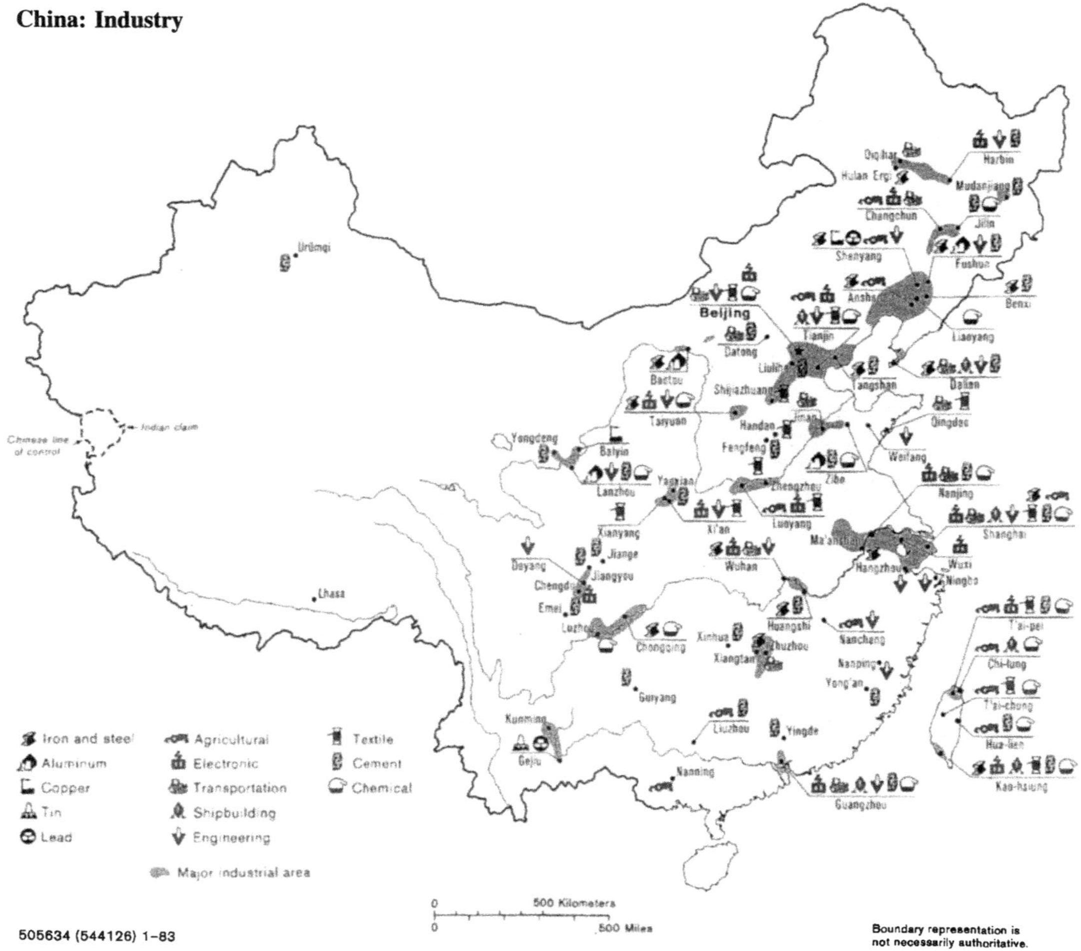

People's Republic of China: Major Industrial Areas

Electric Power Infrastructure

Gas Infrastructure

The Paracel Islands

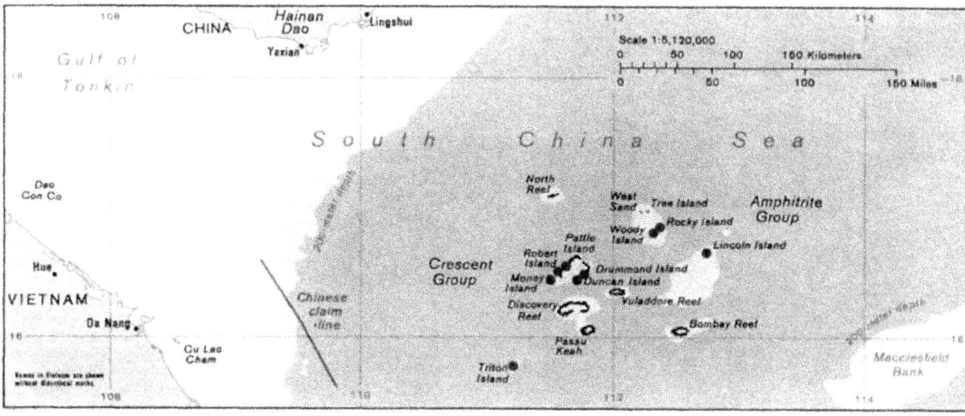

The Spratly Islands

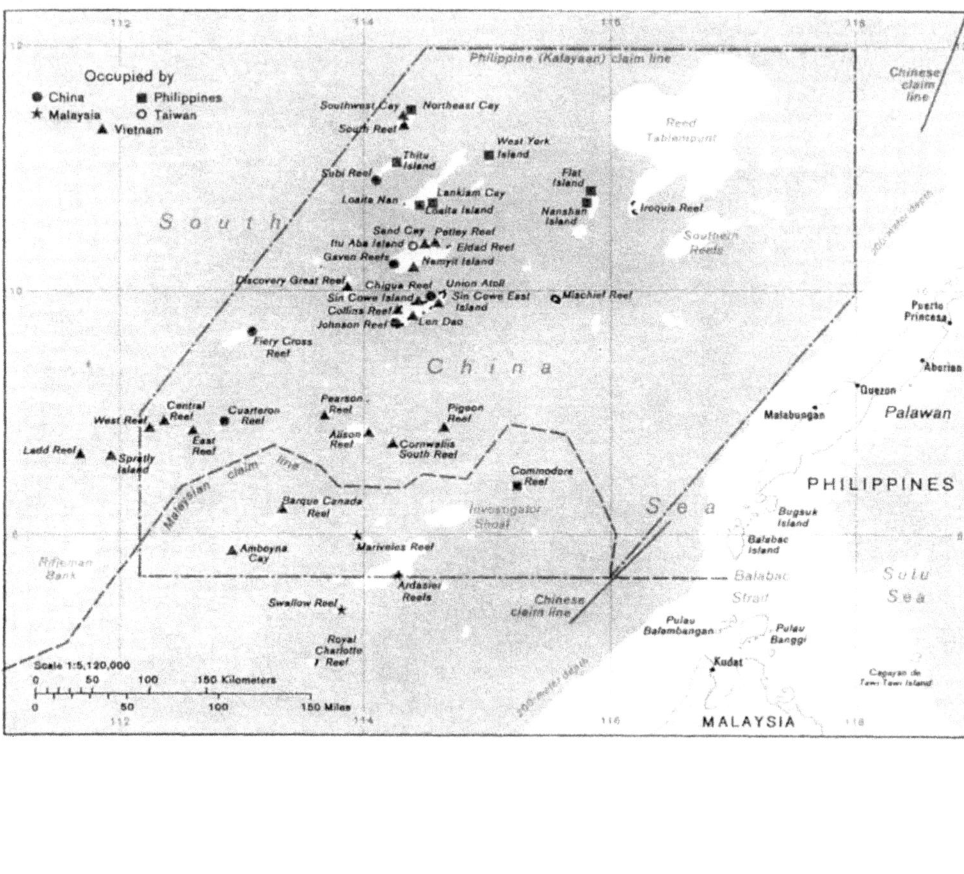

Taiwan

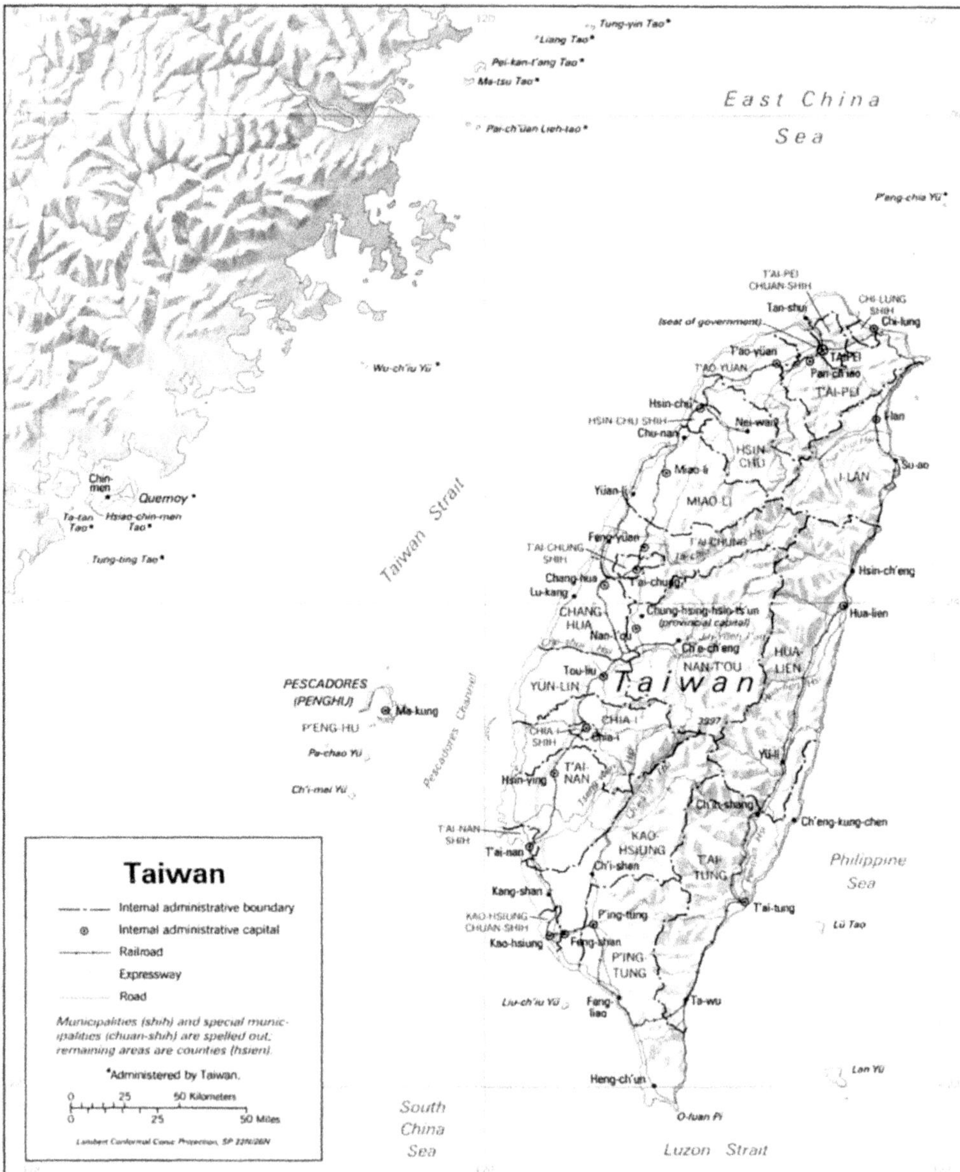

Taiwan
Internal administrative boundary
Internal administrative capital
Railroad
Expressway
Road
Municipalities (shih) and special municipalities (chuan-shih) are spelled out; remaining areas are counties (hsien).
*Administered by Taiwan.
0 25 50 Kilometers
0 25 50 Miles
East China Sea
Taiwan Strait
Pescadores Channel
Philippine Sea
South China Sea
Luzon Strait
Tung-yin Tao*
Liang Tao*
Pei-kan-t'ang Tao*
Ma-tsu Tao*
Pai-ch'üan Lieh-tao*
P'eng-chia Yü*
Wu-ch'iu Yü*
Chin-men
Quemoy*
Ta-tan Tao*
Hsiao-chin-men Tao*
Tung-ting Tao*
PESCADORES (PENGHU)
Ma-kung
P'ENG-HU
Pa-chao Yü
Ch'i-mei Yü
T'AI-PEI CHUAN-SHIH
CHI-LUNG SHIH
Tan-shui
Chi-lung
(seat of government)
T'ao-yüan
TAIPEI
T'AO-YÜAN
Pan-ch'iao
T'AI-PEI
Hsin-chu
HSIN-CHU SHIH
Nei-wan
Chu-nan
I-lan
HSIN-CHU
Su-ao
Miao-li
I-LAN
Yüan-li
MIAO-LI
Feng-yüan
T'AI-CHUNG SHIH
T'AI-CHUNG
Chang-hua
T'ai-chung
Hsin-ch'eng
Lu-kang
CHANG-HUA
Chung-hsing-hsin-ts'un (provincial capital)
Hua-lien
Nan-t'ou
Ch'e-ch'eng
HUA-LIEN
NAN-T'OU
Tou-liu
YÜN-LIN
Taiwan
CHIA-I
3997
CHIA-I SHIH
Chia-i
Yü-li
T'AI-NAN
Hsin-ying
Ch'ih-shang
Ch'eng-kung-chen
T'AI-NAN SHIH
T'ai-nan
KAO-HSIUNG
T'AI-TUNG
Ch'i-shan
Kang-shan
T'ai-tung
Lü Tao
KAO-HSIUNG CHUAN-SHIH
P'ing-tung
Kao-hsiung
Feng-shan
P'ING-TUNG
Liu-ch'iu Yü
Feng-liao
Ta-wu
Lan Yü
Heng-ch'un
O-luan Pi

China: Military Regions

Asia

RUSSIA
CHINA
INDIA
KAZAKSTAN
MONGOLIA
IRAN
AFGHANISTAN
PAKISTAN
UZBEKISTAN
NEPAL
BHUTAN
BANGLADESH
BURMA
LAOS
THAILAND
VIETNAM
CAMBODIA
MALAYSIA
INDONESIA
PHILIPPINES
JAPAN
SRI LANKA
MALDIVES
Arctic Ocean
Arabian Sea
Bay of Bengal
Indian Ocean
South China Sea
Philippine Sea
Bering Sea
Moscow
Beijing
New Delhi
Tokyo
Scale 1:48,000,000
Azimuthal Equal-Area Projection
Boundary representation is not necessarily authoritative.
AUSTRALIA

Probable Militia Organization

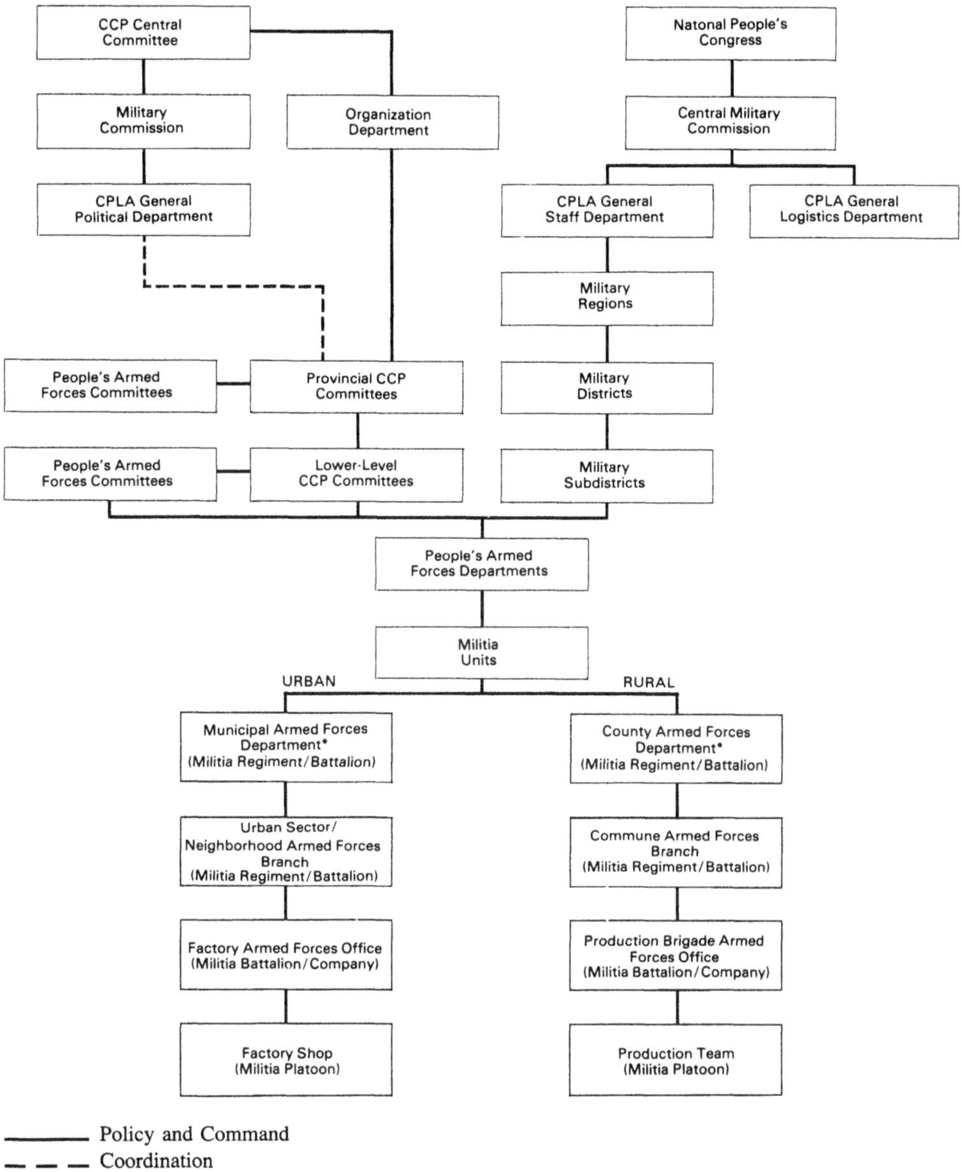

*Militia activity is supervised by county or municipal "People's Armed Forces Department." These departments are probably shared by a mixture of regular CPLA and civilian personnel and are subject to dual control as military bodies. They are subject to the next higher military echelon, the military subdistrict; as political organs, they are subject to the county or municipal party committee, which they serve as a military staff section.

NOTE: Within the various militia units, elements of each of the three militia categories may be found. The armed and basic militia serving primarily as cadres.

INTRODUCTION

The military history of China is best understood in a broader context that explains the geographical setting and the social, economic, and cultural conditions in which Chinese society is based. One of the most prolific American historians on China in this century, John King Fairbank, has used three major differences between China and the United States to help Americans understand the most basic factors that both define and limit China. The first difference is what Fairbank characterizes as China's poverty per capita; the second is the weight of China's own cultural tradition; and the third is a conservative fear of foreign technology and encroachment. The combination of the latter two of these differences, the sheer weight of a long cultural tradition combined with a near xenophobic fear and dislike of things foreign, together forms the conceptual underpinnings for the way that the Chinese emperors reacted to the West in the mid-nineteenth century, when the West and China come into direct conflict.

Almost every introductory briefing given by Chinese hosts to visiting foreign delegations starts the same way, using a stock phrase that would be merely a cliché were it not so true: "Zhongguo shi renkou da de yige guojia" (China is a country with a large population). Watching Chinese television from my apartment in Beijing in 1997, I noted that a commercial for women's beauty products used the same words to explain why women needed to concentrate on their appearance to be competitive in the marketplace. The existential fact of the largest population of any country in the world confronts the Chinese Communist leadership today and has been a factor in Western policy toward China since the early nineteenth century. For Chinese leaders the population is a challenge that must be kept under control at the same time that it must be kept fed, housed, and clothed. Indeed, by the year 2020, 20 percent of the population of China is projected to be over age 60; the state, therefore, will probably have to ensure that these people are cared for. For the West, the Chinese population represents a market of almost fabled proportions. "Oil for the lamps of China" is a thought that drove the Yankee traders, while later the British were dumping opium on the Chinese populace, making the Chinese coast a magnet for foreign ships and merchants.

Given its physical environment and the population problems with which China's leaders must wrestle, simply feeding and educating the population sometimes seems an insurmountable task. China's territory is about 3.7 million square miles, only about 100,000 square miles larger than the United States. However, only about 10 percent of China's land is arable. The United States, by comparison, supports one-fifth of the population of China on roughly the same-size landmass. About 20 percent of the landmass of the United States is arable, however. Fully two-thirds of China's land is either mountain or desert. But the population of China in 1998 was in the area of 1.3 billion people, about five times greater than that of the United States. In fact, population has always been a major factor in China, not just in the late 1990s. An imperial census in 1910, just before the Republican Revolution, put China's population at 330 million. In the sixteenth century, by which time China's borders approximated those of today, the population was about 150 million, and it doubled to 290 million by the close of the eighteenth century. Moreover, because of the pressures of industrialization and urban sprawl, agricultural land is shrinking by a small percentage annually. The challenges of feeding, let alone governing, such a huge population with so few available resources are daunting, to say the least. The Chinese are barely capable of feeding themselves, let alone exporting large quantities of food, as does the United States.

The ability of the Chinese population to migrate internally is also much more limited than for the population of the United States. Much of North-Central China is dry and punctuated by jagged mountain ranges. China's traditional bread baskets, in provinces like Sichuan in central China, are already heavily populated, as are the coastal region and the south. There are large tracts of unsettled forestland that is fertile and has mineral deposits in China's northeast, but a combination of infrastructure problems and the need for buffers against a traditional enemy, Russia, kept the land from being fully utilized. The Qing dynasty (1644–1911), which swept into, and conquered, the Ming dynasty from the northeast (Manchuria; hence, it is often called the Manchu dynasty), actually prohibited entry into northeast China by Han Chinese. Tibet (Xizang) and the Tibetan plateau areas, including Qinghai Province, in the southwest are so mountainous and high as to make large-scale colonization or development impractical, if not impossible. Not only do altitude and relief complicate development on the Tibetan plateau, but the soil and the mountains are made of a loose shale containing a lot of volcanic ash. As a consequence, roadbeds collapse constantly, and there are regular landslides blocking roads. Tunneling through the mountains is also difficult and subject to regular collapses. In the northwest, the area comprising Xinjiang Province is semidesert and high plateau steppes at the periphery and true desert in the center. China has succeeded in making part of this land arable through irrigation, but there are limitations on the size of the population that can be supported, limiting migration into the region. Thus, we see a country that must limit the rate of population increase while, at the same time, it must somehow squeeze annual increases of return per acre in food from a limited area that has been intensively farmed for hundreds of centuries. Whereas in the United States and Europe one simply mechanizes to increase productivity, mechanization in China creates massive dislocations for the agricultural population, which would find itself unemployed. While the World Bank and China's leaders talk about the challenge of raising China's income per capita, John Fairbank's image of China's "poverty per capita" is a moving one.

The second major difference referred to by Fairbank is China's "continuity in the same place, creating the weight of entrenched cultural tradition." The cradle of Chinese civilization in the time of about 1850 B.C. was along the Yellow River in China's heartland. About 4,000 years of recorded Chinese history is really a chronicle of the expansion of that civilization through a succession of some 14 different dynastic periods, some of which saw dozens of rulers in charge. Interestingly, only during the Yuan dynasty (1271–1368), a foreign dynasty imposed by Mongol conquerors, did China achieve the borders that it occupies again today. In the United States, one can talk about a Manifest Destiny of expansion from an east coast, settled primarily by Europeans, successively westward to the Pacific Ocean. In China, one must grasp that the very name for the country means central or middle kingdom and that this expansion took place from the central heartland of the Yellow River eastward to the coast and westward into Chinese Turkestan, now the area of Xinjiang and the Central Asian republics. Under successive dynastic orders, ruled by a military-administrative and priestly class, the cultural tradition expanded outward to control and organize all of the inhabitants who came under their rule. To the east, the expansion took place primarily along rivers and waterways and then along the coast. To the west, the expansion followed established, but ancient, trading routes, which came to be known as the Silk Road.

China was a country—or more appropriately, a kingdom or state—that built itself on military expansion through a government-managed or -controlled bureaucracy that conscripted mass labor, controlled population movement and mobility, and extracted wealth in the form of food and taxes. China administered itself through an elite, all of whom mastered through rote memorization a single form of ideograph script. These scholar-leaders could repeat or write from memory the Confucian classical texts that could be traced back to the central heartland and the first dynastic order. Individualism, a quest for freedom, and creative initiative were not valued qualities in China. Mastery of the writing and replication of the Confucian texts were the road to wealth and elite status.

The importance of tradition and the ancient classics led to a deep respect for established ways of doing things. This only strengthened the Confucian ethic, which envisioned a society that functioned like a family headed by a benevolent elder patriarch. Likewise, in the Confucian order, interpersonal and familial relationships were conducted in rigid ways. Because of the burden of the population and the need to control it by the emperors, certain forms of technology and laborsaving devices that were developed in the West were unnecessary, or perhaps unwanted, in China. Keeping thousands of peasants working the land with hoes was a good way to control the masses and to keep them occupied. China's two main river systems, the Yangtze (Changjiang) and the Yellow River (Huang He), which both flow west to east, from out of China's heartland, flooded over the centuries, creating natural disasters. Ancient dam projects and dikes built from corvée labor by thousands of peasants were an economically effective way to address the problem. What mass transportation links were required, such as roads or canals, were built by conscripted labor. The industrial age, however, was late coming to China, and it was introduced by the West. China's population was tied to near-subsistence agriculture involving rice transplanting and local or regional economics. The canal and river systems in China made railways and major road networks less important than across the more sparsely populated Europe or the United States. Therefore, railroads and the laborsaving production lines of factories were viewed with suspicion by the Chinese people. The dynastic leaders and bureaucrats resisted these Western technical innovations because not only did they radically change social and spatial relationships, but they represented a form of foreign encroachment and penetration into China.

Still, a few transitional figures emerged in the mid-nineteenth century who were directly responsible for facilitation of China's movement into the industrial age. Lin Zexu, the official sent by Emperor Daoguang to Guangzhou in 1839 to suppress opium trade and compel foreign powers to stop trading, represents the catalyst who precipitated the military confrontation between China and the West. Lin burned the British opium stocks in Guangzhou in an attempt to stop the penetration of China's market by the opium trade, leading to the **Opium War** (1839–1842).

The Opium War represents a radical departure from China's earlier history and opens the period during which parts of China were subdivided into trading concessions by the West. This is the period of contemporary history where this book starts. After the Opium War, the emperor's control over the society weakened, and corruption increased. Popular unrest also grew. Millennial movements and peasant rebellions such as the **Taiping Rebellion** and the **Nian Rebellion** threatened national unity and stability, but the Qing dynasty, also weakened by rebellions of Muslims in western China and having for too long depended on established ways of fighting and producing war materials, was unable to handle the challenge. Foreign assistance, often from private mercenaries but also from foreign quasi-official and official sources, assisted the Qing bureaucrats. This also propelled forward the industrialization and mechanization of China.

Between the mid-1850s and the **Sino–Japanese War** (1894–1895), China fought France and England and nearly went to war with Japan over Taiwan. These events led some Chinese bureaucrats to think about modernization. Within China, the **self-strengthening movement** saw Qing administrators like **Li**

Hongzhang and **Zuo Zongtang**, who recognized the value of Western technology and weaponry, create modern shipyards and arsenals. They tried to harness industrialization to China's military and developmental needs. Many of the more detailed entries in the book deal with the period between the Opium War and the **Republican Revolution** (1911). This period, the late nineteenth through the early twentieth centuries, forms the framework underpinning the security posture of the People's Republic of China. This period also reflects the Communist Party's Marxist understanding of how to approach other nations and how the West has treated China. The period also saw the foundations laid for warlord control of the country and its division into armed camps and spheres of influence. Before reading the encyclopedic entries of this book, a brief review of China's earlier history is useful, including some mention of the more important classical military texts.

THE GEOGRAPHICAL REALITIES CONFRONTING CHINA

China's land frontiers of more than 20,000 kilometers are shared with nearly all of the other countries on the mainland of East Asia. Many of these borders are not well demarcated and only recently are being settled. The Sino–Russian border, now about 5,000 kilometers long with the breakup of the Soviet Union, still has several areas under dispute, but surveys and border negotiations have reduced the tensions there that led to the **Zhenbao Island** (Damansky Island) **Clash** between China and the Soviet Union in 1969. In addition, parts of the Sino–Indian border remain in dispute, China and India having fought a war over the borders in 1962. The Sino–Vietnamese border, the focus of China's invasion of Vietnam in 1979, still has sections to be completely and accurately demarcated. Chinese military leaders often use these borders and the long history of invasion from the north by Mongols, Manchus, Russians, and Japanese as a justification for retaining a large ground force in the military. Arguably, however, part of the justification for the size of China's ground forces is also the central government's desire to maintain control over the population.

Eastern China is nearly all seacoast, and China has about 18,000 kilometers of coastline, along which there are many good harbors and offshore islands. However, except for a short period in the fifteenth century, when the Ming dynasty (1368–1644) sent maritime trading missions led by Admiral Zheng He to the South China Sea, the Middle East, and Africa, China's strategic and developmental orientation has been primarily continental. This continental orientation is probably a function of the need to organize and control a large hinterland to the west as much as it is the continued need to defend against invasions from the north. Unquestionably, economic development and trade also contributed to a continental orientation for a defense strategy. At the height of imperial dynastic power, China had the world's largest economy (from the thirteenth or fourteenth through the nineteenth centuries). However, much of the trade in which China was engaged moved along the land route of the ''Silk

Road,'' which stretched from Xi'an, Shaanxi Province, along the north rim of the Tarim Basin, across Central Asia and into Persia (Iran), Turkey, and the Arabian Peninsula. Trading and economic interests affect defense orientation, and until Vasco DaGama found a sea route around the Cape of Africa (Cape of Good Hope), almost all trade from Europe and the Middle East with China was by land. When the European powers began to move goods by sea, China still had little need to develop a navy because merchant ships came into the harbors along that extensive coast seeking to trade, and China's extensive bureaucracy managed interaction with the foreigners. What Fairbank called the ''weight of China's cultural tradition'' also had an effect on strategic orientation. Even when a shift in strategic orientation to a focus on maritime matters was probably called for in the eighteenth or early nineteenth centuries, the constancy of some 4,000 years of a land-based strategic orientation functioned as a brake on developing a maritime strategy. When the Western powers finally sent military expeditions to defend their interests in China, beginning with the Opium War, the emperors had no fleet with which to defend their national interest. More seriously, when they tried to build a fleet during the self-strengthening movement, they ended up with a mix of weaponry and equipment that could not be supported by an indigenous industrial base. The Nationalist government also failed to establish a full industrial base that could support a strong indigenous arms industry, relying on arsenals established by **warlords** such as **Yan Xishan**'s Taiyuan Arsenal in Shanxi or the older **Jiangnan** and **Fuzhou Arsenals and Dockyards**, established with foreign help during the self-strengthening movement in the late nineteenth century. After 1949, the Communist leadership took over these places. Although they established a comprehensive military–industrial complex complete with a dispersed **Third Line** industrial base in case of foreign invasion or nuclear attack, much of the system was initially reliant on help from the Soviet Union. The management of national defense industries, through the **Commission of Science, Technology, and Industry for National Defense** (COSTIND), its predecessor, National Defense Industry Office, and its successor, General Armaments Department (the fourth **General Department of the People's Liberation Army** [PLA]) and State COSTIND, remained a Soviet-based, batch-production system. China in 1998 depended on purchases and licensed production of modern combat of aircraft from Russia for its military.

The border with Russia, to the north, has been in dispute for several hundred years and in 1998 still was not fully demarcated to the satisfaction of both sides. China still claimed parts of Siberia as late as 1954. The western portions of the border in Inner Mongolia and Xinjiang are being surveyed by aircraft for demarcation. On the eastern side of China, the Sino–Russian border includes the areas of the Argun, Amur, and Ussuri Rivers, along Heilongjiang Province, which were the focus of the Zhenbao Island dispute in 1969. The former states of the Soviet Union in the west, especially Kazakhstan and Kyrgystan, also have border areas in dispute with China. The 1997 agreements between Beijing and the leaders of the five Central Asian republics (Turkmenistan, Uzbekistan, Kyrgystan, Kazakhstan, and Tajikistan) formed a ''strategic partnership'' with these countries designed to promote mutual economic and social development. However, during the Yuan dynasty the Mongols swept into these areas and incorporated them under the emperor's suzerainty as what was once called "Chinese Turkestan." After the Taiping Rebellion, Zuo Zongtang, the governor of Fujian Province who established the Fuzhou Dockyard, was sent to Turkestan to put down rebellious Muslims in the Nian Rebellion. Also, there are still territorial disputes in the Pamir Mountains, where some sections of the glacial landscape separating China and Afghanistan remain undemarcated.

Two areas of the Sino–Indian border are still in dispute. The Aksai Chin area, northeast of Jammu and Kashmir on the Pakistan–India border, is under Chinese control but is claimed by India. Another area near Bhutan, along the Bhramaputra River, is claimed by China as part of Tibet but is still under Indian control. In a proposal to the Indian government in June 1980, Beijing suggested that if India would cede the Aksai Chin area (through which China has built a highway) to China, Beijing would recognize the McMahon Line, resolving the **Sino–Indian border war** in the east.

In the South China Sea, Beijing seized the Paracel Islands (**Xisha Islands**) in 1974 from Vietnam and lay claim to the Spratly Islands (**Nansha Islands**). The claims must be resolved in international law, but their legal basis dates to the reign of Khublai Khan, the Mongol leader who became emperor of China in 1271. During the Mongol expansion, which became the Yuan dynasty (1279–1368), the Yuan sent out fleets to Java (Indonesia) in 1292. Earlier, the Yuan conducted naval and land campaigns against what is now Vietnam and invaded Burma, Laos, and Thailand. In 1274, the Yuan dynasty, which controlled China and the Korean Peninsula, sent a naval expedition against Japan. In 1281, the Yuan attacked again and also attempted to take control of the Liuqiu Islands (the Ryukyus, or Okinawa). Vietnam, Malaysia, the Philippines, and even Taiwan also lay claim to the Spratlys, the scene of several conflicts with Vietnam and, in 1995, between China and the Philippines over China's seizure of Mischief Reef. China and Japan have ongoing territorial disputes over the Senkaku Islands, and Vietnam and China still have not agreed on the demarcation of the Gulf of Tonkin, west of Hainan, which has undersea petroleum deposits. There is a major Chinese diaspora around Southeast Asia, where ethnic Chinese, who make up less than 10 percent of the population in many countries, own more than 70 percent of the wealth. These Chinese trading colonies around Asia date back to the seven trading voyages of the Muslim eunuch Zheng He. Between 1405 and 1433, fleets under the command of Zheng He traveled the South Pacific, the Malacca Strait, the Indian Ocean, and the Arabian Sea. Zheng He carried enough men to fight land battles but, for the most part, engaged in trade and the exchange of imperial tribute.

A BRIEF HISTORY OF DYNASTIC SUCCESSION

Two of the finest modern Sinologists, Kenneth Leiberthal in *Governing China* and John Bryan Starr in *Understanding China*, suggest that it is better to focus on the continuities of Chinese culture than to concentrate on the details of each dynastic order. They place great weight on the enduring qualities and characteristics that make Chinese culture and society unique. Among these characteristics are the dominant, authoritarian political institutions based on Confucianism; the ideographic written language that makes rote imitation a necessity to be considered educated; the overwhelming size of the population; and, as Fairbank stated, the seemingly insurmountable task of feeding that population. Despite the advice of Leiberthal and Starr, I will review the succession of dynasties briefly in this Introduction and point out some of the salient features reflected in contemporary Chinese culture of a few of the dynasties.

Although there is no archaeological evidence that the dynasty existed, after a mythological beginning, the first Chinese dynasty is said to be the Xia (2200 B.C. to 1500 B.C.). The existence of this ancient dynasty is part of legend.

The Yellow River valley, however, is the general area where the first historical and archaeological evidence of the existence of an early dynasty can be found. The Shang dynasty was an agrarian-based society that can be dated to between 1766 B.C. and 1027 B.C. Evidence of the dynasty is found in some bronze vases, which demonstrate that metallurgy was part of the culture, and ancient forms of ideographs found on oracle bones. Tombs yield evidence of the existence of royalty and of burial rites. Hundreds of people, possibly slaves, were also buried alive with royal corpses.

From approximately 1122 B.C. to 249 B.C., the Zhou dynasty rose to power, with a capital near what is now the city of Xi'an, Shaanxi Province. The Zhou was the dynasty that seems to have existed longer than any other. Its official histories are also the first to reflect the concept of a "Mandate of Heaven," meaning that the ruler had legitimacy only if the power that was "divinely" invested in him could be authoritatively exercised. Natural disaster, famine, and invasion were later taken as evidence that a ruler had lost the "Mandate." For the purposes of this book, it is important to remember that Zhou kings ruled through an aristocracy that was hereditary and were a warrior class. Their rule was based on interpersonal bonds. In this, one can see the basis for the regional nature of leadership in China and the elements of personal loyalty that were part of "warlordism."

The Zhou dynastic period is divided into a Western Zhou (1122 B.C.–771 B.C.) and an Eastern Zhou (771 B.C. through about 256 B.C.). The period of the Eastern Zhou is most remarkable from a historical standpoint. The period from 722 B.C. to 481 B.C. is known as the "Spring and Autumn Period." It is the subject of one of China's most important historical chronicles, which takes that name. The latter half of the Eastern Zhou (roughly 402 B.C. to 221 B.C.) is known as the "Warring States" period. The incessant strife in the period produced some of the most enduring institutions of Chinese society, including the development of large standing armies, the development of a broad system for tax collection, and the development of literate professional and trading classes. To stabilize agriculture and increase economic growth, major flood control, canal, and irrigation projects were also undertaken. These sorts of projects led both Karl Marx and, later, the Marxist scholar Karl Wittfogel to characterize China as a "hydraulic society" (one that is built around the bureaucracies and people necessary to manage large-scale irrigation projects).

THE PHILOSOPHICAL BASIS OF CHINESE SOCIETY DEVELOPS

Another major feature of the Zhou dynasty was the development of a series of contending schools of thought about government and the nature of man. Six successive scholars stand out, whose thought can be found reflected in the political institutions of China today. Confucius (Master Kung, 551 B.C.–479 B.C.) is probably the most famous of the six. He believed that each man or woman had a specified role in a hierarchical social system and that each person must perform that role for an ordered society. Confucius concentrated on interpersonal relationships—ruler and ruled, father and son, teacher and student, husband and wife, older and younger brother. Filial piety and the maintenance of these relationships, according to Confucius, kept society harmonious.

Mencius (Mengzi, 372 B.C.–289 B.C.) developed Confucian thought but modified it by saying that the basic nature of man was good. In adding this humanistic element to Confucian thought, Mencius also reemphasized the concept of the "Mandate of Heaven," meaning that a ruler governed with the tacit consent of the people.

Hsun-tzu (Xunzi, 300 B.C.–237 B.C.) was another Confucian thinker, but he differed seriously with Mencius. He believed that human nature was inherently selfish and evil. Only education could produce self-cultivation and goodness, Hsun-tzu believed. Another important tenet of his thought was that neither moral nor ethical persuasion would produce a good government, only strong authoritarian control. This emphasis on authoritarianism was further developed in the doctrine of legalism, formulated by Han Fei (Hanfei, who died in 233 B.C.). Like Hsun-tzu, Han Fei believed that man was selfish. To preserve social order, Han Fei insisted that only the strict enforcement of laws and rituals of social order could impose the necessary discipline on society.

An alternative stream of Chinese thought also coexisted with Confucian-based systems. Daoism (Taoism, or "the belief in the way") is attributed to the sage Lao-tzu. Lao-tzu is supposedly older than Confucius. Taoism focuses on the individual and on his or her place in nature instead of his or her place in an ordered society. Each person must find his or her own way to adjust to the natural patterns and rhythms of the universe. This sense of a person's place in a rhythmic universe did not run counter to Confucianism and, in one sense, complemented Confucian moral beliefs. Although some Western scholars, such as Waley and Legge, have interpreted Daoism using Christian terms, the sense of a personal, instrumental supreme being is not part of Daoist beliefs.

Finally, the naturalist Mo-tzu (Mozi, about 479 B.C.–381 B.C.) developed an earlier philosophy, Mohism, dating to the period of Confucius, which held that the mutual nature of opposites in the universe (hot and cold, moon and sun, light and dark, male and female, positive and negative) is basic to all forces. He emphasized universal love, peace, and moderation. Mohism did not emphasize the strong sense of filial piety embedded in Confucianism. The importance of the unity of opposites, however, resonated well with Marxist dialectics and can be found to have blended into some of the thought of Mao Zedong.

CHINA IS UNIFIED FOR THE FIRST TIME

The central area of what we know as China was finally unified by the first Qin dynasty emperor, Qin Shi Huang Di, in 221 B.C. In that year the warring states of China came under the control of a single emperor. Qin Shi Huang Di emphasized the legalist tradition (embodied in Han-Fei-tzu) and relied on scholars as advisers. He also burned the books of those ideologies that competed with his own and killed or drove away Confucian scholars. During the Qin dynasty forms and styles of writing were formalized, legal codes were developed, and ritualized thought was standardized. The Qin emperor also undertook the development of the walls of the contiguous warring states into what is now known as the "Great Wall of China." The project, however, required such a vast pool of impressed labor that when the emperor died in 210 B.C., revolts against the dynasty broke out. The "terra cotta warriors" outside the city of Xi'an are part of a vast, buried clay army that guards the tomb of the emperor Qin Shi Huang Di.

By 206 B.C., the Han dynasty developed out of the collapse of the Qin and the ensuing chaos. The Han emperors adopted Confucian ethics, eschewing the harsher legalist order favored by the Qin. However, the Han leaders still retained the administrative structure and major public works of the Qin. Most importantly, the Han dynasty used Confucian scholars as the core of their civil service, institutionalizing Confucian classics as the standard for an educated person. The dynasty also began an expansion. Han armies moved westward toward the area of Xinjiang to take control of cities and trade along the "Silk Road" with Central Asia and Asia Minor. Toward the northeast, Han influence and control spread onto the Korean Peninsula, carrying with them Confucianism and its bureaucratic infrastructure. Han military forces also invaded and took control of the northern part of Vietnam, turning that kingdom into a tributary state. The system of tributary states was also an important phenomenon that evolved out of the period of the Han dynasty. For the most part, the areas invaded by the Han remained essentially autonomous states with their own kings. However, the kings were expected to generally accept nominal Han control, give preferential trade treatment to China, and exchange gifts and goods at the ruling level. Intermarriage was also an important way of solidifying the hegemonic bonds between China and the tribute state. As the reader of this book examines the interaction of the Qing dynasty with the West in the period that preceded the

Opium War and around the time of the Boxer Rebellion, the remnants of the tribute system can be found. The Western nations, however, never accepted Chinese hegemony.

After about two centuries of rule, which saw the introduction of Buddhism from India into China, the Han dynasty collapsed. By A.D. 221, China was heavily involved in a series of internal civil wars among rival warlord factions competing for power. Over the course of 400 years, China saw only a short period of stability, between the years of 265 and 420, during the Jin dynasty. The Jin, however, had to move the capital from the heartland of China to Nanjing under military pressure in 317, leading to its own demise. The period of conflict after the Han dynasty is immortalized in the dramatic fiction *Romance of the Three Kingdoms* and is known as the "Three Kingdoms" period. The strength of Buddhism in China grew during this time, and gunpowder was invented.

The Sui dynasty, which ruled between A.D. 589 and 618, is known for its massive civil works projects. The Grand Canal was built during this time, and the Great Wall was restored and improved. Most of this work was accomplished with corvée labor, however, and financed with high taxes. This weakened the regime, leading to popular revolts. A combination of natural disasters and the pressures of more military campaigns against the Korean Peninsula led to the overthrow of the Sui, as it lost the "Mandate of Heaven."

After the demise of the Sui, another dynasty, the Tang, established a capital in Xi'an. Ruling between A.D. 618 and 907, the Tang expanded the territory controlled by the emperor, saw a flourishing of literature and art, and, through trade, increased contacts with the Middle East. Over the Silk Road, Islam found its way into China and coexisted with Daoism, Confucian learning, and Buddhism. The government bureaucracy was more formally institutionalized during the Tang dynasty through a system of competitive written examinations required to hold office that were based on mastering the Confucian classics. As the tenth century opened, the Tang dynasty was weakened by a combination of rebellion and invasion, leading once more to China's fragmentation.

By the latter half of the tenth century, the Song dynasty (A.D. 960–1127) rebuilt the centralized bureaucracy of China and supported scholar-officials with regional military governors. China began to urbanize as the population grew, and maritime commerce developed. This made the coastal region important as an engine for development and led to the rise of a stronger merchant class. The term "gentry," as applied to a landed, educated elite, is applied to the middle-class merchants and trading people who made up the artisans and commercial people forming a middle class in the cities. This period also saw the resurgence of Confucian doctrine, which spread into the tributary states of Korea and Vietnam and was introduced into Japan.

The Great Wall, built over the years as a defensive works to keep out invaders from the north, did not hold off the Mongols. Genghis Khan spread Mongol rule out to the Middle East and into Europe toward the end of the Song. Genghis' grandson, Kublai Khan, drove south into China, defeated the Song dynasty forces, and established Mongol rule over China, establishing the Yuan dynasty (A.D. 1271 through 1368). If one looks at the extent of the rule of the Yuan dynasty, for the only time until the occupation of Tibet and Xinjiang by the Communist regime of the People's Republic of China, the map of China looks very similar to that which we see today (minus independent Mongolia). The Yuan began maritime commerce and controlled a vast land area, controlled a flourishing maritime commerce, and developed canal and riverine lines of transportation in China. Islam strengthened in western China, moving in from Central Asia along the Silk Road. Like preceding dynasties, however, the Yuan eventually succumbed to a combination of natural disasters and peasant uprisings.

After centuries of rule by a foreign people, the Han Chinese finally established their own new dynasty in A.D. 1368 with the Ming, which ruled until A.D. 1644. The Ming dynasty initially had its capital in Nanjing, but it moved the capital to Beijing, building the city, its temples, palaces, and walls. Today, leaders of the People's Republic of China also like to point to the maritime trade that took place during the Ming dynasty, pointing out that Admiral Zheng He did not establish colonies but engaged in commerce around the South China Sea, the Indian Ocean, and across to the coast of Africa. Zheng He did that, but if one reads records of the voyages he made, it is clear that he sailed with a large military contingent. His voyages included up to 350 ships and some 15,000 fighting men. When a cranky eunuch from China pulled up to a port with that kind of military force (larger than the Spanish Armada) in the fifteenth century and asked to begin a trading relationship, rulers of smaller kingdoms decided to become tributary states—the only questions asked were, trade what and how much?

The Ming dynasty was weakened by continuous fighting with the Mongols and more serious fighting with the Japanese over control of the Korean Peninsula. It fell to the Manchurian tribes in 1644, with the establishment of the Qing dynasty (A.D. 1644–1911). The interesting features of Qing rule are that Confucianism was revived once more, and the Manchu emperors supported the continuation of Han scholar-officials in administrative positions. Ethnic Manchus were put into military leadership positions. Intermarriage between Manchus and Han Chinese, however, was forbidden by the Qing dynasty, and Manchus were not permitted to engage in trade or manual labor. This led to the development of a strong merchant class among the Han Chinese. The Qing emperors were content to accept nominal suzerainty over Tibet and conquered Mongolia, focusing on preventing rebellion and expanding their rule over the continent. They also relied on the tribute system to control peripheral states.

Here we reach the period of the mid-nineteenth century, the focus of major sections of this book. China encounters the West and in doing so relies on time-tested rituals of tribute and imperial control. However, the old ways fail, and the centrist view of the Qing dynasty is interrupted by the Opium War. Rather than continue with a review of history, the reader can follow the threads of the fall of the Qing through the encyclopedic entries that follow.

ROADS, RAIL LINES, RIVERS, AND CANALS: THE MILITARY SIGNIFICANCE OF LINES OF COMMUNICATION

For centuries, China's north-south lines of communication were restricted by the nature of the mountainous terrain and the river systems, which primarily run east to west. North-south transport and commerce took place primarily along the coast, with goods then moving inland from ports along rivers. The major exception to this was the Grand Canal, a construction project linking North and South China that ranks with the Great Wall, undertaken by the Sui dynasty (A.D. 581–617). By the time of the Yuan dynasty (A.D. 1279–1368), the Grand Canal was extended to Beijing and linked that city with the seaport of Tianjin to the east and Shanghai to the south. In fact, it was along the Tianjin–Beijing branch of the Grand Canal that the **Eight Foreign Armies** marched and transported their supplies during the **Boxer Rebellion**.

The main river systems of China—the Huang He, or Yellow River, in North-Central China; the Yangtse, or Changjiang, in Central China; the Pearl River, or Zhujiang, in the south; and the Amur (Heilingjiang), Ussuri, and Sungari Rivers (Songhuajiang) in Manchuria—for centuries were the primary means of transportation from the coast inland. These river systems became critical during periods of Western incursion and trade in China, since they generally served as the coastal terminus of trade routes and became the treaty ports after the Opium War. Along the Yangtse River the American Asiatic Fleet sailed its gunboats, from Wuhan to Shanghai, protecting U.S. commercial interests. The **Panay Incident** in 1937 also took place along the Yangtse. To the south, the British sailed up the Pearl River to establish the trading posts at Guangzhou (Canton), precipitating the Opium War.

There was no railroad in China until 1876, and this was built by Western powers. The Woosong Railway was completed in Shanghai that year. The railway workers' strike and the "Railway Protection Movement," a protest in Sichuan Province against Qing dynasty attempts to nationalize locally owned railroads built primarily with foreign help, started the turmoil in 1911 that eventually became the **Wuchang Uprising** and precipitated the fall of the Qing and the Republican Revolution of October 10, 1911.

Most of the rail lines, however, were in northeast China, especially in the more heavily industrialized areas of Manchuria. These were important in the **Nine One Eight (918) Incident** of September 18, 1931, when Japanese forces marched into northeast China.

During the **Siping Campaign** and the **Liaoning-Shenyang Campaign** of the **Civil War**, the Nationalist forces were defeated by the **People's Liberation Army** (PLA) primarily because the KMT Army, on the orders of **Chiang Kai-**

shek, concentrated its forces along the rail networks, giving the Communists the chance to mass superior forces at decisive points against a dispersed, but more numerous, Nationalist force. The Japanese army had made the same mistake in eastern China toward the end of the war, as their own supply lines weakened. Between 1949, when one could travel by rail no farther west than Shanxi or Wuhan, and the 1960s, rail networks had been improved to the point that all of eastern China was linked to Lanzhou in Gansu Province, Sichuan, and Xinjiang. The French-built rail lines linking Kunming, Yunnan Province, to Hanoi and Guangzhou on the coast to Hanoi were also operating. Only Tibet and the Qinghai Plateau lacked rail access. This is extremely important for PLA mobility, since the military has the capability to take over the rail lines in China and can rapidly transport its forces around the country if necessary.

Road networks were built more quickly than rail but were based principally on a spatial distribution around central marketplaces. Therefore, until the 1990s, China lacked an integrated, hard-surface national highway system. That system is presently under construction. Like the railways, the road system proved to be a blessing and a curse during the **Anti-Japanese War** and the Civil War. In each case, Japanese, then Nationalist, forces made the mistake of attempting to protect long stretches of roads between towns, leaving their forces thinly deployed and exposed to attacks by numerically inferior forces that concentrated at the decisive points. During the **Beiping-Tianjin Campaign** of the Civil War, the decisive battle at Xinbaoan, west of Beijing between the Great Wall and Zhangjiakou, the PLA managed to first force the Nationalist Army to protect long corridors of road and then to concentrate their forces in a tiny, walled town where Communist artillery decimated the Nationalist units.

NATURAL RESOURCES

China has a good base of natural resources on which to build its economy and industry: The country has adequate supplies of iron and coking coal on which to establish a steel industry, but it must still import steel, pig iron, copper, and aluminum to meet domestic demand and to enable its transportation infrastructure to meet distributed demand created in the mid-1980s and the 1990s. China's proven coal reserves total more than 700 billion tons, and its estimated reserves are over 3,000 billion tons. However, much of it is low-quality bituminous coal, creating a serious pollution problem. Onshore and offshore oil reserves are estimated at around 5.3 billion tons, most of which are not tapped. Like coal, however, much of the oil is of low quality. In fact, China has committed as much as $12.5 billion to the development of oil fields in neighboring Kazakhstan to the west and the construction of a pipeline stretching across the northern rim of the Tarim Basin in Xinjiang to Lanzhou. China also has natural gas reserves, but the size of these resources is not known with any precision. It is estimated at between 129 billion and 24 trillion cubic meters. The major limitation of China's energy industry has been extraction and distribution, with few major pipelines built up to the 1990s. With regard to nonmetallic materials, China has adequate supplies of barite, fluorite, salt, talc, sulfur, graphite, magnetite, and phosphates to meet domestic needs. Although deficient in chromium, platinum, and gold, there are large reserves of rare earth materials such as beryllium, tungsten, molybdenum, barium, manganese, mercury, zirconium, and titanium. China also has abundant uranium deposits, supporting the military use of nuclear weapons and a power industry. But despite these resources, industrial development has been spotty, and many of the industries still rely on the technologies of the 1950s. Central bureaucratic planning is partially responsible for the uneven exploitation of the resources, but geography is also a contributing factor.

THE CHINESE MILITARY-STRATEGIC CULTURE

If one accepts the thesis of Professor Alistair Johnston of Harvard University, "there are consistent and persistent historical patterns in the way particular states (or state elites) think about the use of force for political ends." Johnston maintains, in his book *Cultural Realism: Strategic Culture and Grand Strategy in Chinese History*, that states have "strategic preferences that are rooted in the 'early' or 'formative' military experiences of the state . . . influenced to some degree by the philosophical, cultural, and cognitive characteristics of the state . . . and these develop through time."

This is certainly true of China, and whether one reads the texts of **Mao Zedong**, the ancient philosopher of war Sun-Tzu, or the contemporary strategic texts published by the People's Liberation Army Academy of Military Science, a few constants jump out. Chinese strategic culture defines itself, and is often defined by Westerners or outsiders, as primarily defensive in nature. The strategic view is primarily designed to preserve the state, maintain security, and respond to external threat. Such defensive cultures are characterized by a preference for warfare with limited aims, the establishment of defensive works such as the Great Wall and the common form of walled city or compound in China, and an emphasis on building coalitions and alliances, even by intrigue, over the complete subjugation of an adversary. Thus, in earlier periods of Chinese history, kingdoms like Vietnam, Korea, and Tibet become tributary states of China rather than being subjected to destruction in war. Indeed, Sun-Tzu in the *Art of War* places his primary emphasis on subduing an enemy without fighting as the consummate test of the general or strategist. States, therefore, either accommodate the pressures of other states by alliances or diplomatic strategies, establish strong defenses supported by military expeditions when necessary to ensure security, or conquer other states. The Chinese preference has been to establish defenses while either deterring attack or dominating other states with limited attacks. To quote Sun-Tzu, "the best military policy is to attack strategies; the next to attack alliances; the next to attack soldiers [armies]; and the worst to attack walled cities [states]."

This basic strategic orientation was reinforced by the events of the nineteenth and early twentieth centuries, since the invading Western powers between the Opium War and the Boxer Rebellion put the Qing dynasty on the defensive and forced it to respond by compromise and diplomacy. Indeed, China's own histories reflect the period between the Opium War and the recovery of Hong Kong from British colonial rule on July 1, 1997, as a period of 150 years of humiliation at the hands of foreigners.

Another trend in Chinese behavior, captured in the diplomatic histories of John Garver (*Foreign Relations of the People's Republic of China*), emphasizes what can be characterized as suddenly violent, preemptive behavior designed to shape the outcome of disputes. China exercised this aggressive, violent behavior toward other states more than 16 times in the period between 1949 and 1993, according to Garver. In each case, despite the rapid escalation to violence, which Alistair Johnston calls "parabellum" behavior, China couched its own action in terms of some form of self-defense against a state that allegedly threatened its territorial integrity or sovereignty. The reinforcement of North Vietnam with up to 50,000 Chinese troops during the U.S. war with Vietnam, therefore, is characterized by Beijing as self-defensive in nature and as only assistance to a neighbor. Likewise, the entry of the People's Liberation Army into the **Korean War**, not as the Chinese armed forces but as **Chinese People's Volunteers**, is characterized as assistance to a neighbor by a friendly state. The 1979 attack on **Vietnam** by China, in the same manner, is characterized as a self-defensive counterattack.

This tendency to escalate quickly to preemptive violence while resorting to stratagem and diplomatic means characterizes Chinese strategic culture. It explains Beijing's actions in the **Taiwan Strait Crisis of 1995/1996**, as readily as it does China's actions in the South China Sea when the PLA Navy reacted strongly against Vietnam and the Philippines. Notably, in almost all cases, China declares its own actions to be defensive in nature. **People's War under Modern High-Technology Conditions** also fits this "parabellum" paradigm of the "active defense."

REFERENCES

Roger Ames, trans., *Sun-Tzu: The Art of Warfare* (New York: Ballantine Books, 1993); Jack Beeching, *The Chinese Opium Wars* (New York: Harcourt Brace Jovanovich, 1976); Keith Buchanan, *The Transformation of the Chinese Earth* (New York: Praeger, 1970); Edmund O. Clubb, *Twentieth Century China* (New York: Columbia University Press, 1964); John K. Fairbank, *China: A New History* (Boston: Belknap Press of Harvard University Press, 1992); John W. Garver, *Foreign Relations of the People's Republic of China* (Englewood Cliffs, NJ: Prentice-Hall, 1993); Alistair Iain Johnston, *Cultural Realism: Strategic Culture and Grand Strategy in Chinese History* (Princeton: Princeton University Press, 1995); D. C. Lau and Roger Ames, trans., *Sun Pin: The Art of Warfare*

(New York: Ballantine Books, 1996); James Legge, *Confucian Analects* (Oxford: Clarendon Press, 1938); Kenneth Leiberthal, *Governing China: From Reform through Revolution* (New York: W. W. Norton, 1995); Christopher J. Salter et al., *Essentials of World Regional Geography*, 2d ed. (New York: Harcourt, Brace, 1998); Christopher J. Smith, *China: People and Places in the Land of One Billion* (Boulder, CO: Westview Press, 1991); Jonathan D. Spence, *The Search for Modern China* (New York: W. W. Norton, 1990); John Bryan Starr, *Understanding China* (New York: Hill and Wang, 1997); T. R. Treagar, *A Geography of China* (Chicago: Aldien, 1970); Harro Von Senger, *The Book of Stratagems: Tactics for Triumph and Survival* (New York: Penguin Books, 1991); Arthur Waley, *Analects of Confucius* (London: Allen Unwin, 1938); Arthur Waley, *The Way and Its Power* (London: Allen Unwin, 1934).

A

AGRARIAN REVOLUTIONARY WAR (1927–1937) This is the general term for the period of the Communist Party-led, anti-imperialist, class-based war against landlords, capitalists, and the Guomindang (KMT) government. It encompasses the period 1927–1937, after the period of the **United Front** and the **Northern Expedition**. The period is marked in Communist histories as beginning with the Shanghai Incident on April 12, 1927, when **Chiang Kai-shek** took military action against Communist-organized labor and military groups. The organization and establishment of the Red Workers' and Peasants' Army and the period of guerrilla warfare started by the August 1, 1927, **Nanchang Uprising** are part of the Agrarian Revolutionary War. This period encompasses the **Autumn Harvest Uprisings**, the **Guangzhou Uprising**, and the establishment of the **Revolutionary Base Areas** such as the **Jinggangshan Soviet** and the **Eyuwan Soviet** by the Communist forces. (The Revolutionary Base Areas were also called soviets.) The histories of the Agrarian Revolutionary War also include the successful countercampaigns against the encirclement attempts by Nationalist forces. The retreat from the Eyuwan Soviet by the **Fourth Front Army** and the entire period of the **Long March**, which saw countless clashes between Communist and Nationalist forces, are also part of the history of the Agrarian Revolutionary War. This period in PLA historiography ends only with the **Xi'an Incident**, in December 1936, which began the period of United Front cooperation between Nationalist and Communist forces against the Japanese (the **Anti-Japanese War** or **World War II**). The Agrarian Revolutionary War has also been called the ''Second Revolutionary War Period'' in PLA and Chinese Communist Party histories (the First Revolutionary War is the Communist term for the October 10, 1911, Nationalist revolution).

REFERENCES

James Bertram, *First Act in China: The Story of the Sian Mutiny* (New York: Viking Press, 1938, reprinted 1973); Chen Tinglong, *The Army of the Republic of China* (Taipei:

Army Press, 1993); Jean Chesneau, Francoise Le Barbier, and Marie-Claire Bergere, *China from the 1911 Revolution to Liberation* (New York: Pantheon Books, 1977); ZGDBKQS, vol. 2; ZGRMJFJZS, vol. 1.

AIGUN, TREATY OF (1858) In an effort to counter Russian trade and exploration, the Qing dynasty emperors were able to block Russian penetration into their ancestral homeland in the northeast (Manchuria) by securing claim to the Amur River area in the Treaty of Nerchinsk in 1689. In return, the Chinese granted rights for Russian religious missions to operate in Beijing (on the same ground occupied today by the Russian Embassy). Later, through the Treaty of Kiakhta, in 1727, Russia and China agreed to regulate trade along the common border. In the second half of the nineteenth century, Russian expeditions explored and established towns along the Amur River and, weakened by the **Opium War** and other internal uprisings, despite protests from Beijing, China was powerless to react. The Qing emperor was already involved in handling the **Taiping Rebellion** and the Anglo–French military campaign against Tianjin and Beijing. In a move to secure Russian support against the British and French, a Qing Court official negotiated the Treaty of Aigun on May 16, 1858, granting Russia control of the sea north of the Amur River and agreeing to joint control of the area between the Ussuri River and the sea. Although the Qing Court representative signed the Treaty of Aigun in 1858, the emperor delayed ratification and rejected the treaty in 1859. The seizure of Beijing by France and Britain in 1860 so weakened the emperor that the Treaty of Aigun was finally confirmed by the Qing in 1860, with the Sino–Russian Treaty of November 1860, signed in Beijing. In a supplementary treaty in 1861, Russia gained control of the area east of the Ussuri River.

REFERENCES

Gerald Graham, *The China Station: War and Diplomacy, 1830–1860* (New York: Oxford University Press, 1978); Rhoads Murphy, *The Treaty Ports and China's Modernization: What Went Wrong?* (Ann Arbor: University of Michigan Press, 1970).

ALPHA FORCE After the recall of **General Joseph W. Stilwell** to the United States by President Roosevelt on October 18, 1944, **Lieutenant General Albert C. Wedemeyer** was sent to China as the commanding general of U.S. forces in the theater. Wedemeyer arrived in China on October 31, 1944. In the face of a major Japanese offensive that threatened allied Chinese–American control of southwest China, including the major air and ground base areas of Chongqing, in Sichuan, and Kunming, in Yunnan Province, Wedemeyer suggested ALPHA Plan to organize 36 Chinese infantry divisions into a single field force commanded by a Chinese general but staffed jointly by Chinese and American officers. Under the plan, the United States was to train, equip, and supply the force, which was to be known as ALPHA Force. The 36-division commitment to the force amounted to about 15 percent of the total Nationalist Chinese army. Despite his early opposition to the establishment of such a force, Generalissimo **Chiang Kai-shek**, whose forces were threatened by the Japanese, reluctantly accepted the plan. Nonetheless, because he was afraid that Chinese Communist forces might attack the Nationalists, Chiang kept some of the best Nationalist troops in reserve in the Chongqing area, providing only grudging support and mediocre forces to Wedemeyer.

In January 1945, Wedemeyer, acting in his capacity as the chief of staff to Chiang Kai-shek (he was dual-hatted as commander of American forces and Chiang's chief of staff), established two subordinate commands to man and train the Chinese forces. The first, the Chinese Combat Command, was designed as an advisory group, placing American officers in positions to give advice to Chinese commanders at all echelons, regiment and above. The Chinese Combat Command was led by U.S. Army major general Robert B. McClure. The second command created by Wedemeyer was the **Chinese Training Command**, led by Brigadier General Joseph W. Middleton. The Chinese Training Command eventually operated seven separate training centers and schools, most of which were located in the vicinity of Kunming, Yunnan Province.

ALPHA Force trained, developed, and concentrated itself in the area surrounding Kunming. It was commanded by Nationalist Army general He Yingjin, who had previously held the position of chief of staff of the Nationalist Army and for whom General Stilwell had expressed great respect.

Taking advantage of air support from Major General **Claire Lee Chennault**'s 14th Air Force, strategic bombing support from the U.S. Army Air Force's 20th Bombardment Group, and the U.S.-established supply and sustainment system, General He Yingjin finally began an offensive against the Japanese in spring 1945. Responding to a Japanese advance toward Kunming from the area of Guilin, in the southeastern province of Guizhou, ALPHA Force divisions began a counteroffensive on April 14, 1945. This ALPHA Force campaign moved east from a locus around Zhejiang, in Hunan, near the Guangxi border. In all, General He Yingjin committed forces of six Chinese armies, the 94th Army, the New Sixth Army, the 74th Army, the 100th Army, the 18th Army, and the 73d Army. Between April 18 and June 7, 1945, ALPHA Force armies and divisions forced Japanese troops to retreat to the positions they had occupied before the offensive. However, Chinese losses in the campaign were heavier than Japanese losses.

The ALPHA Force Plan never had time to reach fruition before **World War II** ended, but the advisory system, combined with the leadership of General He Yingjin, succeeded in blunting the Japanese advances in the Zhejiang campaign. A subsequent American-conceived campaign, called the BETA Plan, designed to attack Canton and Hong Kong using the ALPHA Force, was never implemented. The Japanese surrender after atomic bombs were dropped on Hiroshima and Nagasaki ended the war.

REFERENCES

John H. Boyle, *China and Japan at War: The Politics of Collaboration* (Stanford, CA: Stanford University Press, 1972); Ch'i Hsi-sheng, *Nationalist China at War: Military Defeats and Political Collapse, 1937–45* (Ann Arbor: University of Michigan Press,

1982); F. F. Liu, *A Military History of Modern China, 1924–1949* (Princeton: Princeton University Press, 1956); U.S. Army Center for Military History, *China Offensive, 5 May 1945–2 September 1945* (Washington, DC: U.S. Government Printing Office, 1995).

ALPHA PLAN. See ALPHA FORCE

AMERICAN VOLUNTEER GROUP. See CHENNAULT, CLAIRE LEE; FLYING TIGERS

ANFU CLIQUE The Anfu Clique was one of the many **warlord** armies that existed around China at the end of the Qing dynasty and the beginning of the Republic of China, circa 1911. The clique was composed of a faction of warlords centered around the area of Anhui Province, with influence extending into the North China Plain to the area around Tianjin, who allied with warlord leaders from Fujian (hence, the acronym *An-Fu*). The leading warlord figures of the Anfu Clique were Duan Qirui and Xu Shuzheng. The Anfu Clique allied with the armies of Manchuria and with forces of Feng Guozhang, who also operated in Anhui Province. In 1920, the **Zhili Clique** fought a series of engagements with the Anfu Clique, the **Fengtian–Zhili War**. Zhili was allied at that time with Zhang Zuolin's Fengtian Clique and his **Fengtian Army** from Manchuria.

REFERENCES

Donald Gilin, *Warlord: Yen Hsi-shan in Shanxi Province, 1911–1949* (Princeton: Princeton University Press, 1967); Donald Jordan, *The Northern Expedition: China's National Revolution of 1926–1928* (Honolulu: University Press, 1976); Lucian Pye, *Warlord Politics: Conflict and Coalition in the Modernization of Republican China* (New York: Praeger, 1971); James E. Sheridan, *Chinese Warlord: The Career of Feng Yu-hsiang* (Stanford, CA: Stanford University Press, 1966).

ANTI-JAPANESE MILITARY AND POLITICAL COLLEGE The successor to the **Red Army College**, the Anti-Japanese Military and Political College (Kang-Ri Hongjun Daxue, or *Kangda*, in its Chinese acronym form) was established at Wayaobao in Shaanxi Province on June 1, 1936. The college moved to the base camp for the **People's Liberation Army**, **Yan'an**, in 1937. The curriculum offered about 30 weeks of military and political instruction to cadres. **Lin Biao** was the first president and political commissar, and **Luo Ruiqing** was dean of instruction. During the **Anti-Japanese War** in China a total of 12 branch schools were established and operated in other Communist base areas outside Yan'an. These colleges ceased operation in 1945, after having trained over 100,000 military and political cadres.

REFERENCES

Warren Kuo, *A Comprehensive Glossary of Chinese Communist Terminology* (Taipei: Institute of International Relations, 1978); William W. Whitson with Chen-Hsia Huang, *The Chinese High Command* (New York: Praeger, 1973).

ANTI-JAPANESE WAR. See WORLD WAR II

ARROW WAR (1856–1860) Chinese police arrested 12 Chinese crewmen on a Chinese-owned trading vessel, the *Arrow*, on October 12, 1856. The police suspected the crewmen of piracy and smuggling. In the effort to arrest the crew members, the British flag, flown by the vessel because it was registered in Hong Kong, was torn. The British consul in Canton, Harry Parkes, demanded of Chinese authorities an apology for the damage to the British flag and the release of the Chinese crew members. When Chinese authorities released the crew but refused to apologize for the damage to the British flag, Parkes ordered British naval vessels to bombard the city.

The Chinese responded by burning foreign-owned factories and businesses in Canton. Meanwhile, a French priest was murdered in Canton. The British government dispatched a military expedition to China under the command of Lord Elgin. A French military mission was concurrently dispatched under the command of Baron Gros to avenge the death of the French priest. The Anglo–French forces seized Canton and moved north, up the coast of China, attacking ports and shipping until they reached Tianjin. The Chinese signed a treaty in Tianjin (**Tianjin, Treaty of**) in June 1858 but took no action to ratify the treaty until the British and French governments took renewed military action with forces still under the command of Lord Elgin and Baron Gros. The combined French and British forces occupied parts of Peking, burned the emperor's Summer Palace (the Yuanmingyuan) in the western suburbs of the city, and drove the emperor out of Peking. The Arrow War ended with the acceptance and ratification of the **Convention of Peking** by the emperor in 1860.

REFERENCES

John K. Fairbank, ed., *The Chinese World Order: Traditional Chinese Foreign Relations* (Cambridge: Harvard University Press, 1968); Sean Glynn and Alan Booth, *Modern Britain: An Economic and Social History* (London: Routledge, 1996); Immanuel C. Y. Hsu, *The Rise of Modern China*, 5th ed. (New York: Oxford University Press, 1995); Douglas Hurd, *The Arrow War: An Anglo–Chinese Confusion, 1856–1860* (London: Collins Press, 1967); Charles S. Leavenworth, *The Arrow War with China* (London: S. Row Marston, 1901); Peter J. Marshall, ed., *The Cambridge Encyclopedia of the British Empire* (New York: Cambridge University Press, 1996).

AUGUST 6 NAVAL BATTLE (1965). See DONG SHAN ISLAND NAVAL BATTLE

AUTUMN HARVEST UPRISINGS Throughout the years 1926 and 1927, peasant and labor unrest in China had been growing, stirred by conditions created by the death of the 1911 revolution leader Sun Yat-sen, the **Northern Expedition** against **warlords**, and Communist organizers in opposition to the Nationalist government. On March 21, 1927, the Chinese Communist Party

(CCP) helped to organize a strike in Shanghai, led by the General Labor Union. Gangs of the "Society for Common Progress," supported by industrialists and armed by the Nationalist Army, responded by attacking the labor union headquarters on April 12. Meanwhile, in the Hunan countryside and around the city of Wuhan, there was also peasant and labor unrest.

On May 21, 1927, General Xu Gexiang (Hsu Keh-hsiang), the Guomindang garrison commander in Changsha, Hunan Province, began to attack unruly peasant forces in that area, eventually killing thousands of mobilized peasants who had recently seized land and killed the landowners. Peasant forces led by Communist organizers responded by conducting an armed attack on Changsha on May 31, but they were stopped by the intervention of Comintern agents bearing a cable from Stalin and by CCP leader Chen Duxiu.

By this time, the CCP had decided to launch a series of land appropriations by armed peasants in the countryside, supported by the seizure of urban areas and Guomindang military garrison areas. These "Autumn Harvest Uprisings" took place after the August 1, 1927, **Nanchang Uprising** and essentially ended with the **Canton Coup** in December 1927. The Autumn Harvest Uprisings by peasant forces took place in Jiangxi, Hunan, and Guangdong Provinces. They were coordinated by the CCP with uprisings among armed coal miners and with urban uprisings.

After their defeat by the Nationalist Army, the decimated Communist forces, led by **Mao Zedong**, retreated into base areas in the Jinggang Mountains area in Jiangxi, forming the **Jinggangshan Revolutionary Base Area**. Nationalist forces began a series of **Encirclement Campaigns** designed to isolate Jinggangshan and other CCP base areas and destroy the Communist military forces. These Nationalist campaigns were eventually successful in forcing the Communist **People's Liberation Army** to make the **Long March** and settle in new base areas in North-Central China to avoid being wiped out by the Nationalists.

REFERENCES

F. F. Liu, *A Military History of Modern China: 1924–1949* (Princeton: Princeton University Press, 1956); JGDBKQS, vol. 1; William W. Whitson with Chen-Hsia Huang, *The Chinese High Command* (New York: Praeger, 1973); ZGRMJFJDSD.

B

B-57 BOMBER DOWNING OVER BEIJING (October 7, 1959) According to Chinese **People's Liberation Army** Air Force (PLAAF) accounts, Nationalist forces on Taiwan began a reconnaissance program against the mainland in early 1959, using U.S.-made RB-57 reconnaissance aircraft. Because the PLA Air Force had just organized its own surface-to-air missile units, the PLAAF records the October 7, 1959, shoot-down of a reconnaissance model B-57 (RB-57) over Beijing as a successful operation by its newly established air defense missile forces. The PLAAF missile defense forces had become operational only in late September 1959. As a Taiwan-launched B-57 reconnaissance aircraft entered the air defense zone around Beijing around noon on October 10, the PLAAF fired three SA-2 surface-to-air missiles, bringing down the aircraft and ending high-altitude reconnaissance against the mainland with B-57 aircraft. This forced the Nationalist Air Force, with U.S. support, to use U-2 aircraft for military reconnaissance flights over mainland China.

The PLAAF had received its first SA-2 missiles from the Soviet Union only in October 1958, just after the **Taiwan Strait Crisis** of that year.

REFERENCES

Kenneth W. Allen et al., *China's Air Force Enters the 21st Century* (Santa Monica, CA: Rand Corporation, 1995); ZGDBKQS.

BA YI The Chinese ideographs for the numbers ''eight'' and ''one,'' Ba Yi, are used to signify the month and day commemorated as the date of the establishment of the **People's Liberation Army**, August 1, 1927. The ideographs are displayed inside a red star as the insignia of the People's Liberation Army (PLA) on uniform buttons as well as on vehicles, ships, and aircraft. On some PLA flags, the ideographs appear in yellow on a red background. The date commemorates the anniversary of the Communist-led

Nanchang Uprising in Jiangxi Province on August 1, 1927. On that date, as part of the **Autumn Harvest Uprisings**, the Communist Party Central Committee decided to attack the Nationalist arsenal in the city of Nanchang in an attempt to mobilize the proletariat in the region to begin an urban revolution, but the uprisings failed. By October 1927, the Communist Army retreated into the Jinggang Mountains to form the **Jinggangshan Revolutionary Base Area** (''shan'' means ''mountain'' in Chinese). The date August 1, 1927, was actually established as the PLA anniversary by the Military Committee of the Central Committee of the Chinese People's Soviet Areas, in the Jinggang Mountains, on June 30, 1933.

REFERENCES

Tong Bai, ed., *Junren Shouce* (Beijing: Changzheng Chubanshe, 1987); William W. Whitson with Chen-Hsia Huang, *The Chinese High Command* (New York: Praeger, 1973).

BANDIT EXTERMINATION CAMPAIGNS. See ENCIRCLEMENT CAMPAIGNS

BARRETT, DAVID D. (1892–1977) Colonel David Dean Barrett served in China for 23 of his 35 years on active duty in the U.S. Army. He was born in Central City, Colorado, in 1892, and served at different times as both assistant military attaché and military attaché in Beijing. He headed the U.S. Military Observers Mission, known as the **Dixie Mission** to the Communist Army base area at **Yan'an**, in 1944. The Dixie Mission reached the **Eighth Route Army** Yan'an base area on July 23, 1944, but Barrett was withdrawn at the request of **General Wedemeyer** in November 1944 because, in Wedemeyer's view, he had exceeded his instructions by offering U.S. Army assistance and weapons to a Communist-led guerrilla force to operate behind Japanese lines in China.

Barrett, like the rest of the members of Military Observes Mission, was impressed with the organization, discipline, and honesty of the Communist forces. Barrett had spent most of his 23 years in China cultivating his language skills and contacts with the Chinese people. He was an assistant to General **Stilwell** on several assignments, and at the start of **World War II** in China, July 7, 1937, actually visited the site of the **Marco Polo Bridge Incident** on the next day, only to be fired on by the Japanese. His career was effectively ended after he was pulled out of the Dixie Mission by Ambassador **Patrick J. Hurley**. Barrett wrote and lectured on China until his death in 1977, but he retired from the army bitter that he was never promoted to brigadier general.

REFERENCES

David Barrett, *Dixie Mission: The United States Army Observer Group in Yenan, 1944* (Berkeley: Center for Chinese Studies, University of California Press, 1970); John N.

Hart, "The Making of an Old China Hand: A Memoir of Colonel David D. Barrett" (Unpublished manuscript, Hoover Institute, Stanford University).

BEIFA (1926–1928). See NORTHERN EXPEDITION

BEIPING-TIANJIN CAMPAIGN (Ping-Jin Campaign) (November 21, 1948–January 31, 1949) This 64-day campaign, known alternatively as the Ping-Jin Campaign, the Peiping-T'ian-chin Campaign, and the Beijing-Tianjin Campaign, secured the strategic cities, road, rail, and sea lines of communication on the North China Plain. The forces of the **People's Liberation Army** (PLA) suffered almost 40,000 casualties in the effort to secure the cities of Zhangjiakou (then called Kalgan), Beijing (called Bei-p'ing at that time, hence the "ping" in Ping-Jin), and Tianjin, with its port and garrison at **Dagu**. The Communist forces inflicted 520,000 casualties on the Nationalists.

After the **Liaoning-Shenyang Campaign**, which secured Manchuria for the Communists, the PLA commander, **Lin Biao**, was preparing to reconsolidate and rest his forces. However, the Communist Party's **Central Military Commission** met and decided that, from a strategic standpoint, it was important that **Chiang Kai-shek**'s Nationalist forces, led by General **Fu Zuoyi**, be prevented from regrouping farther south to defend Nanjing and the Yangtze River. The Central Military Commission therefore directed Lin to initiate a new offensive, earlier than planned, to secure the cities of Zhangjiakou, Beijing, and Tianjin, as well as the rail lines linking those cities. Lin Biao held a planning conference in Shenyang, outlining a campaign strategy for the seizure of the North China Plain.

The Nationalist forces under Fu Zuoyi, about 500,000 strong in 50 divisions, occupied the key cities on the plain, Zhangjiakou, Beijing, and Tianjin. In anticipation of the PLA offensive, Fu Zuoyi also deployed Nationalist forces to protect the major mountain passes between Manchuria and the North China Plain. These passes, between Qinhuangdao near the coast and Gubeikou north of Beijing, were important choke points where the attacking Communist forces could be slowed or stopped. By the time that the Ping-Jin Campaign was initiated, having incorporated Nationalist divisions that surrendered in the Liaoning–Shenyang Campaign, the Communist forces had grown to about 800,000 personnel.

Because **Liu Bocheng**'s **Second Field Army** was deployed south of the North China Plain, Lin Biao was confident that Fu Zuoyi's forces would not be able to withdraw and avoid a decisive battle. However, Lin also feared that if he first attacked the more important cities of Tianjin or Beijing, Fu Zuoyi would withdraw his forces west toward Zhangjiakou, into more defensible, mountainous terrain. Lin therefore planned to first secure the most defensible objective, Zhangjiakou (Kalgan), and its rail links to Beijing before Fu's forces could move into that area. The tactical plan, from that perspective, was similar to the one adopted by Lin in the Liaoning-Shenyang Campaign. Lin Biao sought to prevent any withdrawal, to secure the west flank, and then to destroy in detail (i.e., piecemeal) any Nationalist forces between Zhangjiakou and the sea.

As the Communist forces began to move south and west from Manchuria, marching along three axes, right (west), central, and left (east), Fu Zuoyi dispatched three Nationalist corps westward to defend the Beijing-Zhangjiakou rail line. The Nationalist 104th Corps sent a division (the 258th) by road from Huairou, north of Beijing, to defend the rail line. The remainder of the Changping-based 104th Corps established defensive positions around Huairou. Meanwhile, the 16th Corps dug in, in the mountains between Nankou and Changping. The Nationalist 35th Corps took positions in the area around Zhangjiakou, while the 105th Corps defended sections of rail line.

What induced the Nationalist units to take static positions defending the rail line and Zhangjiakou were forces of the Communist Huabei (North China) First Army Group and Third Army Group. These two units, which had been involved in an attack on Taiyuan, Shanxi Province, were sent marching east, toward Zhangjiakou, as part of the plan to force Fu Zuoyi to defend that area. In total, about 26 divisions from the Huabei (North China) Field Army deployed in seven columns. Led by **Nie Rongzhen**, the Communist forces were used to assist Lin Biao's Northeast Field Army in the attack. The first attack against Zhangjiakou was led jointly by Cheng Zehua and **Luo Ruiqing**.

Meanwhile, as Fu's forces moved on Zhangjiakou, another major Communist force, the Central Route Army, named for the central axis of the campaign along which it marched, moved against Chengde, the eighteenth-century summer capital about 150 miles north of Beijing. The attack drove the Nationalist 13th Corps out of Chengde, forcing it to withdraw south to the ancient defensive line along the Great Wall at Gubeikou.

On the east coast, the Left Route Army, part of the attack under the control of Lin Biao, moved against Tianjin and the port at Dagu. It had a strength of about 20 divisions. By November 24, on the approach march to Tianjin, these troops had secured Qinhuangdao.

In one of the most notable battles of the Ping-Jin Campaign, Nie Rongzhen's Fourth Column isolated the Nationalist 35th Corps at Xinbaoan on December 7 and 8. The 35th Corps was trying to withdraw through a pass in the Great Wall in Badaling to Beijing. As the 35th Corps withdrew from Zhangjiakou under heavy Communist pressure, it ran into the Communist Fourth Column's 12th Brigade. Despite having air support and artillery, the Nationalist 35th Corps took defensive positions in a small, walled market town, Xinbaoan. In one of the most decisive battles of the campaign, the 12th Brigade fixed and isolated the 35th Corps, destroying one division, while the rest of Nie Rongzhen's forces drove the surviving elements of the 35th all the way back to Beijing's outskirts. The Communist 12th Brigade alone killed 1,000 Nationalist soldiers of the 35th Corps. By December 12, coordinated attacks by the Fourth and 11th Columns of the PLA forces under Nie and the Third and Fifth Columns of the route army attacking on this central axis drove the remaining Nationalist forces south into

Beijing. The Nationalist 16th Corps was nearly completely wiped out at Kangzhuang, while the 104th Corps was destroyed at Huairou. By December 14, forces of the Communist **Northeast Field Army**, which attacked along the central axis, seized control of Nanyuan airfield, fifteen kilometers south of Beijing. Having completely enveloped the city, Lin Biao continued to build up forces for an assault, bringing in armor and artillery. Meanwhile, the forces that attacked on the right, or western, axis seized Zhangjiakou, while the forces that attacked in the east, on the left axis of Liu's advance, seized Tianjin after a bloody fight. Fu Zuoyi, the Nationalist commander in Beijing, completely surrounded and cut off, agreed on January 20, 1949, to evacuate the city and to bring his entire army of some 250,000 soldiers over into the **People's Liberation Army**, joining the Communist cause. Fu's soldiers evacuated the city peacefully by January 27, and on February 3, 1949, Communist troops held a victory parade in Beijing. One reason that Fu Zuoyi agreed to surrender his forces was that his daughter was a Communist.

The Tianjin engagements, however, were not so easy. The forces that attacked along Lin Biao's left axis of advance were commanded by Li Tianyu, Deng Hua, and Zhong Wei. They formed three task groups and around December 12, 1948, began to envelop the city. At that time, Tianjin and the port at Dagu were defended by about 130,000 Nationalist troops. Li Tianyu moved into position to attack the city from the west with nine divisions, Deng Hua from the east with eight armor and artillery divisions, and Zhong Wei from the south with three divisions. In all, the Communists massed five corps to carry out the attack. Tianjin is surrounded by water and crossed by canals and waterways, as it is the gate to the sea for the North China Plain. Nationalist forces, by the first week of January 1949, flooded much of the area, causing the Communist forces to gather boats to conduct the attack. The Communists were able to divert some of the floodwaters back into canals, but their attack was delayed until January 14, 1949. A general assault on Tianjin began on January 14, spearheaded by the Communist 38th Army (Corps). The city surrendered on January 17. After initiating the assault on the city, Communist forces attacked Dagu and the port, which fell to them on January 17, 1949. The advance on the city and the port was a series of long, bloody battles, but the victory there probably contributed to Fu Zuoyi's decision to surrender in Beijing.

A parade in Beijing on February 3, 1949, ended the campaign. The third decisive campaign of the Chinese **Civil War**, the **Huai-Hai Campaign**, took place at almost the same time as the Ping-Jin Campaign. After February 1949, People's Liberation Army forces regrouped and recuperated for the campaign into South China. Chinese strategists and military officers study the Ping-Jin Campaign as an example of three forms of military operational art. The western axis of the campaign, against Zhangjiakou and Xinbaoan, is cited as an example of the effective use of maneuver and mobility. The eastern axis and the attack on Tianjin are cited as examples of the costly "direct approach" to an objective. The capture of Beijing after a siege is cited by PLA strategists as the best example of psychological operations, since, through its influence over the daughter of Nationalist general Fu Zuoyi, the PLA took the city without a bloody, debilitating fight. The campaign is also significant because it marked the transition of the PLA from a guerrilla-like force, as in the **Sungari River** and **Siping Campaigns**, to a force that could conduct maneuver warfare in large formations coordinating the supporting armor and artillery captured from the Nationalists in the Liaoning-Shenyang Campaign.

REFERENCES

Li Lu, *Zhongguo Guoxiang Daizhan Zhengzhan Li Jie Cidian* [A Compendium to Understand Trends in Chinese Warfare] (Beijing: Guofangdaxue Chubanshe, 1991); Liu Qi et al., eds., *Ping-Jin Zhanyi Qin Li Ji: Yuan Guomindang Jiangling de Huiyi* [Diaries of the Ping-Jin Campaign: Original Recollections of Guomindang Generals] (Beijing: Wenshi Chubanshe, 1996); Ma Yunpeng, *Huizhan Ping-Jin* [The Meeting Engagement—the Ping-Jin Campaign] (Shenyang: Liaoning Renmin Chubanshe, 1998); Su Zefeng, *Ping-Jin Zhanyi* [The Ping-Jin Campaign] (Shanghai: Renmin Chubanshe, 1959); William W. Whitson with Chen-Hsia Huang, *The Chinese High Command* (New York: Praeger, 1973); William W. Whitson and Chen-Hsia Huang, *Civil War in China 1946–1950*, 2 vols. (Washington, DC.: Office of Military History, U.S. Army, 1967).

BEIYANG ARMY The Beiyang Army was developed in the 1880s by **Li Hongzhang**, then the northern commissioner of trade, as a model for a modern army in China. The term comes from the title ''Beiyang Dazhen,'' or ''the Commissioner of Trade for Northern Ports,'' which was held by Li Hongzhang from 1870 to 1901. The post was held after 1901 by **Yuan Shih-kai**. The Beiyang Army is often associated with Yuan because of his influence over its officers during the **warlord** period. Li established officer training schools staffed by foreign instructors and also set up institutions for higher military learning, such as command and staff colleges for midgrade officers. Like its counterpart naval organization the Beiyang Fleet, also established by Li Hongzhang, the Beiyang Army was a post-**Taiping Rebellion** phenomenon that was part of the **self-strengthening movement**. It is an example of the effort in the late Qing dynasty to modernize China's military along Western lines.

During the **Sino–Japanese War** (1894–1895), the Beiyang Army, like the Beiyang Fleet, was not decisively committed. Thus, the utility of these modern forces was not demonstrated to the Qing Court. By 1901, under Yuan Shih-kai, the Beiyang Army had a strength of seven divisions. However, like many of the regional armies that later contributed to the warlord era, the leadership selection process for the army was intensely personalized. Thus, many of the military leaders developed strong personal ties to individual senior leaders but had weak loyalties to the state or nation.

While Yuan Shih-kai was governor-general of Hebei Province, which included Beijing and Tianjin, he built the Beiyang Army into an effective regional force, perhaps the best such force in China. It was instrumental in propelling

Yuan to power as the provisional president of China, effectively ruling from 1912 to 1915. However, when Yuan tried to declare himself an emperor between December 1915 and January 1916, he lost the loyalty of the Beiyang Army leadership.

Nonetheless, throughout the warlord period, the Beiyang Army dominated the North China area. By the early 1920s, rivalry between cliques and factions in North China caused several localized wars to break out between the warlord armies. The **Anfu** (or Anhui) **Clique**, headed by Duan Zhirui, fought a war against the **Zhili Clique** in 1920, known as the Zhili–Anfu War. The Zhili Clique was led initially by Feng Guozhang, but after Feng's death in 1919, Wu Peifu was the most effective military leader in the group. **Zhang Zuolin**, the warlord in control of Manchuria, fought two wars with his Fengtian Army against the Zhili Clique, known as the **Fengtian–Zhili Wars**.

REFERENCES

Ralph L. Powell, *The Rise of Chinese Military Power: 1895–1912* (Princeton: Princeton University Press, 1955); James E. Sheridan, *Chinese Warlord: The Career of Feng Yu-hsiang* (Stanford, CA: Stanford University Press, 1966).

BEIYANG FLEET After the French defeat of China in the **Franco–Chinese War** of 1884–1885, in which French ships ranged the Chinese coast winning both naval and ship-to-shore engagements, **Zuo Zongtang** asked the Qing Court to create a national fleet. As part of the effort to do this, the **Navy Board** was established. However, this financial management and policy-making institution was subordinated to the imperial household, and the funds it was designed to manage, in large part, were diverted to build improvements at the imperial **Summer Palace** at the edge of the western hills in Beijing. Some of the funds drained from naval construction into the Summer Palace (Yiheyuan) for Empress Cixi were used to build a large marble boat, a symbolic representation of a navy, on the north shore of the palace's Kunming Lake. Despite this drain on the moneys earmarked for naval construction, some of the funds were initially used for improving the fleet that was not badly damaged in the Franco–Chinese War, the Beiyang Fleet.

Like its counterpart organization, the **Beiyang Army**, the Beiyang Fleet derived its name from the official title of the "commissioner of trade for northern ports" (northern commissioner of trade, or *Beiyang Dazhen*). The position, which had military responsibilities, was held by **Li Hongzhang**. As the northern commissioner of trade, he was charged with maintaining internal order on land as well as sea and land defense against foreign powers. Li Hongzhang organized the ships under his control into functional squadrons and established a naval staff organized along German lines.

China had purchased two German battleships, which were the basis for two of its three seagoing combat fleets, or squadrons. Li also established a naval training squadron, a torpedo squadron for close-in coastal and harbor defense, and a transportation squadron for logistics and troop movement. The set of regulations and guidelines for the development of the Beiyang Fleet called for central control and regular exercises with the southern fleet, which was under control of the "southern commissioner of trade."

A dry dock was put into **Dagu** (near the modern port of Tangu), near Tianjin, along with a shipyard; the port of Lu-hsun (later called Port Arthur), which is now Dalian, was home port for the two battleships; and another port and machine shop were put in at Weihaiwei, on the Shandong Peninsula. Li also received some American supplies and ships, but this mixed sourcing only produced serious supply and logistics problems for the fleet. Another problem that hampered the fleet's effectiveness was a penchant among Chinese naval leaders and lower ranks for corruption. This embedded corruption extended to clerks in supply offices, who allegedly were in the pay of Japanese even during the **Sino–Japanese War** (1894–1895).

The Beiyang Fleet made a few "show the flag" voyages to ports in China's vicinity such as in Korea and Japan, including a ship visit to Yokahama Naval Base in Japan in 1890. However, for the most part, the fleet was poorly led, and the weaponry was old and uncoordinated. During the Sino–Japanese War, the fleet was only once decisively committed. The Chinese fleet of 12 ships was soundly outmaneuvered and defeated by the Japanese on September 17, 1894, in the Yalu River Naval Battle. The Chinese Beiyang Fleet scored only hits with about 10 percent of its naval gun volleys, whereas the Japanese fleet scored hits with 15 percent of its own shots. Much of the ammunition used by both sides was faulty.

By 1898, after the Sino–Japanese War, the Beiyang Fleet was rebuilt to a strength of 11 ships, of which 3 were German-built cruisers of 3,400 tons. The Chinese navy never became more than a coastal defense force, and its officers were often more interested in obtaining a Confucian-degree administrative position on land than continuing in naval service.

REFERENCES

Bernard Brodie, *Seapower in the Machine Age* (Princeton: Princeton University Press, 1941); Ralph L. Powell, *The Rise of Chinese Military Power, 1895–1912* (Princeton: Princeton University Press, 1955); John L. Rawlinson, *China's Struggle for Naval Development, 1839–1895* (Cambridge: Harvard University Press, 1967).

BO YIBO (1908–1996) Bo Yibo was born in 1908 in Ding Xiang County, Shanxi Province. He graduated from normal school in Taiyuan, the capital of Shanxi, and attended Beijing University. As a student of Beijing University he joined the Communist Party in 1925 and was an active student organizer.

After college, Bo Yibo seems to have been heavily involved in political work, organizing Communist Party support in "White areas." These were places outside Communist control where the party concentrated on building support among workers, peasants, and intellectuals. He was arrested by the Nationalists in 1932 for organizing an uprising and sentenced to prison for agitation; however, he was released after serving three years of his sentence.

After his release from prison, Bo Yibo returned to Shanxi and founded the ''National Salvation and Sacrifice League'' with Liang Huasheng. The organization was strongly anti-Japanese but was a front for Communist activity. It grew by 1936 to a strength of 100,000 people. The military career of Bo Yibo appears to have started in 1936, with the use of his forces to get General Wang Ruofei out of a prison run by the **warlord** of Shanxi Province, **Yan Xishan**. By 1937, Bo was commander of the ''Dare-to-Die-Corps'' (Juesidui) in southeast Shanxi Province. The ''Dare-to-Die-Corps'' was a guerrilla unit that operated against the Japanese, particularly in southern Shanxi after Japanese forces occupied Taiyuan, and cooperated closely with the Eighth Route Army's 129th Division. In 1939, the ''Dare-to-Die-Corps'' joined the Eighth Route Army, and Bo Yibo was made a division commander in the Shanxi–Hebei–Shandong–Henan Border Region. He later commanded the Taiyuan Military Subdistrict, encompassing areas in southern Shanxi Province from 1939 to 1945. During the **Civil War** period (1946–1949), Bo Yibo was vice chairman of the Shanxi–Hebei–Shandong–Henan Border Region civil government and concurrently the deputy political commissar for the area from 1946 to 1948. As military forces in the Shanxi area shifted north to drive the Nationalists out of Manchuria in 1948, Bo Yibo was appointed political commissar of the North China Military Region.

Bo Yibo was generally associated with the economic management and reform of China after the establishment of the People's Republic of China (PRC) in 1949. Beginning with the early days of the **Cultural Revolution** in 1966, Bo was attacked as a ''counterrevolutionary revisionist'' and was purged from the party work and political posts. At that time he was a vice premier and chairman of the State Economic Commission. After 1978, when **Deng Xiaoping** was rehabilitated and restored his political rights, Bo Yibo also was rehabilitated politically and returned to government and party work. He served as the minister of the state machine-building industry from 1980 to 1982, was made a state counselor from 1982 to 1983, and from 1982 to 1988 served as a vice minister on the State Commission for Restructuring the Economy. He was one of those who supported the opening of China to the outside world, one of the main policies associated with Deng Xiaoping. He also supported Deng on the use of the military to end the pro-democracy demonstration in 1989 and defended the **Tiananmen Square Massacre**.

REFERENCES

Wolfgang Bartke, *Biographical Dictionary and Analysis of China's Party Leadership, 1922–1988* (New York: K. G. Saur, 1990); Xiao Chaoran, *Zhongguo Dangshi Jianming Cidian* [A Concise Dictionary of the Chinese Communist Party's History] (Beijing: PLA Press, 1986).

BOXER REBELLION (1900) A popular movement that developed in North China, the Boxer Rebellion was directed against both Manchu officials and foreigners, especially missionaries. The Boxers (Yihequan or Fist of Righteous Indignation) began to organize as a secret society in Shandong to oppose extortion by local officials of the Manchu dynasty. The movement was a reaction to economic depression resulting from heavy flooding and did not have a specific ideology. The Boxers practiced traditional Chinese martial arts and believed that they could not be harmed by the bullets of foreigners.

Originally a collection of secret societies and self-defense groups in Shandong in 1898, the Boxer movement spread northwest toward Beijing. The Boxers sought an end to foreign domination and turned their hatred especially toward Chinese converts to Christianity, who were receiving extra food rations during the famine. By 1899, the Boxers had stolen property from Christian converts and killed a number in the Hebei region. The groups grew more militant and more antiforeign until, by spring 1900, Boxer groups drifted toward Tianjin and Beijing, attacking Chinese Christians who were in possession of objects made by foreigners. A few European citizens were also killed. In early June 1900, the western legations in Beijing were reinforced by 400 troops from Tianjin as a defense against the Boxers. However, after destroying rail lines and the telegraph, the Boxers were able to turn back another contingent of 2,000 foreign troops trying to reach Beijing to further reinforce legations and embassies.

Western forces seized the **Dagu forts** protecting Tianjin on June 17, 1900, to support a buildup of more troops landing to reinforce the legations. In Beijing, meanwhile, the German minister was killed in the street, and the Boxers began a siege of the foreign legation district, the Russian school, and the Catholic churches in the city. On June 21, 1900, the empress Cixi issued a declaration supporting the Boxers and condemning foreign powers for impinging on China's sovereignty and oppressing the Chinese people.

The foreign legations in Beijing set up a defensive perimeter centering around what is now Zhengyilu in Beijing, composed of the old Russian, German, British, Japanese, and American compounds. In Taiyuan, Shanxi Province, 44 foreign men, women, and children were killed by the Manchu governor.

On August 4, 1900, a combined column of foreign expeditionary forces started to fight their way from Tianjin to Beijing. The U.S. Ninth Infantry earned its nickname, ''the Manchus,'' in this action. The regiment's motto, ''Keep up the Fire,'' also derives from the Boxer Rebellion. These were the last words of the regimental commander after he was hit by Chinese fire as the regiment stormed the imperial armory in Tianjin. The silver from the armory was cast into a huge punch bowl with handles in the form of a Chinese dragon and is still in the Ninth Infantry Regiment's headquarters in the United States. The foreign relief force arrived in Beijing on August 14, 1900, where it broke the Boxer siege of the legations. Outside the legation area, about 500 Russians had been killed in the Russian church and school in the northeast corner of the city wall, now the site of the Russian Embassy in Beijing, while Italian and French troops had held out at the Northern Cathedral (Beitang). Another U.S. infantry regiment, the 14th Infantry, played a significant role in this action. Pinned down by fire from the walls of the Forbidden City, the regimental commander asked for a volunteer to climb the wall under fire and attempt to open a gate. Private Titus of the 14th Infantry yelled, "I'll try, sir" and climbed up the wall under fire by the Manchu troops, for which he was later awarded the Medal of Honor. Titus went on to attend West Point and become an army officer, and the 14th Infantry motto became his words outside the Forbidden City. The foreign relief forces razed the Forbidden City, the Summer Palace, and the empress Cixi's villas in the Fragrant Hills west of the city. Cixi, meanwhile, fled to Xi'an (where she remained until January 1902). German troops conducted another punitive campaign until the Boxer protocol was signed in September 1901. Histories of the People's Republic of China usually refer to the Boxer Rebellion foreign relief force as the **Eight Foreign Armies Invasion of China**.

REFERENCES

Mary Hooker, *Behind the Scenes in Peking* (Hong Kong: Oxford University Press, 1987); Joseph E. Sherick, *The Origins of the Boxer Uprising* (Berkeley: University of California Press, 1987); Jonathan D. Spence, *The Search for Modern China* (New York: W. W. Norton, 1990).

BROAD OCEAN MISSILE LAUNCH (May 1980) After a series of short, preliminary deployments by Space Event Support Ships (SESS) and oceanographic research vessels over a period of about a year, the **People's Liberation Army** (PLA) Navy deployed a naval task force consisting of 18 ships, including three destroyers, support ships, and SESSs, to the South and Western Pacific in support of its successful test of a CSSX-4 intercontinental ballistic missile (ICBM). The ICBM splashdown was at about 6,000 kilometers from the Chinese coast, on the equator, about 1,000 kilometers from the Solomon Islands. The Chinese naval task force was on site to observe and retrieve the missile.

The ships replenished at sea, displaying a new capability for underway replenishment by the PLA Navy, and the crews apparently carried out their assignments in a professional manner. After splashdown, the Chinese navy retrieved the capsule. This remains the largest Chinese naval deployment to date (1999) and probably illustrates the top end of their naval capability to support a deployment at sea for an extended period. Among the factors limiting naval deployments at great distances for extended periods for the Chinese navy are an inability to make freshwater at sea, a shortage of underway replenishment ships, and a lack of air defense systems of PLA ships.

Vance H. Morrison

REFERENCES

Paul H. B. Godwin, *The Chinese Communist Armed Forces* (Montgomery, AL: Air University Press, 1988); David G. Muller, Jr., *China as a Maritime Power* (Boulder, CO: Westview Press, 1983); Sun Zhen, ed., *PLA Forces* (Hong Kong: CONMILIT Press, 1986).

BURGEVINE, HENRY A. (1836–1865) Henry Burgevine was born in North Carolina and was the second-in-command to **Frederick Townsend Ward** in the **Ever-Victorious Army** against the **Taiping Rebels**. Burgevine's father was French and taught the language in Chapel Hill, North Carolina. Burgevine's family was abandoned by his father, and his mother moved to Washington, D.C., where Henry served as a Senate page. While in Washington, Burgevine met Anson Burlingame, who was subsequently ambassador to China. In search of adventure, at the age of 19, Burgevine fought in the Crimea as a private in the French army, returning to Washington in 1856. He may have met Frederick Townsend Ward in Crimea. Burgevine later relocated to New York, where he decided to accompany Ward to China in 1859. He arrived in Shanghai on October 18, 1859, and assisted Ward in organizing the **Foreign Arms Corps**, serving at different times as its deputy and its commander. When the Ever-Victorious Army was created by Qing dynasty edict out of the Foreign Arms Corps, Burgevine served as the second-in-command to Ward. After Ward's death in 1862, Burgevine's earlier affiliation with Anson Burlingame assured his appointment to replace Ward as the army commander. However, on January 4, 1863, Burgevine struck a Shanghai banker in an argument over pay for the army and took 40,000 Chinese dollars. Burgevine was dismissed as army commander and disappeared, with a price on his head under threat of execution.

Burgevine was later captured and arrested by army authorities on May 13, 1865. While under escort to Suzhou from Fujian, he was reportedly drowned. However, when his body was found, it allegedly had a large portion of flesh removed from it by flaying.

REFERENCES

Caleb Carr, *The Devil Soldier; The Story of Frederick Townsend Ward* (New York: Random House, 1992); H. B. Morse, *In the Days of the Taipings: An Historical Retrospect* (Salem, MA: Essex Institute, 1927); Ssu-yu Teng, *The Taipei Rebellion and the Western Powers* (Oxford: Clarendon, 1971).

BURMA ROAD In September 1931, following up on the defeat of local Chinese forces in Manchuria after the **Mukden Incident** (see **Nine One Eight Incident**), Japan invaded and occupied all of Manchuria, turning it into the puppet republic of **Manchukuo**. Japanese forces later used the **Marco Polo Bridge Incident** on July 7, 1937, as an excuse to invade and occupy large parts of the Chinese mainland. Japan's attacks forced Chinese industry to move inland, away from the coasts, and cut off coastal commerce and lines of communication that would support the Nationalist government and a war effort against Japan.

China depended on three major supply lines to bring in aid and supplies for the war against the Japanese military in the early part of **World War II**. Across the Pacific Ocean, supplies could reach China over rail lines leading from Haiphong, Vietnam (Indochina), crossing through Hanoi, and branching to reach Nanning, Guangxi Province, in southeast China, and Kunming, Yunnan Province, in southwest China. From the northwest, through the Soviet Union, supplies moved into China from the trans-Siberian railroad and Turkestan, where they were then moved by road through Xinjiang to Lanzhou, Gansu Province. From Lanzhou, supplies could be moved by rail through Sichuan Province to southwest China, where the Nationalists had moved the government to Chongqing. The third main supply route (MSR) for war materials destined to support the Chinese war effort against Japan ran from Rangoon, Burma, where material entered at the port, to Lashio, Burma, by rail. Finally, at the end of the rail line in Lashio, the supplies flowed into southwest China along the 700-mile Burma Road to Kunming.

With the assistance of its Axis ally Germany, after Germany occupied France in June 1940, Japan pressured the Vichy French government, which controlled Vietnam and Indochina, to close the Haiphong–Kunming and Haiphong–Nanning rail links to China. This shut down one of the three main MSRs available to China. Then, when Germany attacked the Soviet Union in June 1941, all Soviet supplies coming into China were diverted to the Eastern Front to support the Soviet war effort against the Nazis. The effect was that, even though the MSR was open, no matériel flowed from the Soviet Union to China from the far West. The consequence of these two German military campaigns was that China was left with only one MSR available to transport war matériel—the Burma Road.

The basic plan to construct a road and rail link between Kunming and Rangoon as a means to develop southwest China was conceived by Sun Yat-sen at the establishment of the Republic of China in 1911. By 1938, after the Japanese main attack into China closed coastal commerce, the highway between Kunming and Lashio, Burma, was still not complete. Between early 1938 and late 1939, however, a workforce of over 200,000 Chinese, working mostly by hand, carved a usable road into 688 miles of mountainous slopes and valleys. Because of the U.S. neutrality act, American aid to China had to come in from ''nongovernmental sources.'' The Nationalist government, therefore, established the Southwest Transport Corporation to manage the road and the supply line through it that began to flow from the United States. The average transit time on the road from Lashio, Burma, to Kunming, China, was about five days.

When the U.S. Congress approved Lend-Lease aid for China in March 1941, American transport specialists working in China studied how to improve the traffic flow on the road. The foremost of these specialists were David Arnstein, a trucking expert from Chicago, and John Baker, who already had considerable experience working in China in the transportation industry. However, Japanese air attacks on the Burma Road proved to be a significant factor limiting the resupply effort. To defend against the Japanese air forces, the Nationalist government recruited U.S. aviators, who formed the **Flying Tigers** (formally called the American Volunteer Group, or AVG), led by **Claire Lee Chennault**. For the entire period of World War II, the Allied war effort in the China–Burma–India Theater depended on maintaining a free flow of traffic on the Burma Road.

The flow of supplies on the road was also supplemented by flights over the "Hump," which crossed the mountains between Burma and China.

REFERENCES

Charles R. Bond, *A Flying Tiger's Diary* (College Station: Texas A&M University Press, 1984); Eric R. Craine, *Burma Roadsters* (Tucson, AZ: Western Research, 1992); Lincoln Li, *The Japanese Army in North China, 1937–1941: Problems of Political and Economic Control* (London: Oxford University Press, 1975); F. F. Liu, *A Military History of Modern China, 1924–1949* (Princeton: Princeton University Press, 1956); Charles F. Romanus, *Time Runs Out in the CBI* (Washington, DC: Office of the Chief of Military History, Department of the Army, 1959); Barbara W. Tuchman, *Stillwell and the Americal Experience in China, 1911–45* (New York: Macmillan, 1971).

C

CANTON COUP On March 20, 1926, **Chiang Kai-shek**, the military commander of Canton (Guangzhou), acted against the Chinese Communist Party (CCP) military cadre in the Nationalist Army. Using cadets from the **Whampoa Military Academy** and Guomindang units, Chiang arrested Communist political commissars and groups of their Soviet military advisers at the Whampoa Military Academy. In addition, Chiang's forces disarmed the "Workers' Guard," a paramilitary cadre of the CCP.

The head of the Soviet Military Mission in Canton at the time, General Victor Rogacheff, fled to Beijing. Two other Russian advisers were arrested. They were General Galen (Vassily K. Bluecher), the Guomindang chief military consultant, and Mikhail Borodin, who was in charge of reorganizing the Guomindang along Leninist lines. (Borodin was so successful that the Guomindang and the Nationalist government remained essentially a Leninist organization through the 1990s).

To carry out the coup, Chiang allied with graduates of the Japanese Military Academy who served at Whampoa and with graduates of the Baoding and Yunnan Military Academies. Their goal was to end the role of the Soviets in the Guomindang and to suppress the Communists in the Nationalist Army. The Guomindang leaders were alarmed that the Communist commander and Soviet naval adviser of the warship *Zhongshan* had moved the vessel to Canton to support Communist uprisings in the area. Realizing that the CCP and the Soviet Comintern advisers were secretly plotting against the Nationalist government, Chiang acted to arrest the Communists. The CCP claimed that there was no plot to which Chiang needed to respond, charging that Chiang had acted solely to seize control of the Whampoa Military Academy and to end the influence of Wang Jingwei, the CCP senior representative at Whampoa. See **Nanchang Uprising** and **Autumn Harvest Uprisings**.

REFERENCES

Arif Dirlik, *The Origins of the Chinese Communist Party* (London: Oxford University Press, 1989); F. F. Liu, *A Military History of Modern China: 1924–1949* (Princeton: Princeton University Press, 1956); Michael Y. L. Luk, *The Origins of Chinese Bolshevism, 1920–1928* (London: Oxford University Press, 1990); ZGRMJFJZSTB.

CAO GANGCHUAN (1935–) Lieutenant General Cao was one of the deputy chiefs of the General Staff Department (GSD) of the **People's Liberation Army** (PLA) until 1996. At that time he was responsible for oversight of weapons development, production, and sales, as well as logistical matters. He was later promoted to director of the **Commission of Science, Technology, and Industry for National Defense** (COSTIND), responsible for all military research, acquisition, and development for China and China's space program. In a reorganization of military procurement in July–August 1998, Cao was promoted to general and became the first director of the fourth General Department of the PLA, the General Armaments Department. In this capacity he took over responsibility for all research, development, and procurement of weapons and equipment for the PLA.

Cao Gangchuan was born in December 1935 in Wugang, Henan Province. After entering the PLA, he studied at the Third Artillery Ordinance Technical School in 1954. Cao studied at Russian-language school from 1956 to 1957 and at the Military Engineering College of the Soviet Artillery Corps the same year. Cao later graduated from Advanced Military Engineering School, Soviet Artillery Corps, in 1963. He served as assistant at the Ordnance Department, Military Equipment Department, PLA General Logistics Department from 1963 to 1975. Cao also served in leadership positions in the Military Affairs Department of the General Staff Department between 1979 and 1990. From 1990 to 1992, Cao was in charge of arms sales abroad and of weapons purchases as the director, Office of Military Trade of the Central Military Commission.

REFERENCE

Liao Gailong and Fan Yuan, eds., *Who's Who in China: Current Leaders* (Beijing: Foreign Languages Press, 1994).

CARLSON, EVANS F. Evans F. Carlson, a U.S. Marine Corps officer, was the assistant naval attaché in China during the early phases of World War II. He was a captain at that time and had served an earlier tour in China from 1927 through 1929 as an enlisted marine and in 1933–1935 as a language student in Beijing and Shanghai. Carlson also commanded the Marine Security Guard Detachment at President Roosevelt's personal retreat in Warm Springs, Georgia, and corresponded regularly with Roosevelt. In 1937, in his capacity as a naval attaché, Carlson traveled to **Yan'an** to observe how the Communist **Eighth Route Army** was organized and trained. Carlson came away impressed with the Communist emphasis on egalitarianism and teamwork, as well as with their emphasis on honesty and sharing hardship. Later, when he commanded the U.S. Marine Corps' First Raider Battalion, he used a slogan that he had learned from Chinese, "Gung Ho" (*gong he* in Pinyin transliteration), or "Work Together." Actually, the pronunciation today in Marine Corps parlance is not the same as the original Chinese but is an American-accented rendering of the German-language-based transliteration system in practice at the time Carlson studied in China.

REFERENCES

Evans Fordyce Carlson, *The Chinese Army* (New York: American Council, Institute of Pacific Relations, 1940); Barbara W. Tuchman, *Stilwell and the American Experience in China, 1911–45* (New York: Macmillan, 1971).

CENTRAL MILITARY COMMISSION OF THE CHINESE COMMUNIST PARTY The Central Military Commission (CMC) is the highest military policy- and decision-making organization in China. Its origins go back to the Military Affairs Department of the Chinese Communist Party Central Committee, which was formed in 1925. **Zhang Guotao** was the first director of the Military Affairs Department, but he was replaced by **Zhou Enlai** in September 1926. By 1927, the name of the organization was changed to the Central Military Commission (Zhongyang Junshi Weiyuanhui). Zhou Enlai was appointed the secretary-general. Zhou's title, however, was later changed to "chairman." The CMC moved from Shanghai to the Jiangxi Soviet Base Area in 1932. The chairman became **Xiang Ying**, with **Zhu De** and **Mao Zedong** appointed vice chairmen. By 1934, Zhou Enlai once more became chairman. Later, Zhou was succeeded by Mao Zedong when, at the **Zunyi Conference** on the **Long March**, Mao gained the title, which he kept for 41 years.

Although there is a Ministry of National Defense under the State Council of China, the CMC still has the power to direct the policies and operations of the **People's Liberation Army** (PLA). This is part of the Communist Party's tradition that "the Party controls the gun." The secretary-general of the Communist Party, Jiang Zemin in 1996, continued to concurrently hold the position of chairman of the Central Military Commission.

REFERENCES

Jaques Guillermaz, *A History of the Chinese Communist Party, 1921–1949* (New York: Rand House, 1972); Warren Kuo, *Analytical History of the Chinese Communist Party* (Taipei: Institute of International Relations, 1968); Mao Zedong, *Selected Works of Mao Tse-Tung*, vol. 1, "Struggle in the Chingkang Mountains," pp. 73–104 (Beijing: Foreign Languages Press, 1975); William W. Whitson with Chen-Hsia Huang, *The Chinese High Command* (New York: Praeger, 1973).

CHANG HSUE-LIANG. See ZHANG XUELIANG

CHEN GENG (1903–1961) Chen Geng was one of the 10 **People's Liberation Army** (PLA) officers promoted to general in 1955. He was born in XiangXiang County, Hunan Province, in 1903, where his father was a landowner. From 1922 to 1923, he attended a military school in Guangzhou and was a graduate of the first class at the **Whampoa Military Academy**, 1925. He took part in the **Nanchang Uprising**. Later, during the **Autumn Harvest Uprisings**, Chen Geng was also instrumental in organizing units for the **Huangan-Macheng Uprising**.

Chen Geng commanded the 38th Regiment, 13th Division of the **Forth Front Army** and later the Army's 12th Division during 1931 and 1932. In August 1932, as the Nationalist **Encirclement Campaigns** surrounded Fourth Front Army forces in the **Eyuwan Soviet** area, Chen's unit was among the forces that took part in the Fourth Front Army's **Long March** west to Sichuan Province. He was wounded in the Battle of Xiangyang during the march and had to be left behind with local peasants. Chen Geng was moved to Shanghai, where he received medical care and recovered from his wounds. He returned to be commandant of the **Red Army College** in Ruijin and then took part in the Long March with the First Front Army. From 1937 to 1947, Chen was commander of the 386th Brigade, 129th Division, of the Eighth Route Army in the Taiyue Military District (in the Shanxi-Heibei-Henan area). He worked closely with **He Long**'s guerrilla forces in the area and concurrently served, after 1940, as commander of the Taiyue Military District.

In 1947, during the **Civil War**, Chen Geng took part in the campaign to secure the Central China Plains, commanding the Fourth Corps of the Central Plains Field Army. He fought in the Dabie Mountains (**Dabieshan**) area against the Nationalists in familiar terrain around the old Eyuwan Soviet. In November 1948, Chen's leadership was instrumental in the Battle of Xuzhou in the **Huai-Hai Campaign**. His 4th Corps then swung south, marching on the city of Guangzhou, which he seized on October 26, 1949, as commander of the 4th Army Corps, **Second Field Army**.

Chen Geng then continued to march west, into Yunnan Province, where he was instrumental in driving forces from the Nationalist 93d and 193d Divisions out of China and into Burma. He stayed in Yunnan as second secretary of the Yunnan Province Communist Party and the chairman of the Kunming Military Control Commission until 1951. During the **Korean War**, Chen Geng was commander of the Third Army and a deputy commander of the **Chinese People's Volunteers**.

In 1954, Chen was made a deputy chief of the General Staff Department of the People's Liberation Army. He was among the officers appointed to the rank of general in 1955. He also served as a Chinese Communist Party member, a member of the State Science Planning Commission, and a vice minister of National Defense. In 1960, Chen Geng was appointed commander of the Beijing Military region. Chen Geng was identified throughout his military career with the development of a regular, professional army that carried out infantry maneuver warfare. He understood the factors of time, space, and movement in a campaign and used his political commissars to organize militia and logistics support to integrate **People's War** into his plans. He died on March 16, 1961.

REFERENCES

Wolfgang Bartke, *Biographical Dictionary and Analysis of China's Party Leadership, 1922–1988* (New York: K. G. Saur, 1990); Mu Xin, *Chen Geng Tongzhi zai Shanghai: Zai Zhongyang Teke de Douzheng Jingli* (Beijing: Wenhua Lishi Chubanshe, 1980); William W. Whitson with Chen-Hsia Huang, *The Chinese High Command* (New York: Praeger, 1973).

CHEN XILIAN (1913–) Chen Xilian was closely identified with both **Mao Zedong** and **Lin Biao**. He played a central role in the power struggles in the early formation of the **Eighth Route Army** during and after the **Long March**. In subsequent years he developed a strong power base among orthodox Marxists, and for the period 1959 through 1974, Chen was the commander of the strategically important Shenyang Military Region in Manchuria. In January 1974, Chen became commander of the Beijing Military Region until he was removed from his post by **Deng Xiaoping** and the Central Committee in February 1980.

Chen Xilian was born in 1913 in Huangan, Hubei Province. During the Communist **Autumn Harvest Uprisings** of 1926, he became a member of a "young arsonist group" in East Hubei. In 1930, at the age of 17, Chen joined a local Communist group that called itself the "Peasant Self-Defense Army" and that was later integrated into the Communist forces of the Fourth Front Army. In autumn 1931 Chen joined the **Jinggangshan Revolutionary Base**, where he was trained under Lin Biao. In 1933 Chen was made political commissar of the 263d Regiment of the **Fourth Front Army** and became commander of the 88th Division at the age of only 20. During the Long March, Chen's forces were subordinate to **Zhang Guotao** and had separated themselves from the main forces commanded by Mao Zedong. Not until 1936 did Chen join the main Communist forces in Shaanxi. Once there, he underwent further military training at the **Anti-Japanese Military and Political College**, again studying under Lin Biao. In 1940 Chen commanded the 385th Brigade, which took part in the **Hundred Regiments Campaign**. In 1944, Chen was appointed deputy for the **People's Liberation Army** to the first National People's Congress; he was also appointed member of the National Defense Council (where he continued to serve until approximately 1967). From 1946 to 1949 Chen served in the **Second Field Army** commanded by **Liu Bocheng**. In 1946 he commanded the Third Column, Shanxi–Hebei–Shandong–Henan Military Region, and in October of that year became commander of the West Anhui Military Region. In October 1948, during the decisive **Huai-Hai Campaign**, Chen commanded the Army Group West in the Battle of Xuzhou-Pangbu. In February 1949 Chen was made commander of the Third Army Corps, Second Field Army, which in November of that year liberated Chongqing.

From 1951 through 1959, Chen served as commander of the artillery forces of the People's Liberation Army. In 1959 Chen was appointed commander of the Shenyang Military Region (he held that post during the 1969 Sino–Soviet border conflict).

In 1975 Chen returned to Beijing, where he served as commander of the Beijing Military region until 1980.

In February 1980, probably as a result of conflict with Deng Xiaoping, he was removed from all party and state posts at the fifth plenum of the CCP Central Committee.

REFERENCE

Wolfgang Bartke, *Who's Who in the People's Republic of China* (Armonk, NY: M. E. Sharpe, 1981).

CHEN YI (1901–1961) Chen Yi was a deputy commander of the **Chinese People's Volunteers**, the **People's Liberation Army** (PLA) volunteer forces in Korea, and is a former foreign minister of China. He was born in 1901 in Sichuan Province but attended both Shanghai University and Peking College of Law and Commerce. He studied in France at the same time as **Zhou Enlai, Nie Rongzhen**, and **Deng Xiaoping** but was deported to China in 1921. After the **Nanchang Uprising** in 1927, Chen Yi served as a political commissar in the same regiment as **Lin Biao** and **Su Yu**. Later, in 1928, Chen commanded the 12th Division of the Red 4th Corps.

During the **Long March**, Chen Yi remained behind in the Jiangxi area, where he carried out guerrilla operations in East China for three years. After the **New Fourth Army Incident**, Chen Yi's troops survived, reorganized, and continued to operate in East China, where he conducted a campaign with Liu Shaoqi and Su Yu, mobilizing peasants against the Japanese through 1945.

During the **Civil War** (1946–1949), Chen Yi commanded the Shandong and Central China Field Armies. He conducted successful campaigns against Nationalist (KMT) lines of communication on the Lung-Hai Railroad and took part in the **Huai-Hai Campaign**. Chen was later appointed mayor of Shanghai and chairman of the Shanghai Military Control Committee.

From 1954 to 1968, Chen Yi was China's foreign minister. However, in 1969, he appeared in public far less frequently and also became a victim of the **Cultural Revolution**. He sank into relative obscurity until his death in 1972.

REFERENCES

John W. Garver, *Foreign Relations of the People's Republic of China* (Englewood Cliffs, NJ: Prentice-Hall, 1993); William W. Whitson with Chen-Hsia Huang, *The Chinese High Command* (New York: Praeger, 1973).

CHENBAO ISLAND INCIDENT. See ZHENBAO ISLAND CLASH

CHENNAULT, CLAIRE LEE (1890–1958) Lieutenant General Claire Lee Chennault led the **Flying Tigers** (also known officially as the American Volunteer Group, or AVG) during the early part of **World War II** in China, before the United States entered the war. The Flying Tigers was a group of primarily American, volunteer pilots who flew combat and transport aircraft for the Chinese armed forces in the early part of China's **Anti-Japanese War**.

Claire Lee Chennault was born in Commerce, Texas, on September 6, 1890. He grew up in Franklin Parish, Louisiana, where his father was a cotton farmer and locally elected sheriff. Chennault attended Louisiana State University and Louisiana State Normal School, from which he graduated with a teaching degree. After graduation from college, Chennault taught high school in his hometown, later accepted a teaching position at a business college in New Orleans, and also taught physical education in Ohio. He married the former Nellie Thompson, a Louisiana native, with whom he had four children.

Chennault entered the U.S. Army near the end of World War I and underwent officer training at Fort Benjamin Harrison, Indiana. He was initially commissioned as a reserve officer, a second lieutenant of infantry but later transferred to the Signal Corps. Chennault was trained as an army aviator and became a rated military pilot on April 7, 1919. With the end of World War I and the ensuring drawdown in the size of the military, Chennault was discharged from the army on April 9, 1920, only a year after finishing his pilot training. After his discharge from the army, Chennault attempted to farm cotton in Louisiana but quickly applied for a commission as an active-duty army officer in the newly organized Army Air Service, in which he was commissioned a regular officer on September 14, 1920.

During the years between the two wars, Chennault served in a variety of flying assignments, including on the Mexican border in support of patrolling U.S. Army infantry units. He also excelled in aerobatics and was a highly successful pilot of pursuit aircraft. After promotion to captain, on April 12, 1929, he was assigned primarily as a pursuit instructor. Chennault served the remainder of his active-duty army career as a flight instructor at Maxwell Field (now Maxwell Air Force Base), Montgomery, Alabama, until his retirement from the Army Air Corps in 1937.

Because Chennault had made a name for himself in aviation circles as an expert in aerobatics and air combat tactics, he was sought after by U.S. aviation companies after his retirement. The Soviet air force also tried to hire him as an adviser; however, he refused that employment. Chennault accepted an offer in summer 1937, after the Japanese armed forces attacked China, to become an aviation adviser to the Chinese armed forces, then under the leadership of Generalissimo **Chiang Kai-shek**. When Chennault arrived in China, the Chinese air force comprised about 600 aircraft, some from the United States; some from Germany, which had provided considerable aid to the Nationalist government; and some from Italy. In fact, Italian aviation firms were the strongest competition for U.S. industry for aircraft sales to China.

For a number of years prior to Chennault's arrival in China, dating to early 1932, an unofficial U.S. air mission had functioned in that country. The Chinese had already been under attack by the Japanese in Manchuria since the **Mukden Incident (Nine One Eight (918) Incident)** of September 18, 1931, and were working to develop their armed forces. A former American Army Air Corps colonel, John H. Lovett, had established an aviation training school in Hangzhou, Zhejiang Province, which helped to develop China as the world's largest export market for American aviation equipment at that time. Before Colonel Lovett's return to the United States at the expiration of his own contract in 1935, exports of U.S. aircraft and equipment to China had reached over $9 million.

After his arrival in China in 1937, General Chennault developed a close and carefully cultivated relationship with Chiang Kai-shek's wife, American-educated Soong Mei-ling, and her brother, **T. V. Soong**, who functioned as an emissary to the United States for Generalissimo Chiang during World War II. During the period before the war when he served as an adviser to the Chinese air force, Chennault concentrated on refining fighter aircraft tactics for the Chinese pilots. He also coordinated a number of notable, although not tactically effective, bombing raids against the Japanese armed forces, which were already at war with China. The most spectacular of these raids was a May 1938 propaganda leaflet drop on the Japanese ports of Sasebo and Fukuoka (the raid was originally intended for Tokyo, but the aircraft were diverted).

In May 1941, the United States began to deliver Lend-Lease supplies to China over the **Burma Road**. Lend-Lease aid had been extended to China in March 1941, and the aerial defense of the Burma Road, as China's remaining supply line from the west, became critically important to the war effort. Chennault organized the "American Volunteer Group," or Flying Tigers, with the tacit assent of the U.S. government. From offices in the United States, he recruited pilots to fly against the Japanese. From October 1941 to July 5, 1942, Chennault led the American Volunteer Group against the Japanese air force. He held the rank of colonel in the Chinese air force at that time. After the United States formally entered the war, and the mission of General **Joseph Stilwell** was sent to China, Chennault turned into one of the strongest adversaries of Stilwell, clashing with him over the primacy of the air war against Japan versus the vigorous prosecution of the ground war. Chennault's strong relationship with Chiang Kai-shek's wife, Soong Mei-ling, and brother-in-law, T. V. Soong, as well as Stilwell's vocal distaste for Chiang as a leader, combined to eventually undermine Stilwell's own effectiveness in China.

On June 28, 1942, Chennault was given command of the China Air Task Force (CATF). This task force was subordinate to the 10th U.S. Air Force, based in India under the command of Major General Clayton Bissell. By the end of World War II, Chennault was a U.S. Army Air Forces major general, with command of the 14th U.S. Air Force in China. Major General Chennault returned to the United States after the surrender of Japan, arriving on August 24,

1945. He soon returned to China in a private capacity to organize the China National Relief and Rehabilitation Administration (CNRAA) Air Transport Service (known as CAT), with 20 C-46 transport aircraft and two C-47s. CAT went on to ferry Nationalist troops into Manchuria to fight the Communist forces, to evacuate the Nationalist Army out of Manchuria when it was taken over by Communist forces, and to assist in the evacuation of the Nationalist government to Taiwan in 1949. It eventually became a U.S. proxy contract air service. Under Chennault's direction, CAT also flew supplies for Far Eastern air forces in the **Korean War** under contract for the U.S. government and also flew supplies into Dien Bien Phu in support of French forces in Indochina. Chennault returned to the United States in July 1958 and died of throat cancer.

REFERENCES

Martha Byrd, *Chennault: Giving Wings to the Tiger* (Tuscaloosa: University of Alabama Press, 1987); Anna Chennault, *Chennault and the Flying Tigers* (New York: P. S. Eriksson, 1963); Claire Lee Chennault, *Way of a Fighter: The Memoirs of Claire Lee Chennault* (New York: G. P. Putnam's Sons, 1949); Charles F. Romanus and Riley Sunderland, *The United States Army in World War Two. The China-Burma-India Theater: Stilwell's Mission to China* (Washington, DC: Government Printing Office, 1953); Roland Sperry, *China through the Eyes of a Tiger* (New York: Pocket Books, 1990).

CHI HAOTIAN (1929–) Chi Haotian served as China's minister of defense from 1993 through 1997, after which he advanced in seniority to become the number 3 man on the **Central Military Commission** while concurrently holding the position of defense minister. From 1987 until becoming defense minister, Chi was the chief of the General Staff Department (GSD) of the **People's Liberation Army** (PLA) (see **General Departments of the PLA**). Chi Haotian was born in July 1929, in Zhaoyuan, Shandong Province. He joined the **Eighth Route Army** in 1944 during the war against Japan. He served in the East China Field Army and as a company political commissar in the **Third Field Army**. Chi took part in a number of critical campaigns in the Anti-Japanese and Revolutionary War (**Civil War**), including the Laiwu, **Huai-Hai**, Yangtse River Crossing, and Shanghai Campaigns. He was awarded the title of "East China's People's Hero" in 1949. During the "War to Resist U.S. Aggression and Aid Korea" in 1950, Chi served as a battalion political commissar of the **Chinese People's Volunteers**. After returning to China, he served as a regimental political department director and later the political commissar for a division. Chi graduated from the PLA Military Academy in 1960. In the critical Beijing Military Region, Chi was the deputy political commissar. Chi Haotian later served as director of the Political Department of PLA GSD Headquarters and political commissar of PLA Jinan Military region. In 1988 Chi became a member of PRC Central Military Commission. As minister of national defense he concurrently held two other positions: member of the PRC Central Military Commission and state councillor. Chi was promoted to general in 1988.

REFERENCE

Liao Gailong and Fan Yuan, eds., *Who's Who in China: Current Leaders* (Beijing: Foreign Languages Press, 1994).

CHIANG KAI-SHEK (JIANG JIESHI) (1887–1975) Chiang Kai-shek, whose name in most Taiwan publications appears in the form of the Wade-Giles transliteration used here, was born near Ningbo, Zhejiang Province, in 1887. He trained at the Baoding Military Academy, south of Beijing, in 1907–1908, as the New Army was formed under the influence of **Yuan Shih-kai**. Chiang also studied in Japan and met Sun Yat-sen in Tokyo, where he joined the Anti-Manchu (Qing) dynasty alliance, the Tongmeng Hui. Chiang associated with other officers from the Shanghai area, and in 1911 he fought in Shanghai in the **Republican Revolution**. He was made a regimental commander and led his troops against Imperial forces in Hangzhou. It is alleged that Chiang personally assassinated a dissident member of the Revolutionary Alliance during the formative period of the republic. Other allegations concerning Chiang's early life indicate that in the late 1910s and early 1920s, Chiang remained in Shanghai, where he was connected with the powerful underground criminal and political organization the "Green Gang." Chiang was alleged to have helped Sun finance his activities through Chiang's own contacts with smugglers, financiers, and local Shanghai military strongmen. In 1923, Chiang Kai-shek was made chief of staff in Sun's military headquarters in Canton (Guangzhou). Chiang was sent to the Soviet Union from September to November 1923, where he studied the Soviet military system. On his return to China, Chiang was elected to the Military Council of the Guomindang Party (KMT, or Nationalist Party), which was organized along Leninist lines. He also was made head of the **Whampoa Military Academy**, where he built the Nationalist Army as a KMT Party Army with Soviet help. After the death of Sun Yat-sen (March 12, 1925), Chiang worked to reduce Communist penetration of the KMT and the army, while retaining a Leninist organization and structure. Chiang led the **Northern Expedition** in 1926, consolidating KMT power over the country. In April 1927, he worked together with his former underworld "Green Gang" associates to destroy the infrastructure of the Chinese Communist Party organization in Shanghai. Chiang led the April 1927 Shanghai Coup, breaking the back of the Communist-controlled labor movement, which allegedly had used assassins to force workers to submit to union control. Chiang later married the sister-in-law of Sun Yat-sen, Soong Mei-ling, the American-educated daughter of a powerful Chinese family. To court Soong Mei-ling, Chiang first had to divorce his wife, by whom he had a son. Chiang also converted to Christianity, at Soong Mei-ling's insistence. Soong Ching-ling, Mei-ling's older sister, was the second wife of Sun Yat-sen. His marriage to Soong Mei-ling linked Chiang Kai-shek to a wealthy and influential Shanghai family that had been Christian for 300 years. Mei-ling's father had been raised and educated in the United States.

In 1928, Chiang was appointed the chairman of the National Military Council of the KMT. As generalissimo of the Nationalist Army he led the second stage of the **Northern Expedition**, consolidating power over Beijing, financed by Soong Mei-ling's brother, T. V. Soong. Through the 1930s Chiang pursued the Communists at each point where they tried to organize their forces. He worked to revitalize the KMT, weakened by the Japanese invasion of Manchuria in 1931. In December 1935, Chiang was appointed president of the Executive Yuan, a position equivalent to prime minister. As a result of the **Xi'an Incident** in December 1936, Chiang Kai-shek grudgingly entered a period of cooperation with the Chinese Communist Party (CCP) against the Japanese.

Despite the agreement to cooperate with the Communists against the Japanese made by Chiang Kai-shek after the Xi'an Incident, neither side really trusted the other or complied. This period is known as the "Second **United Front**." Chiang's subsequent imprisonment of **Zhang Xueliang** was a sign that the Nationalists could not be taken at their word. Nevertheless, **Zhou Enlai** made several trips to Xi'an and Hangzhou, near Nanjing, to negotiate with the Nationalists on the conduct of the war. Several times during these negotiations Zhou met directly with Chiang. However, the negotiations took the same turn as those in the first United Front, in 1926, as Chiang demanded the disbandment of the Red Army and the incorporation of its Communist forces into the Guomindang Army as part of the National Revolutionary Army.

Japanese forces landed in Shanghai in August 1937, after a Japanese Imperial Marine lieutenant was killed in an incident there. After invading and capturing Shanghai, Japanese forces drove toward Nanjing, which at the time was Chiang's headquarters and the capital of China. The Japanese reached the city by December, and the bloody occupation of that city by Japan, the infamous **Rape of Nanjing**, drove Chiang Kai-shek and the Nationalist government west along the Yangtze River to Chongqing.

By 1938, Chiang Kai-shek not only was the head of the political leadership of the Guomindang Party but was also formally made commander in chief of the military forces of China. Fighting a holding action against the Japanese, Chiang focused for some time on the growing Communist forces in the Shaanxi–Ningxia–Gansu border area, seeking to cut off the Communist headquarters at Ya'an. Although Chiang got some financial and military support from the United States, such as the way that the Roosevelt administration permitted the **Flying Tigers** to form and fly for the Chinese, Chiang's forces were essentially on their own until 1941, when the United States entered the war, and an Allied war effort formed. In fact, Chiang was initially made supreme allied commander of the entire China Theater, putting allied military operations in Thailand and Indochina under his control. Meanwhile, Chiang became known for corruption and currency speculation. He was thought to have allowed the four leading families of Nationalist China—the Kungs, the Soongs, the Chens, and the Chiangs themselves—to enrich themselves by currency manipulation and by padding the army payroll. Much of this money came from U.S. Lend-Lease and war loans to China.

Chiang Kai-shek clashed often with the U.S. commander in China, General **Joseph Stilwell**. From the time that he had served as military attaché in Beijing (then Peiping), Stilwell had known both Generalissimo Chiang and Mrs. Chiang. However, Stilwell had no stomach for the corruption he saw as endemic in the Chiang family and the Guomindang. Stilwell's own rival, the commander of U.S. Army Air Forces in China, Major General **Claire Lee Chennault**, got along very well with both Generalissimo and Madame Chiang, creating greater friction among all parties.

Chiang got his biggest ally when **Patrick J. Hurley** was nominated to Congress as American ambassador to China on November 30, 1944 (Hurley actually presented his credentials to Chiang on January 8, 1945). Hurley was a strong anti-Communist whose stated goal on assuming his position was to "sustain Chiang Kai-shek as President of the Chinese Republic and the Generalissimo of the Armies." Although Chiang was not pleased with the secret agreements at Yalta, which gave the Soviets control of parts of Manchuria after the war, he acquiesced because he still needed American aid. The northeast part of China came to haunt Chiang as the war closed, since he had to contend not only with the Soviets there but with a drive by the Communists to control the heavily industrialized region, which was rich in petroleum deposits. Chiang's closest advisers counseled that the "Young Marshal," Zhang Xueliang, who had conspired to kidnap Chiang in Xi'an in 1936, be released from prison to consolidate Nationalist control of Manchuria. These advisers argued that as the former **warlord** of the area, Zhang would command the most loyalty there against the Communists. However, Chiang Kai-shek refused to allow Zhang to be freed.

Although he was advised not to do so by the commander of U.S. forces, General **Wedemeyer**, after the surrender of Japan in August 1945, Chiang attempted to move Guomindang forces into Manchuria. His actions in moving forces north were prompted by the fact that in the **Yalta Agreement** in February 1945 Roosevelt consented to allow the Russians a role and ports in Manchuria. Only after Roosevelt's death, on April 12, 1945, was Chiang finally told of the agreements between the United States and the Soviet Union at the Yalta Conference. Ambassador Hurley reported the Yalta agreements to Chiang on June 15, 1945. The Chinese finally agreed to let the Soviet Union use Port Arthur as a naval base and to declare Dalian (Darien) a free port. In return, Chiang's negotiator in Moscow, foreign minister Wang Shijie, got an agreement from Stalin not to supply material aid to Mao Zedong and the Communists. By the time of the Japanese surrender on August 14, 1945, the Soviets were in control of Manchuria.

On August 28, 1945, Mao Zedong and Chiang Kai-shek finally met in Chongqing. Ambassador Hurley had personally escorted Mao to the Nationalist wartime capital, since Mao would not entrust his own safety to Jiang's guarantees. Mao and Chiang agreed to cooperate and to cease hostilities, but both the Nationalist and the Communist armies maneuvered to get to Manchuria, each side attempting to block the other's progress. The Russians allowed the Communists to set up local administrative headquarters in Manchuria and Inner Mongolia, while Chiang sent his own force to occupy the major cities of South and Central China, Nanjing, Shanghai, Wuhan, and Beijing. An immediate battle in Manchuria was stopped by pressure from General Marshall (see the **Marshall Mission**), who pressured Chiang to stop his advance north of Beijing. Meanwhile, Marshall got two U.S. Marine divisions into China. On October 6, 1945, American naval forces assisted Jiang in this effort, attempting to land Chinese troops at Port Arthur (Dalian). The Soviets, who already occupied the area, refused to allow the Nationalist troops to disembark, giving the Communists a chance to build their forces in Manchuria.

Mao and Chiang drafted a joint communiquè on October 10, 1945, the national day of Republican China, which agreed to cooperation between the two parties (Communist and Nationalist) to achieve national reconstruction. However, the Nationalist government refused to recognize Communist control of the "liberated areas." Mao agreed to reduce his forces from a strength of between 80 and 100 divisions to a strength of about 24 divisions and to station these Communist divisions in zones assigned by Chiang. However, despite these agreements, Nationalist and Communist troops started skirmishing. By November 27, 1945, Ambassador Hurley resigned, and General George Marshall became President Truman's special representative in China. A cease-fire agreement between Mao and Jiang was reached on January 10, 1946, with hostilities to cease on January 13. But the Communists immediately took advantage of the cease-fire to move more troops into Manchuria.

At least through summer 1946, it appears that Stalin had kept his agreement with Chiang and did not provide significant material to the Communists. In fact, using Chiang Kai-shek's eldest son, Chiang Ching-kuo, as an intermediary, Stalin invited Chiang to Moscow in 1946. The son had been sent to Moscow as a student by Chiang Kai-shek and had married a Soviet wife. The father refused Stalin's invitation to visit Moscow, however; despite this fact, General Marshall encouraged Chiang to travel to Moscow to discuss a peaceful resolution to the Nationalist–Communist dispute. By the end of the year, **Li Lisan**, who had been in Moscow, was sent back to China to coordinate the transfer of 100,000 North Korean troops into **Lin Biao**'s forces. As the situation in Manchuria deteriorated, Marshall requested his own recall to the United States, and he returned to become secretary of state on January 7, 1947.

Through the first half of 1948, the Communists secured their foothold in Manchuria and fought their way into Central China. Chiang traveled to Taiyuan, Shanxi Province, in July 1948, hoping to boost the morale of his troops, which concentrated on holding strong points and populated areas. All of this was to no avail, as the Nationalist forces slowly collapsed. The Communist forces continued to gain ground, crossing the Yangtze River on April 21, 1949. They captured Shanghai on May 27 and continued to drive south. In the west and northwest, meanwhile, the **First Field Army** captured Lanzhou, Xining, and Yinchuan. Seeing the end, Chiang finally fled to Taiwan, reaching the island on December 12, 1949. He retained his best forces, principally those trained by the United States, which fled with him. Chiang had some 800,000 troops, 600 airplanes, and 70 ships with him, plus 40,000 marines.

For the next 26 years, until his death from a heart attack at the age of 87 on April 5, 1975, Chiang Kai-shek remained president of the Republic of China and vowed to recapture the mainland. He kept the island of Taiwan under martial law the whole time and created a number of crises that he hoped would weaken mainland China. His son from his first marriage, Chiang Ching-kuo (Jiang Jingguo), who was educated in the Soviet Union, inherited the presidency of Taiwan (and the Republic of China). Chiang Ching-kuo died in 1988.

REFERENCES

Donald Jordan, *The Northern Expedition* (Honolulu: University of Hawaii Press, 1976); F. F. Liu, *A Military History of Modern China* (Princeton: Princeton University Press, 1956); C. Martin Wilbur, *Sun Yat-sen, Frustrated Patriot* (New York: Columbia University Press, 1976).

CHIHLI CLIQUE. See ZHILI CLIQUE

"CHINESE" GORDON. See GORDON, CHARLES GEORGE

CHINESE PEOPLE'S VOLUNTEERS (CPV) On July 7, 1950, **Mao Zedong** initiated preparations in China for a participation in a potential war on the Korean Peninsula. On October 8, 1950, after some debate, Beijing authorities decided to authorize armed intervention to assist the North Koreans in their war against the Nations (primarily the United States) and the South Korean forces. On behalf of the **Central Military Commission** (CMC), Mao Zedong instructed that the Northeast Border Defense Forces (NBDF) be renamed the Chinese People's Volunteers (CPV). The NBDF was hastily assembled from units of the **Fourth Field Army** and other commands. Intensive training began during late August and early September. Marshal **Peng Dehuai** assumed command of the forces on October 5, 1950. Peng served as both the commander and political commissar and commanded the CPV until the armistice was signed in 1953, although he spent much of 1952–1953 in Beijing.

The CPV secretly crossed into North Korea beginning October 18–19, 1951. Fearing the UN/U.S. forces would destroy the bridges over the Yalu River, severing the international border between China and North Korea, Peng pressed for all units to cross ahead of schedule. On directions from Mao Zedong, the CPV moved only at night to conceal its movements and retain the element of surprise.

In late 1950, the CPV consisted of six armies (the 38th, 39th, 40th, 42d, 50th, and 66th), an artillery command (three artillery divisions), and a logistics command and totaled over 250,000 personnel. The CPV deputy commanders were

Deng Hua, Hong Xuezhi, and Han Xianchu. Xie Fang served as the chief of staff.

A CPV Logistical Command was formally established on May 19, 1951, under the command of **Hong Xuezhi**. Hong reportedly assumed this position reluctantly and only under the condition that he would not be forced to serve in logistics work after the war. His selection proved to be critical in the development of Chinese logistics. He introduced effective organization and discipline over logistics forces and effectively maximized all available resources (including combat forces) to support logistics operations. Through his leadership, Hong transformed the logistics of the CPV into a modern organization that could effectively support defensive and offensive operations for the remainder of the war. In spite of his reluctance to be involved in logistics matters, Hong later became the director of the General Logistics Department of the PLA.

The CPV strength grew throughout 1951 until July 1953, when it consisted of 20 infantry armies (60 divisions), 10 artillery divisions, four tank regiments, and 10 air force divisions. Unit rotations began in 1952. Withdrawal of CPV units began in May 1953 and continued until October 1958, when the last unit, First Army, departed North Korea.

REFERENCES

Chen Jian, *China's Road to the Korean War: The Making of the Sino-American Confrontation* (New York: Columbia University Press, 1994); Jürgen Domes, *Peng Te-huai: The Man and the Image* (Stanford, CA: Stanford University Press, 1985); Alexander L. George, *The Chinese Communist Army in Action: The Korean War and Its Aftermath* (New York: Columbia University Press, 1967); Samuel B. Griffith II, *The Chinese People's Liberation Army* (New York: McGraw-Hill, 1967); Nie Rongzhen, *Inside the Red Star: The Memoirs of Marshal Nie Rongzhen*, trans. Zhong Renyi (Beijing: New World Press, 1988); Peng Dehuai, *Memoirs of a Field Marshal*, trans. Zheng Longpu (Beijing: Foreign Languages Press, 1984); Allen S. Whiting, *China Crosses the Yalu: The Decision to Enter the Korean War* (Stanford, CA: Stanford University Press, 1960); William W. Whitson with Chen-Hsia Huang, *The Chinese High Command: A History of Communist Military Politics, 1927–71* (New York: Praeger, 1973); Zhang Shu Guang, *Mao's Military Romanticism: China and the Korean War, 1950–1953* (Lawrence: University Press of Kansas, 1995).

Susan M. Puska

CHINESE TRAINING COMMAND In response to Japanese advances during **Operation Ichigo**, in fall 1944, Lieutenant General **Wedemeyer** had a small force of about 4,800 American officers and soldiers under his command assigned to various training centers and as advisers to the **Y-Force** and **Z-Force**. Wedemeyer reorganized these advisers into two commands: the Chinese Training Command, which was responsible for running training centers for the Nationalist Army, and the Chinese Combat Command, which was actually an advisory force that functioned as a network down to the regimental level of command in the Nationalist Army.

The Chinese Training Command was led by Brigadier General John W. Middleton. It trained both individual soldiers and the cadre and staff of units and divisions that had special roles. This training structure was centered in Kunming, Yunnan Province, at the end of the **Burma Road**, where resupply was easier, and it was protected from Japanese forces. The center operated a major field artillery training center, which concentrated on teaching Chinese officers and soldiers to use and effectively employ American-supplied artillery, and seven other service schools, concentrating on the logistics structure to provide support for the combat forces of the Nationalist Army.

The United States also began to operate a command and general staff course to school midgrade officers for handling positions on division and corps staffs, a war college to train senior officers for regimental command and in battlefield operational art. Modeled on the U.S. military schools system, the Chinese Training Center opened specialized schools for training troops and officers to use heavy mortars, operated an infantry school and a signal school, and established an English-language training center to train interpreters to assist the American advisers. The effort was short-lived and withdrawn after the surrender of Japan and the opening of the **Civil War** between the Nationalists and the Communists. Many of the installations used by the Training Center are still operational **People's Liberation Army** bases in Yunnan today.

REFERENCES

F. F. Liu, *A Military History of Modern China: 1924–1949* (Princeton: Princeton University Press, 1956); Barbara W. Tuchman, *Stilwell and the American Experience in China, 1911–45* (New York: Macmillan, 1971); Charles F. Romanus and Riley Sutherland, *The United States Army in World War Two. The China-Burma-India Theater: Time Runs Out in CBI* (Washington, DC: Government Printing Office, 1959); Albert C. Wedemeyer, *Wedemeyer Reports!* (New York: Henry Holt, 1958).

CHINESE WORKERS AND PEASANTS RED ARMY The Chinese Workers and Peasants Red Army (*Zhongguo Gong Nong Hong Jun*) is the forerunner to the **People's Liberation Army**. Originally formed in late 1926–1927 in the border areas of Hunan and Jiangxi Provinces from local armed organizations of peasants and workers, the local organizations worked with regular forces that had been part of the National Army to conduct the **Nanchang Uprising** and the **Autumn Harvest Uprisings**. They formed the **First Front Army** in the **Jinggangshan Revolutionary Base Area**.

REFERENCES

Mao Zedong, *Selected Works of Mao Zedong*, vol. 1 (Beijing: Foreign Languages Press, 1975); Xiao Zhaoran, *Zhonggong Dangshi Jianming Cidian* [A Concise Dictionary of the History of the Chinese Communist Party] (Beijing: Jiefangjun Chubanshe, 1987); ZGDBKQS.

CHO NAM GI (Zhao Nanqi) (1926–) Cho Nam Gi was the chief of **People's Liberation Army** (PLA) General Logistics Department (GLD) before his retirement. An ethnic Korean, Cho was born in April 1926 in Yongji, Jilin Province, and studied at the Jilin Branch of the Northeast China Military and Political Science University from 1945 to 1946. His name is romanized in Pinyin as Zhao Nanqi. He served as deputy chief of the General Affairs Section of the Yanbian Prefectural Communist Party Committee during 1948–1949. In 1950 he was a staff officer at the Northeast Military Command Headquarters, when he went to North Korea as a member of the **Chinese People's Volunteers** (CPV) and served in logistics positions in the **Korean War**. He studied at the PLA Logistics Academy 1955–1957. From 1963 to 1984 Cho occupied a series of military and civilian leadership positions in Jilin Province, including director of the Political Department of Jilin Provincial Military District in 1977–1978; first secretary of CPC Yanbian Korean Autonomous Prefectural Committee 1978–1980, and vice governor of Jilin in 1980–1983. Cho was the political commissar of the Jilin Provincial Military District in 1984–1985. In 1987 he served as deputy director and deputy political commissar of PLA General Logistics Department, after which he was appointed director of PLA General Logistics Department in 1988. He served as president of the Academy of Military Science from 1992 until he retired in 1995.

REFERENCE

Liao Gailong and Fan Yuan, eds., *Who's Who in China: Current Leaders* (Beijing: Foreign Languages Press, 1994).

CIVIL WAR (1946–1949) At the end of **World War II** in China, some 1.25 million Japanese troops in Central China and 900,000 Japanese troops in Manchuria surrendered to Chinese forces. The Nationalist Army was about 2.7 million strong, and the Communists' **Eighth Route Army** and **New Fourth Army**, combined, had a strength of about 1 million armed troops.

Lin Biao, exploiting the Soviet attack into Manchuria against the Japanese, occupied much of that area before the Nationalists could act to hold terrain. Meanwhile, in Shanxi Province, the **warlord**-turned-Nationalist **Yan Xishan** accepted the surrender of Japanese forces, only to turn them into his own. He used these former Japanese troops to mount his own defense of Taiyuan against the Communist forces.

In August 1945, talks began in Chongqing between the Nationalists and the Communists, attended by **Mao Zedong** and **Chiang Kai-shek**. U.S. ambassador **Patrick J. Hurley** tried to encourage the two sides to form a coalition government. These talks continued through early fall, and Hurley, in frustration over the lack of progress, resigned as U.S. ambassador on November 26, 1945. **Zhou Enlai** had replaced Mao as the Communist negotiator by that time. After a protracted series of debates in the United States, President Truman sent **General George Marshall** to China in mid-December 1945. Although Marshall got the

Nationalists and Communists to agree to a cease-fire, which was to begin on January 10, 1946, and to hold a political consultative congress in Nanjing on January 11, the truce was one in name only. Communist and Nationalist forces continued to fight throughout China, most severely in the north.

Lin Biao's hold on Manchuria was subject to waves of Guomindang assaults between November 1945 and July 1946. Nationalist forces sent the sixth and 52d Corps into eastern Manchuria, driving Communist forces out of the Shenyang area toward the Yalu River; the focus of the Nationalist effort was to secure the strategic road and rail links from Shenyang south through the passes at Shanhaiguan.

In Hebei, Shanxi, and Shandung, from 1945 to 1947, the forces of **Liu Bocheng** and **Deng Xiaoping** maneuvered against the Nationalists. The field army under **Chen Yi** was almost destroyed in detail by a powerful Nationalist offensive, but Liu Bocheng came to Chen Yi's aid. Liu successfully used a series of ambushes to destroy Nationalist forces by surrounding small garrisons and then ambushing Nationalist reinforcing and relief forces. Liu Bocheng then launched another major series of attacks in a campaign along the Long-Hai railroad to control the Yellow River crossing points.

In June 1947, in Central China, the Communists crossed the Yellow River, opening up a new offensive campaign into Hunan and the Dabieshan Mountains. Through August 1947, the forces of Liu Bocheng and Deng Xiaoping moved through the Central China plains, approaching the **Dabieshan** Mountains and the old **Eyuwan Soviet** area. Chen Yi, meanwhile, continued military pressure against the Nationalists in Shandong and on the Long-Hai railroad. Both sets of forces positioned themselves for the **Huai-Hai Campaign**, the decisive series of engagements beginning in November 1948.

In Manchuria, continuous Communist pressure from Lin Biao's forces conducting the decisive **Liaoning-Shenyang (Liao-Shen) Campaign** drove thousands of refugees south. In July 1948, about 5,000 students who had fled Manchuria massed in Beijing, where they marched on the municipal government. The Nationalist forces blocked their way and, using armored cars and automatic weapons, killed a score of students and wounded another 100. This Beijing Massacre of 1948 increased popular dissatisfaction with the Nationalists. Lin Biao's forces pushed the Nationalists out of the key Manchuria cities of Shenyang and Changchun, forcing a major evacuation of Nationalist forces from Manchuria by sea. Death, desertion, and capture whittled a force of over 400,000 Nationalist troops down to 20,000. Many of the Nationalists switched sides. Some, like Xu Huizi, fought for the Communists; he rose to general after a career that saw him fighting in Korea, against India, and against Vietnam. General Xu Huizi retired as the commandant of the PLA Academy of Military Science in 1997. As the Communist forces grew in strength, they also captured Nationalist and older Japanese weapons and equipment, including tanks, artillery, and vehicles. What was essentially a guerrilla force conducting hit-and-run tactics changed into a conventional army able to conduct maneuver warfare between the Liao-Shen Campaign and the **Beiping-Tianjin (Ping-Jin) Campaign** of November 21, 1948, and January 31, 1949. The three campaigns, the Liao-Shen, the Ping-Jin, and the Huai-Hai, are studied at People's Liberation Army military academies as the decisive campaigns of the Civil War.

By spring 1949, the Nationalists were completing their evacuation from the mainland to Taiwan, a process that started in 1947. Indeed, Taiwan residents were angered at the Nationalist evacuation. When Taiwanese protested in Kaohsiung, Taiwan, in February 1947, Nationalist troops massacred some 40,000 Taiwanese.

Nanjing fell to the Communists in April 1949, and the PLA then swept east to Shanghai and Hangzhou. Another Communist column swept into Wuhan in May 1949, and then forces under Peng Dehuai swept northwest, capturing Xi'an and Lanzhou in August. Lin Biao's forces, having secured Manchuria, swept south from Wuhan, capturing Changsha in August 1949 and Canton (Guangzhou) in October. By November 1949, one month after Mao Zedong declared the establishment of the People's Republic of China on the rostrum of the Forbidden City on October 1, 1949, Lin Biao's troops were defeating the last remnants of the Nationalist forces in Xiamen, Fujian Province.

REFERENCES

Herbert Feis, *The China Tangle: The American Effort in China from Pearl Harbor to the Marshall Mission* (Princeton: Princeton University Press, 1953); Jonathan D. Spence, *The Search for Modern China* (New York: W. W. Norton, 1990); William W. Whitson with Chen-Hsia Huang, *The Chinese High Command* (New York: Praeger, 1973); ZGRMJFJZSJB, 1992; ZGRMJFJZS, vol. 3, *Quan Guo Jiefang Zhanzheng Shi Qi* [The Entire History of the Liberation War Period].

COMMISSION OF SCIENCE, TECHNOLOGY, AND INDUSTRY FOR NATIONAL DEFENSE (COSTIND) The Commission of Science, Technology, and Industry for National Defense, often referred to by the acronym COSTIND, was the most important body in China overseeing China's military–industrial complex between August 1982 and July 1998. COSTIND was established by the State Council of China as an independent commission in August 1982, combining the National Defense Science and Technology Commission, the National Defense Industry Office, and the Science and Technology Equipment Commission of the Communist Party's Central Military Commission. COSTIND supervised and coordinated the numbered, state-owned industrial plants of the various machine-building industries of China and the **Third Line**, or Third Front, industries of the defense industrial base. Some of the COSTIND industries today have evolved into major conglomerates that are either collectively owned or state-owned; others have spun off into joint venture industries with foreign companies.

COSTIND was organized in an attempt to centralize rational decision making on military research, development, acquisition, and production. However, it immediately began to have difficulties and organizational rivalries with the Equipment Department of the General Staff Department of the **People's Liberation Army** (PLA). COSTIND not only exercised oversight of the finances for defense research and production but also managed the Chinese space-launch industry, the missile industry, the satellite industry, and nuclear weapons production. The organization was loaded with the children of senior-level PLA cadre, many of them the children of **Long March** veterans. As a consequence of this nepotism and system of political patronage, COSTIND was rife with corruption.

In July 1998, in an attempt to divest the PLA of its business interests and to put the weapons factories under the control of the civil sector of the state, the General Armaments Department of the PLA was created. This "fourth department" (the other three being the General Staff Department, the General Logistics Department, and the General Political Department) was to direct all weapons development. The factories were put under the supervision of the newly created State Commission of Science, Technology, and Industry for National Defense (SCOSTIND). General **Cao Gangchuan**, who had been the COSTIND chief, was made the new director of the General Armaments Department (GAD). One stated goal of creating the GAD was to assist in ensuring that the **Local War Doctrine** for local wars under high technology conditions was supported with appropriate new weapons.

REFERENCES

John Frankenstein and Bates Gill, "Current and Future Challenges Facing Chinese Defense Industries," *The China Quarterly* (June 1996): 394–427; Bates Gill and Taeho Kim, *China's Arms Acquisitions from Abroad* (Oxford: Oxford University Press, 1995); Bruce Gilley, "Stand-Down Order," *Far Eastern Economic Review* (September 10, 1998); Xie Guang et al., *Dangdai Zhongguo de Guofang Keji Shiye* [The National Defense Science and Technology Efforts of Modern China] (Beijing: Dangdai Zhongguo Chubanshe, 1992); John Lewis and Xue Litai, *China's Strategic Seapower* (Stanford, CA: Stanford University Press, 1994).

CONVENTION OF PEKING (1860). See PEKING CONVENTION

COOPERATION PLANS, U.S. AND YAN'AN COMMUNIST FORCES

In 1944, Guomindang recruitment of forces for the Nationalist Army was so poorly managed, and troops were so badly malnourished that the United States attempted a rapprochement with the Communist forces at **Yan'an**. After General **Stilwell**'s recall and replacement by General **Wedemeyer** in October 1944, Wedemeyer proposed to **Chiang Kai-shek** that, in view of the serious Nationalist losses at the hands of the Japanese during **Operation Ichigo**, greater effort be made to incorporate Communist forces into the war. Wedemeyer suggested that a column of Communist troops, about 5,000 soldiers organized in three regiments, be equipped with American weapons and provided American training. Operating under U.S. command and supervised by 10 U.S. liaison officers, these regiments were to carry out combat operations against the Japanese in areas normally reserved for Nationalist operations. As it was conceived, one aspect of the plan that was designed to make it more palatable to Generalissimo Chiang was that the American presence throughout the force would serve to reassure Chiang that the Communist troops would not operate against the Nationalists. Chiang rejected this concept.

A short time later, in December 1944, while General Wedemeyer was in Burma supervising operations during the Salween Campaign, a second plan was prepared by his chief of staff, Major General Robert B. McClure, and by the head of the **Dixie Mission** (the U.S. Army Military Advisory Group at Yan'an), Colonel **David Barrett**. The second plan, which was prepared by McClure at Wedemeyer's direction and with his knowledge, called for sending U.S. airborne regiments into Communist-held areas to operate behind Japanese lines, destroying Japanese installations and supply lines. Jiang's principal representative to the Americans, **T. V. Soong**, ignored this plan.

A third plan, far more controversial, was then put forth in January 1945 by McClure, Barrett, and officers of the Office of Strategic Services (OSS, the wartime forerunner of the U.S. Central Intelligence Agency). As conceived by OSS officers and Wedemeyer's staff, Americans would work together with Communist guerrillas, in a manner similar to the way OSS officers worked with the French and Yugoslav guerrillas in Europe, to destroy Japanese installations, communications, airfields, and troop concentrations. This plan called for arming and equipping as many as 25,000 Communists and assisting in providing small arms for as many as 100,000 Communist people's militia personnel. The OSS also envisioned forming an intelligence network to operate behind Japanese lines using Communists guerrillas. As the concept for the plan was "floated" in Yan'an to test the reaction of the Communists, it also came to the attention of Ambassador **Hurley**. Hurley was furious that such a plan would be advanced without his consent and embarrassed that it had leaked to Chiang Kai-shek. Hurley sent a cable back to President Roosevelt and the State Department accusing some of the officers under Wedemeyer's command, including McClure, Barrett, and Foreign Service officer advisers John Davies and John Service, of being Communist sympathizers and of operating without authority. The result was that Barrett was withdrawn and replaced on the observer mission in Yan'an and that American forces were ordered not to "assist, negotiate with, or collaborate with" Chinese political parties unless they were specifically authorized to do so by Wedemeyer. Since both Wedemeyer and Hurley were solid anti-Communists, this firmly committed the United States to support only the Nationalist government.

REFERENCES

David Barrett, *Dixie Mission: The United States Army Observer Group in Yenan, 1944* (Berkeley: University of California Press, 1970); Albert C. Wedemeyer, *Wedemeyer Reports!* (New York: Henry Holt, 1958); U.S. Department of State (Far Eastern Series 30),

United States Relations with China: With Special References to the Period 1944–49 (The China White Paper) (Washington, DC: Government Printing Office, 1949).

COSTIND. See COMMISSION OF SCIENCE, TECHNOLOGY, AND INDUSTRY FOR NATIONAL DEFENSE

CULTURAL REVOLUTION (1965–1976) The ''Great Proletarian Cultural Revolution'' (Wenhua Da Geming) was a movement created by **Mao Zedong** designed as an anti-Soviet Union, anti-Western mass mobilization campaign. Mao revived the form of violent class struggle against people identified as former landlords and capitalists that characterized the **Agrarian Revolutionary War** period. Arguably, the Cultural Revolution was also a means by which Mao attacked his old enemies, especially **Liu Shaoqi**, and critics within the Communist Party who criticized him because of both his failings during the ''Great Leap Forward'' and his military policies. The Great Leap proved to be an economic debacle of catastrophic proportions, creating mass starvation and disrupting industrial production. The Cultural Revolution officially started in November 1965, with the publication of an editorial in Shanghai denouncing one of Mao's critics who, according to Mao's supporters, had attacked Mao through a play. The play, *The Dismissal of Hai Rui from Office*, by Wu Han, used literary allusion to criticize Mao for his dismissal of **Peng Dehuai**.

Mobilizing millions of young Chinese students into **Red Guards**, Mao set out to use the movement to attack his critics. Mao literally purged and, at times, instigated the death of hundreds of loyal Communist Party members, including some of the leading military figures of the **Anti-Japanese War** and the **Civil War**. Many of these were rivals for power or had been critical of Mao's leadership. By 1967, China was so turbulent that the **People's Liberation Army** (PLA) had to take action against the Red Guard Units. The **Wuhan Incident** is one example of PLA action; a second example is that of the 38th Group Army, which was moved from Manchuria to the city of Baoding, Hebei Province, to stabilize Red Guard fighting in that city. Although Cultural Revolution turbulence essentially ended after 1969, the Cultural Revolution was declared over only after the death of Mao and the arrest of the **Gang of Four** Cultural Revolution leaders, including Mao's wife, Jiang Qing, on October 6, 1976.

REFERENCES

Lee Hong-yung, *The Politics of the Chinese Cultural Revolution* (Berkeley: University of California Press, 1978); Roderick MacFarquhar, *The Origins of the Cultural Revolution*, 3 vols. (New York: Columbia University Press, 1981–1986); Richard H. Solomon, *Mao's Revolution and the Chinese Political Culture* (Berkeley: University of California Press, 1971).

D

DABIESHAN Dabieshan, or the Dabie Mountains, was the location of the **Eyuwan Soviet**, one of the two main **Revolutionary Base Areas** of the **People's Liberation Army** before the **Long March**. The second main base area was **Jinggangshan**.

REFERENCES

Laszlo Ladany, *The Communist Party of China and Marxism, 1921–1985* (Stanford, CA: Hoover Institution Press, 1988); William W. Whitson with Chen-Hsia Huang, *The Chinese High Command: A History of Communist Military Politics, 1927–71* (New York: Praeger, 1973).

DAGU FORTS (TAKU FORTS) The forts are redoubts outside the city of Tianjin on the coast of North China, 90 miles east of Beijing. The Dagu forts were defensive works commanding the approaches to the harbor and city of Tianjin that were intended to defend the port and strategic lines of communication of the North China Plain. These fortifications were repeatedly attacked and breached or captured by foreign invaders in the **Opium War**, the **Arrow War**, the **Franco–Chinese War**, the **Sino–Japanese War**, and the **Boxer Rebellion**. They were the key fortifications that had to be secured by foreign forces, and they never succeeded in doing more than slowing foreign attack.

REFERENCES

Wolfram Eberhard, *A History of China* (Berkeley: University of California Press, 1977); Jonathan D. Spence, *The Search for Modern China* (New York: W. W. Norton, 1990).

DAI LI (Tai Li) (1895–1946) Dai Li was the head of all intelligence and counterespionage services for **Chiang Kai-shek** and the Nationalist government. He organized a guerrilla force to operate against the Japanese during the **Sino–**

Japanese War (1937–1945), using his connections to the Shanghai underworld. He was also the director of the **Sino-American Cooperative Organization** (SACO), in which he cooperated with the U.S. Strategic Services (Office of the Strategic Services or OSS) representative to China, Captain (later Rear Admiral) **Milton E. Miles**. SACO gathered intelligence, destroyed Japanese supply lines, and conducted guerrilla operations against the Japanese.

Dai Li was born in Jianggshan, Zhejiang Province, in 1895. His exact date of birth is not known. His father failed in business (although the father's family contained several successful businessmen and traders) and died in 1900. Dai Li was raised by his mother, but he left school in 1909 to become a military cadet in the ''model regiment'' of the Zhejiang Army. Almost nothing is known of his activities between 1909 and 1926. Dai Li became a member of the Guomindang Party in 1926 and entered the fourth class at the **Whampoa Military Academy**. After graduation from Whampoa in the same year, he became a cavalry officer.

As a cavalry officer during the **Northern Expedition**, he was sent ahead of the main body of troops to gather information on public attitudes toward the **warlords**, on the military situation, and on avenues of attack. It is not clear whether his acumen at intelligence gathering brought him to the attention of Chiang Kai-shek, or whether they had a previous association in Shanghai related to the Green Gang, but Dai Li was sent to Shanghai by Chiang in 1927 to work with the underworld gangs and secret societies (principally, the Green Gang [Qing Bang]), in preparation for Chiang's own move to Shanghai in April 1927.

Dai Li served on Chiang Kai-shek's staff and in 1931 was appointed as chief of the Second Department, Bureau of Investigations and Statistics, Military Affairs Commission. In this capacity he was responsible for the conduct of espionage operations against Japan and Japanese forces in China. He also had responsibility for counterespionage operations against Japanese agents in China. He put together a staff of officers drawn from other graduates of the Whampoa Military Academy. In addition to his intelligence work against the Japanese, between 1931 and 1936, Dai Li also carried out clandestine operations against the Communists and their forces.

After the **Marco Polo Bridge Incident** on July 7, 1937, and the formal start of the war against Japan, Dai Li was sent to Shanghai. He used his earlier connections to the Green Gang and the Shanghai underworld to organize a guerrilla force to fight Japan that was known as the ''Loyal and Righteous Army of National Salvation'' (Zhongyi Qiuguojun). In late 1937, in Nanjing, still the capital of Nationalist China, Dai Li was made deputy director of the successor organization to the Second Department, Bureau of Investigations and Statistics of the Military Affairs Commission. He became the director in 1938. With a powerful network of agents and guerrillas under his control, he penetrated both the Communist **New Fourth Army** and the puppet regime established by the Japanese in Nanjing in 1939, headed by **Wang Jingwei**. Within a few years, Dai Li also took over the Anti-Smuggling Bureau, the Commodity Transport

Control Bureau, and the Transportation (Jiaotong) Control Bureau. This effectively put him in charge of all commodity distribution for the Nationalist government, an extremely powerful position in a very corrupt regime. Meanwhile, in his capacity as the director of espionage and counterespionage, Dai Li managed to penetrate and control the security and police forces in Nanjing.

When the United States assigned Captain Milton E. Miles, an officer of the Strategic Services, to China in May 1942, in cover as the chief on the U.S. Naval Observer Group, Dai Li accompanied Miles on several covert trips into Japanese-held areas. Miles and Dai Li were the directors (Dai Li was director, and Miles his deputy) of the Sino-American Cooperative Organization, established under a bilateral agreement signed by the U.S. Navy secretary Frank Knox and **T. V. Soong** on April 15, 1943. Under the direction of Dai Li and Miles, SACO established 14 weathers stations, guerrilla training bases, and intelligence collection sites throughout China.

After **World War II**, Dai Li undertook the duties of tracking down and arresting those Chinese who had cooperated with the Japanese and the Wang Jingwei puppet government. He is credited with having brought over 3,000 people to trial. Dai Li was elected to the Executive Committee of the Guomindang Sixth National Congress in 1945. As part of an effort to suppress the increasing strength of the Communists as **Civil War** broke out, Dai Li flew to Qingdao on March 16, 1946. After conferring with the commander of U.S. Navy forces in Qingdao, he returned to Shanghai on March 17 on a plane belonging to the China Civil Aeronautics Commission. The plane vanished but was found crashed in the mountains near Nanjing several weeks later. Chiang Kai-shek ordered a period of mourning for Dai Li's death.

REFERENCES

Howard L. Boorman, ed., *Biographical Dictionary of Republican China* (New York: Columbia University Press, 1970); Milton E. Miles, *A Different Kind of War* (Garden City, NY: Doubleday, 1967).

D'AIGUEBELLE, PAUL ALEXANDRE NEVEUE (1831–1879) Paul D'Aiguebelle was a French naval officer, a graduate of the French Naval Academy who, during the **Taiping Rebellion**, took command of the **Ever-Triumphant Army** after Tardif de Mordrey died in March 1863. During D'Aiguebelle's period of command, the Ever-Triumphant Army fought at Shaoxing, Fuyang, Hengzhou, and Huzhou, often in concert with the British-led **Ever-Victorious Army**. Like the original leader of the Ever-Triumphant Army, **Frederick Townsend Ward**, D'Aiguebelle was given the Chinese title of "Tidu" (commander in chief) of the army by the Qing dynasty. After the end of the Taiping Rebellion, D'Aiguebelle worked with the governor-general of Zhejiang and Fujian, **Zuo Zongtang**, to build the **Fuzhou Dockyard**. During the period of the **self-strengthening movement**, D'Aiguebelle continued to advise Zuo.

REFERENCES

Prosper Giquel, *A Journal of the Chinese Civil War, 1864*, trans. Stephen Leibo (Honolulu: University of Hawaii Press, 1985); Franz Michael, *The Taiping Rebellion: History and Documents*, 3 vols. (Seattle: University of Washington Press, 1966).

DALIAN. See PORT ARTHUR

DAMANSKY ISLAND. See ZHENBAO ISLAND CLASH

DEMOCRATIC BASE AREAS (also known as Soviet Base Areas) The "Democratic Base Areas" organized by the Communist Party and its armed forces during the time of the Land Reform Campaign (1927–1936) and through the **Anti-Japanese War** are synonymous with the "Soviets" organized and operated by the Communist forces (e.g., the Jiangxi Soviet, or **Jinggangshan Revolutionary Base Area**). They were areas under the control of the Communist Party, administered by the party using Leninist "Democratic Centralism," and headed by "People's Governments." The Democratic Base Areas all had some form of People's Armed Department, which controlled a **militia** organization, and many of the base areas also supported main force or guerrilla organizations of the **People's Liberation Army**. There were over 45 different base areas during the period when the Communist Party was fighting, whether for its existence against the Nationalists, against the Japanese, or for control of the mainland of China. Some of the better-known base areas include the **Eyuwan Soviet** (or Revolutionary Base Area), the Jinggangshan Soviet, and the Tai-Yue Soviet. The names are taken from the geographic regions where the base areas were located, and the terms Soviet Base Area and Revolutionary Base Area are used more or less interchangeably in the literature on China. The base areas were generally self-sufficient in agriculture and even tried to manufacture their own ammunition and weapons.

REFERENCES

F. F. Liu, *A History of Modern China: 1924–1949* (Princeton: Princeton University Press, 1956); William W. Whitson with Chen-Hsia Huang, *The Chinese High Command* (Praeger, 1973); ZGDBKQS, vol. 2.

DENG XIAOPING (1904–1997) Deng Xiaoping is most closely identified with the reforms of China's economic and social program begun about 1978. After **Mao Zedong**'s death, Deng was the senior leader of the country, directing its political, economic, and security policies. His early party background, however, was in the military, as was the case with many of China's first, post-1949 generation of leaders. Deng Xiaoping was born in August 1904 in Guang'an, Sichuan Province. In 1920 Deng went to Lyon, France, on a work–study program probably funded by the Communist International. He joined the Socialist Youth League of China in 1922 and the Chinese Communist Party in 1924. In

1926, after leaving France, Deng studied at Zhongshan University in Moscow. Deng was later an instructor at the Military and Political Academy run by Feng Yuxiang. Deng moved to Guangxi in 1927, after Feng purged the Communists from the academy. He led uprisings against the Guomindang in Guangxi Province, 1929–1930, during which time he served as political commissar of the 7th and 8th Red Armies. In 1930 Deng served as Peng Dehuai's chief of staff in the **First Front Army**. Deng entered the **Jinggangshan Revolutionary Base Area** in Jiangxi in 1931 and served as secretary of the Chinese Communist Party (CCP) Ruijin County Committee in 1933; after supporting Mao in an inner-party struggle, Deng was removed from his posts. Deng later worked in the General Political Department of the Military Committee of the CCP Central Party Committee and served as editor in chief of *Red Star News*.

In 1935, Deng took part in the **Long March** while serving as secretary-general of the CCP Central Committee. During **World War II** (the "War of Resistance against Japan" or Anti-Japanese War, to the Chinese), Deng served as deputy director of the General Political Department, **Eighth Route Army**, and also the political commissar of the 129th Division. In the Shanxi-Hebei border areas working with **Liu Bocheng**, Deng established anti-Japanese base areas in the Taihang Mountains. During the **War of Liberation**, Deng served as political commissar of the Shanxi-Hebei-Shandong-Henan Field Army, which eventually evolved into the Second Field Army. Again with **Liu Bocheng**, in 1947, Deng led armies across the Yellow River to China's Central Plains. During the **Huai-Hai** and Crossing Yangtze campaigns in 1948, Deng served with Liu Bocheng and **Chen Yi**, participating in the Central Plains Field Army liberation of Nanjing, the seat of the Nationalist (KMT) government. After 1949, Deng returned to southwest China and the Sichuan area, where he served as 1st secretary of Southwest Bureau, CCP Central Committee; vice chairman of Southwest China Military and Administrative Committee; and political commissar of Southwest China Military Area Command. Deng returned to Beijing in 1954, where he served as vice premier of the State Council; secretary-general and head of the Organization Department of the CCP Central Committee; and vice chairmen of the National Defense Council. In 1955, Deng was appointed general secretary of the CCP Central Committee.

At the beginning of the **Cultural Revolution**, in 1966, Deng was condemned as a "rightist, a capitalist roader" and was purged from all posts. He was reinstated as vice premier of the State Council and a member of the Communist Party Central Committee's Political Bureau upon the decision of CCP Central Committee in 1973. In 1975, Deng was elected vice chairman of the CCP Central Committee, vice chairman of the **Central Military Commission**, and chief of the **People's Liberation Army** (PLA) General Staff Department. After Premier **Zhou Enlai**'s illness and death, Deng took care of routine matters of the CCP and government. However, in a power struggle with Mao's wife, Jiang Qing, and the **Gang of Four**, Deng was once more purged from all posts in 1976. In July 1977, after the arrest of the Gang of Four, Deng was reinstated to his former posts. In 1978, Deng out forward a platform that called for the party to "emancipate the mind, seek truth from facts and unite as one in looking to the future" at the Third Plenary Session of CCP 11th Central Committee. After his December 1978 visit to the United States, Deng directed that the PLA conduct its **Self-defensive Counterattack against Vietnam** in February 1979. During the demonstrations in Beijing and around China in 1989, Deng directed that the PLA be used to forcefully suppress student and labor activists who had occupied Tiananmen Square, leading to the **Tiananmen Square Massacre** of June 4, 1989. He resigned from his post of chairman of the CCP Central Military Commission in November 1989 and as chairman of the State Central Military Commission in March 1990. Nonetheless, throughout at least his last public appearance in 1994, Deng remained the most powerful senior leader in China. Deng Xiaoping died on February 19, 1997.

REFERENCES

Wolfgang Bartke, *Who's Who in the People's Republic of China* (Armonk, NY: M. E. Sharpe, 1981); Liao Gailong and Fan Yuan, eds., *Who's Who in China: Current Leaders* (Beijing: Foreign Languages Press, 1994).

DEPARTMENTS OF THE GENERAL STAFF. See GENERAL DEPARTMENTS OF THE PLA

DIXIE MISSION (July–November 1944) As American military operations against Japan expanded in 1944, it became clear that U.S. forces needed more intelligence on Japanese forces in Manchuria to plan their campaign. The U.S. War Department also needed to formulate means by which downed airmen might be aided by local forces if they landed in Japanese-occupied parts of China controlled by the **Eighth Route Army** and Communist forces. In discussing the need to open contacts with Communist forces in China, in an internal memorandum circulated within the American command in China, two other reasons were cited as justification to send U.S. military observers to the Communist headquarters in **Yan'an**: first, to gather information about Chinese Communist forces and potential Russian operations (in the event that Russia entered the Far East war); and second, to break the isolation that the Nationalists had imposed on the Communists in order to prevent total Chinese Communist dependence on the Soviet Union.

From February 9, 1944, through the spring, President Roosevelt, through letters and emissaries, pushed Generalissimo **Chiang Kai-shek** to agree to permit a U.S. military mission to go to Yan'an. Chiang Kai-Shek finally acceded to the mission after a June 1944 visit to China by Vice President Wallace. U.S. Army colonel **David Barrett**, the military attaché at the Nationalist Headquarters at Chongqing, was selected to head the mission. Colonel Barrett had served in China repeatedly since 1924 in varying capacities. He served as head of the U.S. Army Observer Group, Yan'an, from July through November 1944. The group was composed of military intelligence, logistics, meteorological, and communications specialists. Also included was a U.S. State Department representative, John S. Service, who was later vilified by the McCarthy Hearings in the 1950s for his reports on Communist forces at Yan'an.

The Dixie Mission was charged with obtaining intelligence on Japanese forces, gathering meteorological information to support U.S. air operations, and assisting downed pilots with escaping through Japanese lines. It also reported on the fighting capacity of Chinese Communist forces. Members of the Office of Strategic Services (OSS), the forerunner of the CIA, were included in the mission. Eventually, the Dixie Mission also began to coordinate assistance to Chinese Communist forces, but by January 1945, plans for U.S. assistance to the Communists were stopped.

REFERENCES

David Barrett, *Dixie Mission: The United States Army Observer Group in Yenan, 1944* (Berkeley: University of California Press, 1970); Herbert Feis, *The China Tangle: The American Effort in China from Pearl Harbor to the Marshall Mission* (Princeton: Princeton University Press, 1953).

DONGSHAN ISLAND "DEFENSIVE" BATTLE (July 16, 1953) On July 16, 1953, the Nationalist Army sent an attack force based on **Jinmen** Island (Quemoy) estimated in size as over 10,000 men to attack Dongshan Island. Dongshan is the second largest island off the Fuijan coast between mainland China and Taiwan, with an area of 165 square kilometers. The island was defended by a garrison force from a battalion of the 80th **People's Liberation Army** (PLA) Public Security Regiment.

In a coordinated air, naval, and amphibious attack, Nationalist troops landed on the island. Concurrently, two airborne units dropped on the island in supporting attacks. Local Chinese militia forces and PLA personnel of the 80th Regiment defended, allowing a counterattack by forces of the PLA 82d Division and the 122d Division. By People's Republic of China (PRC) accounts, 3,379 Nationalist troops were killed and wounded, two tanks were destroyed, three Nationalist naval vessels were sunk, and two Nationalist aircraft were destroyed. Other PRC accounts say that 400 PLA soldiers were killed and that the Nationalists lost 1,000 dead.

A 1996 story in the Hong Kong press said that the vice chairman of the Chinese Communist Party **Central Military Commission, General Zhang Wannian**, took part in the battle for Dongshan Island. When the PLA conducted amphibious exercises at Dongshan Island in the **Taiwan Strait Crisis of 1995/1996**, Zhang Wannian was publicized as one of the commanders of the exercise.

REFERENCES

JGDBKQS, vol. 1; ZGRMJFJDSD.

DONGSHAN ISLAND NAVAL BATTLE (August 6, 1965) On the night of August 6, 1965, the 1,250-ton Republic of China (ROC) (Taiwan) Navy patrol ship *Chinmen*, with Rear Admiral Hu-Chia-sheng embarked and surreptitiously landed commandos in rubber boats in the southern Taiwan Strait in the vicinity of the border between Fujian and Guangdong Provinces. A second ROC naval patrol craft of about 450 tons was assisting in the operation. The area was under surveillance by People's Republic of China (PRC) coastal radars, and the Chinese may even have had advance knowledge of the operation. The two ROC ships were caught and surprised by between 7 and 10 small PRC Navy combatant vessels, and both were sunk in the ensuing battle. A total of 203 ROC sailors were killed or captured in the engagement, including Admiral Hu. The PRC took advantage of the opportunity to issue another warning against ''U.S. imperialism'' in the region. On August 17, 1965, senior PRC leaders, including **Mao Zedong**, **Liu Shaoqi**, and **Zhou Enlai**, received the crews of the PRC naval vessels as heroes in Beijing.

REFERENCE

David G. Muller, Jr., *China as a Maritime Power* (Boulder, CO: Westview Press, 1983).

Vance H. Morrison

E

EASTERN CAMPAIGN (COMMUNIST) (1936) This was a 75-day campaign conducted by the forces of the **First Front Army** and the Shaanxi-Gansu Soviet Base Area against Nationalist forces in Shanxi Province. The Communist 15th Army and the 28th and 29th Corps began organizing to move east out of the base area on February 18, 1936, under **Peng Dehuai**'s direction. The political commissar was **Mao Zedong**. The First Front Army crossed the Yellow River on February 20, 1936, attacking the forces of Shanxi governor and former **warlord** leader **Yan Xishan**. In a broad sweep north and east, the main forces of the 15th Army moved east to threaten the main north-south rail link, approaching the capital of Shanxi, Taiyuan. They were thrown back, however, by the 70th and 71st Nationalist Divisions. The campaign amounted to a preemptive attack against the Nationalists, securing time and equipment for the First Front Army as it built up a base in neighboring Shaanxi. In all, seven regiments of Nationalist forces were rendered ineffective or destroyed, and the Red Army captured 4,000 rifles and automatic weapons and over 20 artillery pieces. Between defections from Nationalist troops and new recruitment, the Communist forces also increased their ranks by between 5,000 and 8,000 men. This campaign also helped the Red Army to consolidate its base area in Yan'an and perhaps convinced **Zhang Xueliang** and Yang Hucheng, whose forces were south of the campaign operating area in Xi'an, to consider cooperating with the Communists against the Japanese. Zhang and Yang opened discussions with the Communists after this campaign, which eventually led up to the December 1936 **Xi'an Incident**.

REFERENCE

ZGDBKQS, vol 1.

EASTERN CAMPAIGN (NATIONALIST) (1925) This 1925 Nationalist Army campaign was conducted against **warlords** in the Shanghai-Nanjing area to consolidate the power of the Nationalist governments. It is also called the "First Eastern Campaign" because it was divided into two phases, the first in early 1925 and the second in fall 1925. The warlord Chen Qiangming, who had in 1922 refused to join the Nationalist government, began to gather forces in the lower Yangze River Valley, threatening Guangzhou itself. The Eastern Campaign began on February 1, 1925, and led to the recovery of control of the Guangzhou-Kowloon Railway as well as control of eastern China from Guangdong to Anhui. In the first major battle of the campaign, Lake Mian, a brigade of about 3,000 newly trained cadets of the **Whampoa Military Academy**, the "First Whampoa Brigade," drove back a superior force of some 20,000 troops fighting under one of the warlord's leaders, Lin Hu. In another major engagement of the Eastern Campaign, known as the "Second Eastern Campaign," Nationalist forces took the city of Huizhou, again securing the Nationalist position and consolidating control of Guangdong Province and the lower Yangze River valley.

REFERENCE

Chen Tinglong, *The Army of the Republic of China* (Taipei: Army Press, 1993).

EIGHT FOREIGN ARMIES INVASION OF CHINA (1900) In order to relieve the siege of the foreign legations in Beijing by Boxers during the **Boxer Rebellion** (June 20–August 14, 1900), a combined army from eight nations seized Tianjin and Beijing. Initially, after the Boxers began to kill Chinese Christians and foreign missionaries, the legations in Beijing were reinforced with 400 military personnel in May 1900. By June 17, over 2,000 Western troops had seized the forts at **Dagu** in preparation for a potential reinforcement by more forces. These troops attempted to fight their way to Beijing but were repulsed by the Boxer forces supported by troops of the dowager empress Cixi. The empress issued a declaration of war on June 21, 1900, accusing foreign forces of oppressing the Chinese and praising the Boxers. By August 1900, foreign troops from Japan, Russia, Britain, the United States, France, Italy, Germany, and Austria had built their strength to about 20,000. This combined army formed an expeditionary force that began to march on Beijing on August 4. By August 14, 1900, the foreign troops had relieved the legations, seized and burned parts of the Forbidden City and the Summer Palace, and advanced all the way into the Fragrant Hills, west of Beijing, where they sacked the empress' villa.

REFERENCES

Joseph Esherick, *The Origins of the Boxer Uprising* (Berkeley: University of California Press, 1987); Victor Purcell, *The Boxer Uprising: A Background Study* (New York: Cambridge University Press, 1963).

EIGHTEENTH GROUP ARMY OF THE RED ARMY When the Nationalist and Communist forces decided to conduct another **United Front**-like offensive in cooperation against Japan, the **New Fourth Army** was sent to its old operating areas near the pre-**Long March** location of the **Eyuwan Soviet**. The **Eighth Route Army** was designated by the Nationalists as the Communist 18th Group Army, nominally part of the National Army of China. When, in 1944, after President Roosevelt's insistence, General **Stilwell** was to be given command of all Chinese forces, Nationalist and Communist, in order to consolidate efforts in the face of Japanese **Operation Ichigo**, Stilwell insisted that the Communist 18th Group Army be included under his command. However, Stilwell was recalled before this took place. The Communist headquarters at **Yan'an**, despite the designation of their forces as the Communist 18th by the Nationalists, continued to refer to their forces as the Eighth Route Army. The Communist 18th Group Army never really existed or fought as part of the National Army of China.

REFERENCES

Hsi-cheng Ch'i, *Nationalist China at War* (Ann Arbor: University of Michigan Press, 1982); Barbara Tuchman, *Stilwell and the American Experience in China, 1911–45* (New York: Macmillan, 1970); William W. Whitson with Chen-Hsi Huang, *The Chinese High Command* (New York: Praeger, 1973).

EIGHTH ROUTE ARMY After the outbreak of **World War II** in China, marked by the **Marco Polo Bridge** (Lugouqiao) **Incident** of July 7, 1937, Communist and Nationalist forces focused on the fight against Japan, combining into a National Revolutionary Army. The forces of the Communist **People's Liberation Army** (PLA) were organized into the Eighth Route Army, commanded by **Zhu De. Peng Dehuai** was the deputy commander. The Eighth Route Army comprised three divisions: the 115th Division had formed from the former Red **First Front Army** and part of the **Fourth Front Army** (the 15th and 25th Corps). The 115th Division was commanded by **Lin Biao**. The deputy commander was **Nie Rongzhen**. The 115th Division operated primarily in the Shanxi, Hebei, Lianoning, and Shandong regions. The 120th Division was built around **He Long**'s former **Second Front Army**. It was commanded by He Long, and the deputy commander was **Xiao Ke**. The 120th Division operated principally in the Shaanxi, Gansu, Ningxia area (although its operations also extended into northwest Shanxi). The 129th Division was composed of the bulk of the former Fourth Front Army. Its commander was **Liu Bocheng**, and the deputy commander was **Xu Xiangqian**. The 129th Division's operating areas extended into the east and southeast, spanning parts of Shanxi, Hebei, and Henan and reaching Shandong. Each of these divisions was that in name only, since its strength was often in excess of 100,000 personnel, including support and logistics personnel.

REFERENCES

Kataoka Tetsuya, *Resistance and Revolution in China: The Communists and the Second United Front* (Berkeley: University of California Press, 1974); F. F. Liu, *A Military*

History of Modern China, 1924–1949 (Princeton: Princeton University Press, 1956); ZGRMJFJZS, vol. 2.

ENCIRCLEMENT CAMPAIGNS (1930–1934) Between December 1930 and October 1934, Guomindang Forces (Nationalist, or KMT) mounted five distinct campaigns against Communist forces in Jiangxi Province and in the Dabieshan area. The Nationalist forces sought to surround, pin down, and destroy, in detail, Communist forces. In all, the Communist forces never were able to employ more than 150,000 men in their counterencirclement campaign, while Nationalist forces employed against them numbered as high as 600,000 men. The Nationalist effort served as the crucible that forged the military strategy of **Mao Zedong**, which relied on smaller guerrilla units, an active defense that relied on skilled use of familiar terrain to conduct ambushes, and the piecemeal defeat of attacking forces. These tactics characterized what became known as the **People's War** strategy used by Communist forces against the Japanese. These campaigns are collectively called ''counterencirclement campaigns'' in Chinese Communist histories.

In the First Encirclement Campaign, between mid-November 1930 and January 3, 1931, about 100,000 Nationalist troops surrounded some 40,000 Communists of the **First Front Army**, only about 25,000 of whom were armed, in the mountainous area of Ningtu in Jiangxi. Communist First Front Army forces managed to lure divisional elements of the Nationalist Army into narrow valleys in the area of Langgang and Dongqu, block the forces, surround them, and destroy the Nationalists by attacking from concealed positions in the hills surrounding the valleys. Prominent **People's Liberation Army** (PLA) leaders in the campaign were **Lin Biao** and **Peng Dehuai**. In one incident, the commander of the Nationalist 18th Division, Zhang Huican, was captured by the PLA and beheaded after a public trial. Zhang's head was mounted on a raft and floated down the Gan River from Longgun to Zhian to announce the PLA victory. The PLA also captured weapons and material from the Nationalists.

The Second Encirclement Campaign took place in essentially the same area of Jiangxi during April and May 1931. The Nationalist Army commander in chief, He Yingqin, brought about 15 Nationalist divisions (about 130,000–150,000 men) to bear on about 30,000 PLA forces. In the Second Encirclement Campaign, PLA commanders Mao Zedong and **Zhu De**, operating out of their base area in the Jinggang Mountains (**Jinggangshan**), selected Dongqu as the strategic point for an ambush, where Lin Biao's Fourth Corps, Zhu De's First Army, and Peng Dehuai's Third Army ambushed and defeated the Nationalist Seventh, 28th, 43d, and 54th Divisions. By the end of the campaign the Communist Jiangxi Soviet area had tripled in size to about 5,000 square miles.

During July–September 1931, the Nationalist Army under **Chiang Kai-shek** reassembled more divisions for the Third Encirclement Campaign. By this time the Nationalists had about 200,000 men available, of whom 130,000 were employed. The armed strength of the Communist forces in the Jiangxi Soviet area was about 55,000 men. The Nationalists also attempted to engage decisively and destroy the **Eyuwan Soviet**, led by **Zhang Guotao** where the **Fourth Front Army** was based, in the Dabie Shan (Mountain) area. Communist units of the Fourth Front Army took advantage of their knowledge of the terrain, conducted a 400-mile forced march south and west behind four Nationalist divisions over a two-week period, and threatened the Nationalist main supply route on the Gan River after swinging north again. A major night battle in the area of Dongshao was the decisive factor in forcing the withdrawal of Nationalist forces. The Japanese invasion of Manchuria on September 19, 1932 (**Nine One Eight (918) Incident**), forced the Nationalists to redefine their military priorities.

Despite the success of the guerrilla strategy in the first three Encirclement Campaigns, Mao's influence and personal power in the PLA were reduced because of two factors: a desire to regularize the Red Army on the part of some Chinese leaders and the influence of the Comintern's military adviser, Li De, who sought to operate as a conventional force against the Nationalists. The Communist forces initiated a new strategy of defense outside the mountainous Jiangxi Soviet area, after the Third Encirclement Campaign.

The Nationalist Fourth Encirclement Campaign (July 1932–April 1933) concentrated about 600,000 troops against three Communist Base Areas (Soviets), Jiangxi, Eyuwan, and the Central Soviet. After an initial offensive against the Eyuwan Soviet, north of Jiangxi in the Dabie Mountains, causing a withdrawal by the Fourth Front Army, the Nationalists again focused on the PLA forces around the Jiangxi Soviet. The Communists were beaten in an attempt to launch attacks outside the mountains but were successful in isolating and destroying Nationalist garrisons and logistical points. The Nationalists eventually withdrew from action without a decisive victory on either side in the Fourth Encirclement Campaign around Jiangxi.

The Fifth Encirclement Campaign took place between October 1933 and October 1934. Nationalist forces encircled the Communist Jiangxi Soviet area by establishing a system of linked, fortified lines of fortifications and blocking any passage or travel. This reduced Communist mobility. Although the Communists attempted to break out of the encirclement, a Nationalist thrust employing about 800,000 men into the mountains of the Jiangxi Soviet area forced the eventual withdrawal of Communist forces and the decision to conduct a strategic withdrawal, on the **Long March**, south, west, and north into Shaanxi Province.

REFERENCES

Chen Tinglong, *The Republic of China Army* (Taipei: Army Press, 1993); Alexander L. George, *The Chinese Communist Army in Action* (New York: Columbia University Press, 1967); William Wee, *Counter-revolution in China: The Nationalists in Jiangxi during the Soviet Period* (Ann Arbor: University of Michigan Press, 1985); William W. Whitson with Chen-Hsia Huang, *The Chinese High Command* (New York: Praeger, 1973).

EVER-TRIUMPHANT ARMY This French-led, joint Sino–foreign force emerged after the **Arrow War** to combat Chinese rebel forces in the **Taiping**

Rebellion after France and England decided to assist the Qing dynasty against the rebels. The French and English provided this assistance to the Qing to protect their own national and commercial interests in China. The organizer of the Ever-Triumphant Army, **Prosper Giquel**, planned the force and worked with Edourd le Brethon de Caligny (Brethon) to bring it into being. Brethon initially was the commander. Giquel worked on organizing financial and political support from the wealthy Chinese of the city of Ningbo and its suburbs, which the force was established to protect, while Brethon set about training Chinese volunteers to be soldiers with help from foreign instructors. Brethon also received training assistance and, at times, artillery support from another Sino–foreign mercenary group that rose up at the time, the Franco–Chinese Corps of Jiangsu, an artillery unit organized in June 1861 by Adrien Tardif de Mordrey (Tardif). Tardif at one time also commanded the Ever-Triumphant Army.

After Ningbo fell to the Taiping rebels in December 1961, Giquel worked with **Frederick Townsend Ward** and Tardif in training Chinese–foreign contingents. After Ningbo was recaptured by foreign units and units of Ward's **Ever-Victorious Army**, Giquel returned to Ningbo in June 1962. Then he began to cooperate with Brethon to form the Ever-Triumphant Army. Brethon and Giquel received assistance from Tardif, and when Brethon was absent, Tardif at times commanded the army.

The initial contingent of the Ever-Triumphant Army was formed with about 50 Europeans, who served as instructors, 50 Filipinos, and several Chinese. Weapons were often of poor quality and even defective. Even as it organized, the Chinese governor-general of Zhejiang and Fujian, **Zuo Zongtang**, was hostile to the mercenary unit and, in spring 1863, forced it to reduce in size. Part of Zuo's opposition was because the Qing government had authorized only one foreign-led force in China, Ward's Ever-Victorious Army. Other opposition to the Ever-Triumphant Army also came from the British naval commander, Vice Admiral James Hope, who supported Frederick Townsend Ward and wanted to see a British-dominated Sino–foreign army in China. Meanwhile, the French naval commander in China, Charles Louis Faucon, sought to maintain the French-dominated force in Ningbo, arguing that since the men had been taught their commands in French, they could not easily operate with, or under, English control.

On July 31, 1862, the Ever-Triumphant Army, supported by a contingent of Ward's force, attacked a Taiping fort at Yuyao, northeast of Ningbo. The attack was initially beaten back in the face of stiff resistance by an estimated 10,000 Taiping rebels. However, a Taiping counterattack was repulsed with the assistance of the British ship HMS *Hardy* and the Franco–Chinese ship *Confucius*, which was controlled by Brethon. This combined, foreign-led force routed the Taiping's and captured Yuyao. Giquel commanded the ground contingent of the Ever-Triumphant Army in this battle.

On November 17, 1862, the Ever-Triumphant Army, under Brethon and Giquel, marched on the city of Shangyu, 12 miles farther southeast of Yuyao. En route to Shangyu, the troops, supported by a French and an English gunboat, had to fight through some 14 prepared Taiping defensive positions. On November 21, 200 troops led by Giquel attempted to storm Shangyu, but they were beaten back by the Taipings. Giquel was wounded in the attack and did not recover from a shattered right elbow until 1864. He convalesced for 19 months in France. The Taiping fort at Shangyu was captured on November 22, 1862, with naval gun support from the British gunboat.

Brethon, who commanded the *Confucius* during the battle, died in winter 1862. He was replaced as commander of the Ever-Triumphant Army by Tardif. Meanwhile, Frederick Townsend Ward was sent to defend Ningbo by Jiangsu governor **Li Hongzhang**, but in an attack on the town of Qiqi, in eastern Zhejiang, Ward was killed. After Ward's death, the Ever-Victorious Army was weakened, and the French-led Ever-Triumphant Army grew stronger. Brethon, this time with Chinese agreement, commanded the Ever-Triumphant Army, while Tardif was in Shanghai with his Franco–Chinese Corps. By this time **Charles "Chinese" Gordon** had arrived to command the Ever-Victorious Army with the support of the British.

Between January and March 1863, the Ever-Triumphant Army took part in a campaign to capture Shaoxing, commanded by Brethon. Shaoxing was inland, east of Ningbo and northeast of Shangyu. The city of Shaoxing was protected by a 50-meter wide canal, making an attack very difficult. Supported by two 12-pound howitzers and 9-pound guns, Brethon attempted to assault the city with a force of about 1,200 men. However, he was killed by an explosion of one of his guns in mid-January. Brethon was replaced by Tardif in February 1863. On February 19, supported by a critical artillery piece provided by the British navy, Tardif attempted another assault on the city. However, he was shot in the back of the head by one of his own men and killed. The Taipings managed to capture six foreign attackers and beheaded them. Chinese troops, under Qing dynasty control, had refused to take part in the attack.

Tardif was replaced by a French naval officer, Commander **Paul D'Aiguebelle**, who had supported Ward's Ever-Victorious Army in the attack at Fenghua and also took part in the Shangyu engagement. The Taipings, after almost a one-month siege, finally evacuated Shaoxing on March 14, 1863. The new commander of the Ever-Triumphant Army, D'Aiguebelle, got along with Chinese governor-general Zuo Zongtang no better than had Giquel or Brethon. When D'Aiguebelle's demands for pay from the Chinese were not met, D'Aiguebelle occupied a city and demanded the money owed his troops. After a formal protest by the Chinese government to Paris, the resolution of the matter was that French and Qing Chinese forces cooperated more closely.

Governor-General Zuo's own forces were engaged in a campaign to capture Hangzhou but were stalled in an attempt to seize Fuyang, about 30 miles southwest of Hangzhou. On September 8, 1863, Zuo sought the assistance of the Ever-Triumphant Army to capture the city. D'Aiguebelle provided 1,500 men and artillery, and within two days the Taipings fell back from Fuyang to Yuhang.

However, they attempted to recapture the city throughout the fall and winter. Zuo's Qing troops finally moved on Hangzhou in February 1864, assisted by the Ever-Triumphant Army. Between March 3 and March 28, 1864, Hangzhou remained under siege from the guns of D'Aiguebelle. The city walls were breached on the twenty-eighth, and the combined Qing dynasty, Sino–French force stormed the city. Zuo made D'Aiguebelle a lieutenant general, a rank equivalent to that of Charles "Chinese" Gordon, who commanded the Ever-Victorious Army. Giquel returned to China in June 1864, after recuperating from his wounds, in time to see the Ever-Triumphant Army take part in the final campaign to capture the Taiping capital of Nanjing. D'Aiguebelle left for Hangzhou on August 30, 1864, and the Qing authorities began to reduce the size of the Ever-Triumphant Army (the Ever-Victorious Army had been disbanded earlier in the summer). The army was disbanded between September 11 and 28, 1864.

The major significance of the contact with Western military art was that the Chinese later adopted the **self-strengthening movement** so that their own forces could achieve combat efficiency.

REFERENCES

Prosper Giquel, *A Journal of the Chinese Civil War, 1864*, trans. Steven Leibo (Honolulu: University of Hawaii Press, 1985); Ssu-yu Teng, *The Taiping Rebellion and the Western Powers: A Comprehensive Survey* (Oxford: Clarendon Press, 1971); Andrew Wilson, *The "Ever-Victorious Army": A History of the Chinese Campaign under Lt. Col. C. G. Gordon and the Suppression of the Tai-ping Rebellion* (London: W. Blackwood and Sons, 1868, reprinted Novato, CA: Presidio Press, 1991).

EVER-VICTORIOUS ARMY One of the multinational units that emerged in China to assist in the defeat of the **Taiping Rebellion**, the Ever-Victorious Army was first commanded by **Frederick Townsend Ward**, an American, and later by **Charles G. "Chinese" Gordon**, an Englishman.

The Ever-Victorious Army emerged from an earlier foreign mercenary contingent, the **Foreign Arms Corps** (this has also been translated as Foreign Rifle Corps) (Yang Qiang Dui), organized by Ward in Shanghai in 1860. Ward started his force with a group of Filipino mercenaries, whom he hired in spring 1860. At times, the corps was also commanded by **Henry Burgevine**, when Ward was absent. Burgevine eventually succeeded Ward in command of the Foreign Rifle Corps.

Although some references date the establishment of the Ever-Victorious Army to June 1860, when the Foreign Arms Corps was formed, according to biographies of Frederick Ward, that appellation was formally given to the unit by an imperial decree of March 16, 1862, after Ward's marriage to the daughter of his Chinese financial benefactor. When Ward married Yang Fang's daughter, Chien-Mei, the Qing local governor, Xue Huan, conferred the title "Ever-Victorious Army" on the unit because of the earlier successes of the Foreign

Rifle Corps against the Taiping rebels. There was historical precedent for such a name because in the Song dynasty a barbarian (non-Han Chinese) commander led a Song Army by the same name to victory.

The Ever-Victorious Army played a decisive role in a number of battles with the Taiping rebels. In Songjiang, a city about 20 miles southwest of Shanghai, in June 1860, the Foreign Arms Corps fought a small engagement against a group of rebels who had seized positions on the canal leading from Shanghai. Ward faked an attack on Qingbu on July 16 but turned back to attack Songjiang with the gunboats *Confucius* and *Vulcan* and a force of men from the Foreign Arms Corps. On July 17, he took Songjiang, evicting the Taiping rebels and leaving a small garrison force there. For the next two years, in fact, Ward used Songjiang as a forward base of operations and a training area for his force, as he attempted to clear the Taipings from a corridor around Shanghai.

Ward's plan to keep Shanghai secure called for seizing the city of Qingbu, between Shanghai and Suzhou (Suzhou was a Taiping garrison), seizing Jiading, northwest of Shanghai, and holding Songjiang. Ward believed that by doing so, he could secure an area of some 30 miles in diameter, with critical road and canal systems, centered on Shanghai.

Ward next turned to more recruiting and, having gathered and trained more troops, tried to move beyond Songjiang, attempting to take Qingbu. He was beaten back by the Taipings, however, and received five wounds in the attack. All were relatively minor, except for one shot by a musket ball that hit him in the jaw and exited through his cheek. Leaving Burgevine in charge, Ward sought medical care.

Frederick Townsend Ward continued to recover after returning again to Songjiang and recruited and trained more men. In January 1862, he moved to take Guangfuling with a force of 2,000 Chinese from Shanghai, a regiment of 800 of his own foreign-led men, 200 Filipino soldiers, and nine artillery pieces. Ward's forces inflicted severe casualties on the Taipings at Guangfuling, and he continued his campaign. He cleared Pudong and Gaoqiao, outside Shanghai, as well as Minkang, securing the area immediately around Shanghai. By March 1, 1862, Ward's force also secured Nanqiao, at which time he returned to Shanghai to marry Yang Fang's daughter. Prior to the marriage, however, he fought one more battle, finally driving the rebels from Qingbu, where he had earlier been wounded. At this point the Foreign Arms Corps was renamed the "Ever-Victorious Army."

By April 1862, the rebel Taiping force attempted to regain the territory it had lost and moved into the area west of Shanghai, between that city and Songjiang. The Ever-Victorious Army rendezvoused with British forces under Admiral Hope at Tonggadu and, with the forces from the French Ever-Triumphant Army as well as Captain (later General) Charles Gordon, moved against the rebels. These combined forces encountered a strong rebel force at Wangjiazi but were able to drive the Taipings out. The army remained there until April 15, after which the Ever-Victorious Army attacked Qibao.

On April 28, the Ever-Victorious Army attacked a Taiping force at Jiading, which was a walled city protected by a moat. Again, the Ever-Victorious Army, allied with the English fleet, cooperated in the attack with the French-led Ever-Triumphant Army. They cleared out the Taipings and continued on the campaign, seizing Celin, on Hangzou Bay, by May 20. However, the Taipings recaptured Jiading, which had been turned over to Qing forces by the foreign-led armies. Taipings also reseized Songjiang, in the absence of the Ever-Victorious Army.

After once more seizing the cities that he had captured in earlier battles, Ward attempted to recruit and train six more regiments for the army at Songjiang. Meanwhile, in early July, **Li Hongzhang** sent for Ward to confer on finances and a campaign strategy. On August 8, the Ever-Victorious Army marched out of Songjiang again, to fight the Taipings at Qingbu. One regiment was left to garrison Songjiang, while five infantry regiments, a musket sharpshooter regiment and 24 artillery pieces prepared to attack Qingbu. After two unsuccessful assaults, the army drove the Taipings out of Qingbu on its third attack.

In the interim, the French-led Ever-Triumphant Army secured Ningbo. Seeking to capitalize on this victory, Ward's Ever-Victorious Army marched there on September 20, with the objective of attacking Cigei, 10 miles northwest of Ningbo. After driving the Taiping defenders behind the city wall, Ward prepared to attack. He was shot in the abdomen as he stood to give the attack order to his troops. The Ever-Victorious Army took the city, but Ward was evacuated to Ningbo, where he died on September 22, 1862.

Ward's deputy, Henry A. Burgevine, was eventually given command of the Ever-Victorious Army after Ward's death and was later appointed a general. On October 23, 1862, Burgevine, still a colonel, led an allied force of about 4,500 men against Jiading, which had once more been occupied by the Taiping rebels. Included were 1,800 English troops, 400 French-led troops from the Ever-Triumphant Army, and 2,300 troops of the Ever-Victorious Army. With 30 pieces of artillery the allied force succeeded in again capturing the city after three assaults. Burgevine then attacked again at Qingbu, retaking this city from the Taipings on November 19. The Chinese officials, at this point, were prepared to disband the army, which had cleared a 30-mile zone around Shanghai, Ward's original plan. By January 1, 1863, the Ever-Victorious Army had not been paid by the Shanghai merchants, but the Qing officials wanted Burgevine to march on the Taiping capital, Nanjing. Burgevine allegedly took a bodyguard force from the Ever-Victorious Army to Shanghai on January 4 and seized $40,000 from a bank to pay his force. He was dismissed and a 50,000-tael reward was posted for his arrest, to be executed. More than 50 officers of the army signed a public protest, swearing to end their service to Chinese authorities if the execution order was carried out.

Burgevine was replaced as commander of the army by Captain John Holland, a British naval officer, on January 14, 1863. In Holland's first engagement as commander, at Taicang on February 14, he prematurely ordered the withdrawal of the artillery. The army was defeated and lost 190 killed and 174 wounded.

Captain Charles G. Gordon, a British engineer officer, was promoted to major after that incident and given command of the Ever-Victorious Army on March 23, 1863. Gordon continued to experience problems getting pay for the army, however. Gordon went so far as to threaten to defeat the Taipings with the entire army before he secured pay. Between March and July, Gordon kept the 30-mile zone around Shanghai clear of Taipings and advanced outside that perimeter toward Suzhou. The Ever-Victorious Army laid siege to Suzhou through the fall of 1863. He granted amnesty and a pardon to the Taiping leaders of the city and secured its surrender on December 4, 1863. However, Li Hongzhang executed the Taiping leaders in spite of Gordon's promise. Gordon then led the army on the city of Changzhou, which he captured on May 11, 1864. After the fall of Changzhou, the force was effectively disbanded by June 1. Gordon's own journal, however, records four more battles, including the capture of Nanjing from the rebels on July 19, 1864.

REFERENCES

Caleb Carr, *Devil Soldier: The Story of Frederick Townsend Ward* (New York: Random House, 1992). Charles G. Gordon, *General Gordon's Private Diary of His Exploits in China* (London: Sampston, Low, 1885, reprinted New York: Kraus Reprints, 1971); J. S. Gregory, *Great Britain and the Taipings* (New York: Praeger, 1969); Franz Michael, *The Taiping Rebellion: History and Documents*, vol. 1 (Seattle: University of Washington Press, 1966); H. B. Morse, *In the Days of the Taipings: An Historical Retrospect* (Salem, MA: Essex Institute, 1927); Andrew Wilson, *The ''Ever-Victorious Army''* (London: William Blackwood, 1868, reprinted Novato, CA: Presidio Press, 1991).

EYUWAN SOVIET (OYUWAN) With roots in the Red 11th Corps from the **Northern Expedition**, Red Army troops under **Zhang Guotao** combined with **Xu Xiangqian**'s Red Third Army to form a **Revolutionary Base Area** (a ''Soviet'') in the Dabie Mountains (Dabieshan). The Soviet formed in the Anhui, Hubei, Henan border areas and, from 1930 to 1932, was the target of four successive **Encirclement Campaigns** by Nationalist forces. In their own counteroffensive to the first Encirclement Campaign, under Xu Jishen, Communist forces relied on mobility and maneuver, concentrating forces at decisive points to defeat the Nationalists. Zhang Guotao did not take part in this counter encirclement action; he actually reached the revolutionary base area in April 1931. The Second Encirclement Campaign by the Nationalists in spring 1931 was an attempt to completely blockade the Soviet in the Dabieshan Mountains, but it ended indecisively. In January 1932, a third encirclement was attempted by the Nationalists. By this time, Zhang Guotao's army had been designated the **Fourth Front Army**. In this campaign, the Communists defeated the Nationalists because the Guomindang units failed to concentrate their forces and committed units to battle piecemeal. In April 1932, Nationalist forces from Shanghai were moved against the Communists in the Fourth Campaign to encircle the Eyuwan Soviet. Over the summer, Nationalist battlefield successes gradually contracted the maneuver area available to the Fourth Front Army to the point that Zhang Guotao, independently of the Communist Party Central Committee, abandoned the Eyuwan Revolutionary Base Area and marched west toward Sichuan Province, leaving behind a small force in the Dabieshan Mountains. By the time that the forces of the Eyuwan Soviet reached Sichuan, they had dwindled from 16,000 to 9,000 personnel. This 1,500-mile march by the Fourth Front Army took place two years before the more famous **Long March** by the First Front Army.

REFERENCES

William W. Whitson with Chen-Hsia Huang, *The Chinese High Command* (New York: Praeger, 1973); Robert W. McCall, "The Oyuwan Soviet Area, 1927–1932," *Journal of Asian Studies* 27, no. 1 (November 1967): 41–59.

F

FENGTIAN ARMY. See FENGTIAN-ZHILI WAR

FENGTIAN–ZHILI WAR The Fengtian Army, headed by the **warlord Zhang Zuolin**, was based in Manchuria. It fought two wars, in 1922 and 1924, with the North China-based forces of the **Zhili Clique**, headed by the warlord Wu Peifu. The Zhili Clique had fought one war over control of the territory of North China with the **Anfu Clique** (composed of warlords from Anhui and Fujian) in 1920. In this Anfu–Zhili War, Wu Peifu was aided by Zhang Zuolin and the Fengtian faction (or clique). This led to the fall of the Anfu-dominated Chinese central government and created a rough balance of power among China's warlords. Shaanxi Province, at this time, came under the power of one of the Zhili faction members, Feng Yuxiang. As warlords jockeyed for power and the right to control and tax the populace, Hunan Province forces attacked Hubei Province and the forces of the Hubei warlord Wang Zhanyuan. When Wang fled the province to avoid the attacking Hunan Army, Wu Peifu stepped in and moved his own forces south to ''assist'' the people of Hubei. Wu defeated the invading Hunan Army and was appointed the ''Inspecting Commissioner of Hubei and Hunan.'' (On the genesis of the Hunan Army, see **Zeng Guofan**).

The relative strengthening of the position of Wu Peifu angered Zhang Zuolin. In response, Zhang maneuvered to gain control of the central government in Beijing by forcing the resignation of the premier and replacing him with his own man, Liang Shiyi, in December 1921. Liang immediately pardoned some of the members of the Anfu Clique, which angered Wu Peifu. More seriously, Liang issued instructions to the Chinese team attending the **Washington Conference** to yield more land on the Shandong Peninsula to Japan in return for loans from Tokyo. Wu Peifu reacted by issuing a round of telegrams that condemned Liang's instructions as unpatriotic, gathering warlord support against

Liang. Liang resigned his post under pressure ''due to illness,'' creating more friction between Zhang Zuolin and Wu Peifu. During the month of January 1922, Zhang and Wu exchanged a series of telegrams designed to influence other warlords and seek their support. In April 1922, Zhang Zuolin moved his forces south from Manchuria, attempting to invade the North China Plain and the area of Zhili, controlled by Wu Peifu. Wu's Fengtian Army, which had the advantage of having incorporated the older **Beiyang Army**, proved the more effective and disciplined fighting force. At the end of one week of combat, Zhang Zuolin's Fengtian Army retreated into Manchuria. Almost immediately after winning the first Fengtian–Zhili War, however, the Zhili Clique split its coalition into Zhili-, Baoding-, and Tianjin-based factions.

Despite his defeat in the first Fengtian–Zhili War, after retreating into Manchuria in 1922, Zhang Zuolin remained in control of Manchuria. Zhang declared himself to be independent of the central government, hired foreign military instructors, bought new equipment, and worked to develop a weapons manufacturing industry. In 1923, when bandits in Shandong Province captured a train and kidnapped several hundred foreigners, the threat of foreign intervention probably postponed another attempt by Zhang Zuolin to invade Zhili and North China. In September 1924, the warlords of Zhejiang and Jiangsu began a war for the possession and control of Shanghai. Since this once more involved Wu Peifu with the old Anfu Clique that he had defeated in the Anfu–Zhili War, Wu began to support the Zhejiang faction, headed by Lu Yongxiang, while Zhang Zuolin supported the Jiangsu faction headed by Ji Xieyuan. General alignments of the warlord factions at this time were the Zhili Clique, controlled by Wu Peifu with influence in Zhili, Shandong, Jiangsu, Henan, Anhui, Shaanxi, Hubei, Jiangxi, Fujian, Sichuan, and Gansu, and areas opposed to the Zhili Clique were Manchuria, under Zhang Zuolin; Zhejiang, under Sun Quanfang; and Guangdong, controlled by a coalition of warlords. The warlord leaders of Yunnan, Guangxi, Guizhou, and Shanxi (essentially independent under **Yan Xishan**) remained essentially neutral in the factional alignments.

In response to Wu Peifu's support for the Zhejiang faction in the fight over Shanghai, Zhang Zuolin invaded Zhili for a second time in mid-October 1924. Before he moved, however, Wu Peifu had time to position defending forces. Feng Yuxiang, fighting under Wu Peifu's control, moved north to defend the area of Jehol; Peng Shouxin took forces to defend Shanhaiguan, where the Great Wall meets the sea; and Wang Huaiqing moved to control the area north of Beijing. Feng was at Nanyuan (now a military airfield) 30 kilometers south of Beijing, and he moved north of Beijing in late September. However, he never acted, siding with central government at the last minute, and seizing Beijing. The Fengtian Army broke through the Great Wall at Shanhaiguan, approaching Tianjin and forcing Wu Peifu to flee to Shanghai by sea. The Wu Peifu-supported president of China, Cao Kun, resigned on November 2, 1924. Feng Yuxiang kept the ''last emperor'' Pu Yi prisoner in the Forbidden City, while he and Zhang Zuolin negotiated the establishment of a new government. On November 19, Pu Yi fled to the Japanese Legation in Beijing, where he remained a guest until February 1925. Pu Yi lived in the Japanese Concession area of Tianjin until 1931, when he moved north into Manchuria, where the Japanese established Manchukuo in March 1932, after the **Nine One Eight (918) Incident** (Mukden Incident).

REFERENCES

Wunsz King, *China at the Washington Conference, 1921–1922* (New York: St. John's University Press, 1963); James E. Sheridan, *Chinese Warlord: The Career of Feng Yu-hsiang* (Stanford, CA: Stanford University Press, 1966); Jonathan D. Spence, *The Search for Modern China* (New York: W. W. Norton, 1990).

FIELD ARMY SYSTEM The Field Army System of the **People's Liberation Army** (PLA) is a regionally based construct from organizations that incorporate the local combat forces in a specified geographical area, the Military Region, with that area's maneuver forces, called "main forces," and service support organizations. Because most Chinese military officers generally are not rotated out of their military region until they gain high rank, the system turned into a network of strong patron–client relations and political networks that influenced the internal politics of China throughout the first 30 years of the existence of the People's Republic and still has some influence on military promotions today.

The Front Armies (Fangmianjun) of the Red Army were organized into the **Eighth Route Army** and the **New Fourth Army** in 1937. They fought during the period of United Front cooperation with the Nationalists and during the **Anti-Japanese War (World War II)** in that organization, supplemented by localized guerrilla forces, and altered that organization only during the **Civil War** period in order to accommodate more mobile maneuver warfare. In February 1949, toward the end of the Civil War, the Communist military forces were organized into field armies, which were primarily regionally oriented. The **First Field Army** evolved out of the **Second Front Army** and, later, the 120th Division of the Eighth Route Army. It was based in northwest China and dominated the Xinjiang, Lanzhou, and Chengdu regions. The **Second Field Army** evolved from the Fourth Front Army and the 129th Division, Eighth Route Army. It was based around Central China and dominated the Wuhan, Tibet, Yunnan, and southern Sichuan regions. The **Third Field Army** evolved from the First and Fourth Front Army's stay-behind forces in Jiangxi after the base areas were abandoned on the **Long March**. It was the former New Fourth Army, which was centered around eastern China and dominated the Jiangsu, Fujian, and Shandong areas. The **Fourth Field Army** and the **North China Field Army** (the "Fifth" Field Army), evolved out of the First Front Army and, later, the 115th Division, Eighth Route Army. The Fourth Field Army dominated Manchuria, but its influence after the Civil War also extended into Guangdong Province. The North China Field Army was based on the North China Plain, encompassing parts of Shandong, Beijing, Hebei, and Inner Mongolia.

The seminal book on the field army system and its influence is William W.

Whitson's *The Chinese High Command*, which must be read to fully understand the widespread influence of the Field Army System on the PLA. Through at least the 1980s the influence of the senior cadres of these armies was instrumental in military and political affairs in China. The cohort groups and patron–client relationships forged among PLA leaders from years of combat and shared hardship while serving in the same field army extended throughout the PLA and still influence assignments and promotions at the senior levels today.

REFERENCES

Wang Xuepeng, *Zhongguo Dangdai Jiangjun Fengcai* (Beijing: Zhongguo Dangshi Chubanshe, 1993); William W. Whitson with Chen-Hsia Huang, *The Chinese High Command* (New York: Praeger, 1973).

FIRST FIELD ARMY The First Field Army was organized as a distinct military formation in June 1949. It was organized from elements of the **Eighth Route Army**, primarily those operating in the Shaan-Gan-Ning District (comprising forces in parts of Shaanxi, Gansu, and Ningxia Provinces). When it was formed, the First Field Army also drew on forces from the Shanxi Province Pacification District and local force units in these areas. The 18th Army (a U.S. Corps equivalent unit) and the 19th Corps, of the Northeast (Huabei) Field Army, were also put under the control of the First Field Army at the time of its formation. By the end of 1949, the First Field Army had a strength of over 448,000 men, including 15,000 personnel transferred into it from the Northwest Military Region.

In August 1945, the **Central Military Commission** established a Shanxi Military Pacification District, which was commanded by **He Long**. The political commissar for the district was Guan Xiangying. From the time of the defeat of Japan until the end of 1946, the First Field Army took part in the campaigns against the Pingsui rail lines, the campaign to secure northern Shanxi Province, and the capture of Datong (in northern Shanxi Province). In March 1947, when the Nationalist Army sent a force of over 250,000 troops to attack Yunnan, the First Field Army defended that area. The Central Military Commission, after July 31, 1947, established a Northwest Field Army led by **Peng Dehuai**, as part of the Communist Party's Northwest Bureau, which conducted campaigns to liberate Qinghai Province. The Northwest Field Army comprised three columns, two separate divisions, and an independent brigade. Its total strength was about 50,000 troops. At the same time, He Long was given command of the Shaan-Gan-Ning-Jin (Shaanxi, Gansu, Ningxia, Shanxi) Unified Military District. These forces then conducted a series of campaigns to gain control of their respective areas of operation in February 1949. The Northwest Field Theater of Operations was again reorganized, and the forces in the district were redesignated the **People's Liberation Army** First Field Army. Peng Dehuai was appointed Field Army commander, with eight armies (corps) under his control. The forces in the Shaan-Gan-Ning-Jin-Sui district were reorganized into the

Northwest Military Region under the control of He Long. As the war of liberation (**Civil War**) drew to a close, the First Field Army and the forces of the Northwest Military Region conducted a series of campaigns to the west designed to ensure control over Xinjiang and securing all of northwest China (Gansu, Ningxia, and Xinjiang Provinces) for the Communist forces.

REFERENCES

Jin Zha and Xue Qing, *Gonghe Guo Da Junqu Siming Zhuangqi* (Chengdu: Sichuan Renmin Chubanshe, 1995); William W. Whitson with Chen-Hsia Huang, *The Chinese High Command* (New York: Praeger, 1973).

FIRST FRONT ARMY The First Front Army was organized in the Jinggang Mountains (**Jinggangshan Revolutionary Base Area**) of Jiangxi Province after the Nationalist Army reacted to the **Autumn Harvest Uprisings** (1927–1928), crushing the hope of the Communist Party that it could stimulate a general uprising of peasants and workers in South China. Up through the time of the **Long March**, in 1934, the First Front Army often fought independently, often unaware of the activities of its sister units in the **Chinese Workers and Peasants Red Army**. While the First Front Army was defending its **Revolutionary Base Area** in Jiangxi against the Nationalist Army's **Encirclement Campaigns**, designed to destroy all Communist forces, the **Fourth Front Army** was doing the same thing in the **Dabieshan** Mountains of Henan, near the borders of Anhui and Hubei Provinces. Meanwhile, the **Second Front Army**, led by **He Long**, was fighting the Nationalists in the western part of Hunan Province. In Guangxi Province, to the south of the First Front Army's operating area, **Deng Xiaoping** was engaged in a similar defensive effort while trying to organize the Chinese peasantry.

In First Front Army the egalitarian, peasant-based style of leadership and discipline characterized as "Maoist" developed in the **People's Liberation Army**. The First Front Army dates its history to late August 1930, when the Red First Army, led by **Mao Zedong** and **Zhu De**, in Jiangxi Province, combined forces with the Red Third Army, led by **Peng Dehuai. Ye Jianying** served as the first chief of staff of the First Front Army. Mao Zedong's influence over the army is evidenced not only by its egalitarian style but also because the First Front Army developed a strong, active political commissar system. Later, when leaders of the First Front Army met with the leaders of the Fourth Front Army at the **Zunyi Conference**, in 1935, **Zhang Guotao** blamed the political commissar system for fostering mistrust in the ranks of the troops.

By the time of the Third Encirclement Campaign, in December 1931, Mao Zedong was effectively removed from leadership posts in the army and instead operated in the Communist Party structure, where he supervised political commissars in a government leadership post. He was restored to influential leadership positions at the Zunyi Conference. The major rupture between the First Front Army and the Fourth Front Army happened at the **Maoergai Conference** in Sichuan Province. There, in July 1935, Zhang Guotao took his Fourth Front Army farther west, heading for Gansu Province, while Mao Zedong continued north to Shaanxi and **Yan'an** with the First Front Army. When the **Eighth Route Army** formed in 1937, the First Front Army made up the bulk of the 115th Division, commanded by **Lin Biao**.

REFERENCES

William W. Whitson with Chen-Hsia Huang, *The Chinese High Command* (New York: Praeger, 1973); Xiao Zhaoran, *Zhonggong Dangshi Ming Cidian* [A Concise Dictionary of the Chinese Communist Party's History] (Beijing: Jiefangjun Chubanshe, 1986); ZGDBKQS, vol. 1; ZGRMJFJZS, vol. 2.

FIRST REVOLUTIONARY CIVIL WAR. See REPUBLICAN REVOLUTION

FIVE SEVEN ONE (571) PLAN (March 21–22, 1971) The 571 Plan was a secret document allegedly prepared by **Lin Biao** and a group of his associates calling for a coup d'état against **Mao Zedong**. The title of the plan, 571, is pronounced "*wu qi yi*" in Mandarin. The words are a homonym, however, of the term "armed uprising" (*wuzhuang qiyi*). The split between Mao and Lin Biao dated back to the Second Plenum meeting of the Ninth Central Committee of the Chinese Communist Party, August 23–September 6, 1970, where Lin Biao and a group of supporters proposed to establish the position of "state chairman," which would have weakened Mao's authority. By January 1971, in an attempt to weaken Lin Biao's political strength and break apart his base of power, Mao reorganized the Beijing Military Region, replacing some of Lin Biao's **Fourth Field Army** comrades with those of **Chen Yi**'s **Third Field Army**. The attempt to overthrow Mao failed, and Lin Biao was killed on September 13, 1971, when a Trident jet on which he was fleeing China crashed in Mongolia.

REFERENCES

Michael Y. M. Kau, ed., *The Lin Biao Affair: Power Politics and Military Coup* (White Plains, NY: International Arts and Sciences Press, 1975); "Report to the 10th National Congress of the Communist Party of China," *Peking Review*, no. 35–36 (1973); Yao Ming-le, *The Conspiracy and Death of Lin Biao* (New York: Alfred A. Knopf, 1983).

FLYING TIGERS The "Flying Tigers" was a group of aviators, formally known as the American Volunteer Group, or AVG, who were organized by retired U.S. Army Air Corps captain **Claire Lee Chennault**. In the wake of the Japanese attack into China after the **Marco Polo Bridge Incident** (July 7, 1937), which marked the start of **World War II** in China, Chennault accepted an offer from the Chinese government to become an aviation adviser to the Chinese armed forces. Although the United States at that time officially remained neutral in the war, a great deal of support was funneled into China. Between September and November 1940, just after Japan joined the Axis with Italy and Germany, the United States loaned $75 million to China ($25 million in September 1940 and $50 million in November). China also sought from the United States 500 American combat aircraft manned by U.S. pilots to assist in the Chinese war effort.

In January 1941, Chennault, then appointed a colonel as an adviser to the Chinese air force, traveled to Washington to lobby in support of the Chinese request for 500 aircraft. In the end, the U.S. government diverted 100 P-40 fighter aircraft from a lot designated for Great Britain and shipped them to China. The brother-in-law of Chinese generalissimo **Chiang Kai-shek, T. V. Soong**, organized a company called China Defense Supplies, Inc., which handled the shipment of aircraft and supplies. Then, in March 1941, the U.S. Congress approved a law lending money or leasing equipment (Lend-Lease) to "the government of any country whose defense the President deems vital to the United States." This provided about $630 million in Lend-Lease supplies for China, supplemented not long afterward by a $500 million loan to China. A presidential executive order, signed by Franklin D. Roosevelt in April 1941, authorized the release of military pilots from the U.S. armed forces for recruitment as mercenary pilots to fly for the American Volunteer Group, the Flying Tigers, against the Japanese in China. The company that handled the recruitment of these pilots was headquartered in Rockefeller Center, New York, and was called the Central Aircraft Manufacturing Company, or CAMCO. Pilots were attracted to the Flying Tigers by a combination of the desire for combat experience and adventure and a salary that was very high for that time, $750 a month. The contract also gave the pilots a bonus of $500 for each confirmed aerial shootdown of a Japanese aircraft.

In order to avoid violating America's neutrality in the war, the pilots of the Flying Tigers and their aircraft traveled to Burma on ships of Dutch registry. They arrived and went into combat in October 1941, in time to blunt the Japanese military and air effort to block the major surviving supply line into China, the **Burma Road**. The American Volunteer Group was organized into three squadrons and had its own integral medical, maintenance, personnel, and supply support—all recruited from the United States. Functionally, it was an independent air wing.

The Flying Tigers of the American Volunteer Group fought an undermanned, poorly equipped, and outnumbered air war against the Japanese, even after the United States entered the war in December 1941. Finally, on July 4, 1942, the Flying Tigers were disbanded, and many of the pilots reentered the American air forces. Even this, however, was not accomplished with ease. Colonel Chennault had a poor relationship with the American commander in China, General **Joseph W. Stilwell**, and initially balked at serving under Stilwell, agreeing to do so only after a personal interview. Chennault was not trusted by Stilwell, nor was he trusted in Washington by the chief of staff of Army Air Forces, General

H. H. Arnold. The Flying Tigers were initially integrated into the U.S. Army Air Forces as the China Air Task Force, a component of the 10th Air Force, which was based in India and responsible for the air campaign in the China-Burma-India Theater. Discipline was a serious problem as the Flying Tigers were reintegrated into the U.S. military. Many of the pilots had joined for the high pay and the bonuses, which military officers did not earn. A number of the Flying Tiger pilots, even after the personal intervention of Chennault, refused induction into the U.S. armed forces. Because General Chennault's loyalty to the U.S. military leadership was questioned by Stilwell and General Arnold (Chennault had a habit of appealing to Chiang Kai-shek for personal intervention when things didn't go his way), the 10th Air Force commander, Clayton Bissell, was appointed a brigadier a day earlier than Chennault to ensure Bissell's seniority. After operating for a short time as the China Air Task Force, in 1943, the Flying Tigers eventually formed the nucleus of the U.S. 14th Air Force in China and was commanded by Chennault as a major general. However, the theater air commander remained Bissell, who was promoted before Chennault. Even toward the end of the war, when Lieutenant General **Albert C. Wedemeyer** succeeded Stilwell as the commander of American forces in China in 1944, Wedemeyer did not trust Chennault.

REFERENCES

Martha Byrd, *Chennault: Giving Wings to the Tiger* (Tuscaloosa: University of Alabama Press, 1987); Anna Chennault, *Chennault and the Flying Tigers* (New York: P. S. Eriksson, 1963); Claire L. Chennault, *Way of a Fighter* (New York: G. P. Putnam's Sons, 1949); Robert B. Holtz, *With General Chennault* (New York: Coward-McCann, 1943); Roland Sperry, *China through the Eyes of a Tiger* (New York: Pocket Books, 1990); Barbara W. Tuchman, *Stilwell and the American Experience in China, 1911–45* (New York: Macmillan, 1971).

FOREIGN ARMS CORPS (Foreign Rifle Corps) In 1860, **Frederick Townsend Ward**, an American from Massachusetts, with **Henry Burgevine**, as associate of Ward's, organized a group of foreign mercenaries to protect the business interests of his employer, Yang Fang, in Shanghai. The military group was also intended to be used to fight the Taiping rebel forces threatening the area. Ward sought to train only foreigners, and he enlisted European and Filipino mercenaries to fight the Taiping rebels, using Western tactics. However, Ward failed to gather the financial and administrative support needed to sustain the unit, since it offended Chinese officials to have foreign mercenaries operating on their soil (the Chinese were still smarting from their defeat in the **Opium** and **Arrow Wars**). By 1862, after Ward and Burgevine won a number of battles, the Foreign Arms Corps gave way and was replaced by several jointly manned Sino–foreign units. The foremost of these was the English-influenced **Ever-Victorious Army**, which was led by Ward and Burgevine, the French-led Franco-Chinese Corps of Jiangsu, and the French-led **Ever-Triumphant Army**.

REFERENCES

Prosper Giquel, *A Journal of the Chinese Civil War, 1864*, trans. Steven Leibo (Honolulu: University of Hawaii Press, 1985); Franz Michael, *The Taiping Rebellion: History and Documents* (Seattle: University of Washington Press, 1966); Ssu-yu Teng, *The Taiping Rebellion and the Western Powers: A Comprehensive Survey* (Oxford: Clarendon Press, 1971).

FORMOSA CRISIS (1874) A shipwreck in 1871 on Taiwan (Formosa) involving sailors from the Ryukyu Islands (Okinawa) grew in seriousness between that year and 1874, until it developed into a full-blown crisis between China and Japan. Japanese sailors from the Ryukyu Islands were marooned on Taiwan in 1871 as a result of a shipwreck. Some of the sailors were killed by native Taiwanese aboriginals. Japan, arguing that the Ryukyu Islands acknowledged Japanese suzerainty because residents paid tribute to Japan, claimed that the murdered sailors were subjects of the emperor and demanded that China provide some form of compensation or other redress for the actions of the Taiwanese aborigines. The Qing government, which, like Japan, also received tribute from the rulers of the Ryukyu Islands (Liujiu in Chinese), took no action. The dispute, therefore, went unresolved for a number of years.

Under pressure from militarists in Japan to resolve the matter, the Japanese government sent a naval expedition to Taiwan in 1874. At that time, the Qing dynasty had two officials who dealt with maritime and international matters: the commissioner of trade in South China, based in Nanjing, and the northern commissioner of trade in Tianjin. Although both positions had been created by the Qing Court to attend to issues involving foreign relations and to meet China's responsibilities relating to the **Treaty of Tianjin**, neither official had responsibility for coastal defense. At the time the crisis developed, **Li Hongzhang** was concurrently governor-general of Zhili and northern commissar of trade. Li Zongxi was southern commissioner of trade in Nanjing. The Qing Court directed both officials to address the Japanese threats, but it was Li Hongzhang who actually took action.

In early 1874, Japan sent a force of 3,600 troops and three ships to Taiwan to deal with the matter. Li Hongzhang, in response, recommended to the Qing Court on May 10, 1874, that the superintendent of the **Fuzhou Dockyard**, Shen Baozhen, be dispatched to Taiwan with troops and ships in response to the Japanese actions. The Chinese fleet, at the time, was not concentrated in a naval base, nor had it trained for naval action. Instead, the ships were distributed along the coast, where they tried to control piracy. Another reason the fleet was dispersed was to distribute the cost of maintaining the ships and their crews to local officials, reducing expenses for the Qing government. As a consequence of this dispersion, Shen had a great deal of trouble assembling a fleet. Moreover, Shen was hampered by faulty intelligence. He believed that Japan had a fleet of steam-powered ships and two ironclad steamers. Shen's own ships were all made of wood.

By July, Shen Baozhen still had not reacted to the Japanese naval force. He asked for ships and troops from the northern and southern commissioners, seeking to assemble 19 ships and a credible force of troops. By September 1874, however, he had gathered only 6,500 troops from the Anhui Army and seven steamships, all supplied by Li Hongzhang. He sent these forces to the Pescadore (Penghu) Islands, where he had gathered another 6 ships. He also got a battalion of troops from Hubei Province, which he sent to the Pescadores by ship. By November 1874, after six months of effort, Shen assembled a force of 10,000 troops and 16 ships in the Pescadores. However, at no time had he taken any action to intercept the Japanese fleet or any Japanese ships on their way to Taiwan.

By the end of 1874, rather than risk war, the Qing Court settled the matter with the Japanese. China paid a monetary indemnity to Japan, which tacitly recognized Japan's claims to the Ryukyu Islands. Japan, in response, withdrew its forces from Taiwan. The Formosa Crisis of 1874 had the effect of focusing China's attention on its need for a credible, effective fleet unified into a navy. The northern and southern commissioners, from that time on, assisted by **Prosper Giquel**, who figured prominently in the **Taiping Rebellion**, were part of the **self-strengthening movement**. They ordered a number of cruisers and gunboats from foreign shipyards and established a "Sea Defense Fund."

REFERENCES

James P. Baxter, *The Introduction of the Ironclad Warship* (Cambridge: Harvard University Press, 1933); John L. Rawlinson, *China's Struggle for Naval Development, 1839–1895* (Cambridge: Harvard University Press, 1967).

FOURTH CORPS OF THE RED ARMY After the **Autumn Harvest Uprisings** and the withdrawal into the **Jinggangshan Revolutionary Base Area** in June 1928, the Chinese Communist Party (CCP) Central Committee decided on a name change for the Fourth Army of the Workers and Peasants Revolutionary Army. The Central Committee renamed the surviving Red Army unit the Fourth Corps of the Red Army (Hongsijun). The commander was **Zhu De**, while **Mao Zedong** was named party representative. **Chen Yi** assumed the position of secretary of the Party Committee of the Army, and Wang Erzuo was the chief of staff. Four regiments, the 28th, 29th, 31st, and 32d, were subordinated to the army, which was actually smaller than a division in modern terms. The total manpower strength was over 6,000. For the six months after formation the Fourth Army of the Red Army was engaged against Guomindang (Nationalist) forces attempting to encircle and destroy the Jinggangshan Mountain Revolutionary base.

The Fourth Corps joined with the Fifth Corps of the Red Army in December 1928, under the command of **Peng Dehuai**, to effect a defense of the base area. By January of the next year, one regiment, the 32d, was resubordinated to the Fifth Army of the Red Army. The remaining forces of the Fourth Corps were organized into three columns to conduct guerrilla warfare operations against the Nationalists. After two years of operations, mostly engaged in guerrilla actions, the Fourth Army of the Red Army combined with the third and 12th Armies to form the **First Front Army**.

REFERENCES

William W. Whitson with Chen-Hsia Huang, *The Chinese High Command* (New York: Praeger, 1973); ZGDBKQS.

FOURTH FIELD ARMY The military lineage of the Fourth Field Army is traced in the **People's Liberation Army** (PLA) military histories to the First Corps of the **First Front Army. Lin Biao** and **Xiao Jingguang** were the leaders most closely associated with this organization. The Field Army was really first merged together in Manchuria, under Lin Biao's leadership. In the days of the Jiangxi Soviet (**Jinggangshan Revolutionary Base Area**), **Luo Ronghuan**, whose career was closely tied to the fate of **Mao Zedong**, was also an influential leader of the Fourth Field Army.

The **Encirclement Campaigns** around the area of the **Jinggangshan Base Area** by Nationalist forces were all countered by Fourth Field Army antecedent forces. The fifth Encirclement Campaign, which precipitated the **Long March**, drove the Red Army's units out of Jiangxi, and the Long Marchers included many of the units that in 1949 were forged together into the **Fourth Front Army**. While some units split off into the other field armies, the bulk of **Zhu De**'s First Front Army units eventually ended up in the Fourth Field Army. During the **Eighth Route Army** period, the 115th Division of the Eighth Route Army was built from these units.

During the **Civil War**, Lin Biao led his armies in the **Manchurian Campaign** and the **Sungari River** offensive and defensive campaigns. The units then continued south, participating in the **Ping-Jin (Beiping-Tianjin) Campaign**. By the time that the cities and rail lines of the North China Plain were secured, the former Northeast Field Army troops under Lin Biao, in March 1949, were reorganized into the Fourth Field Army. They were also preparing for a major campaign to take South China from the Nationalists.

The Fourth Field Army's 12th, 13th, and 15th Armies crossed the Yangtze River into South China in June and July 1949, behind units of the **Second Field Army**. Lin Biao drove west and south to Hunan and Guangzhou with a force of 10 armies. (In 1949, each numbered army comprised three corps; today a PLA Group Army is the equivalent of one of the corps of a numbered army subordinate to the 1949 Field Army order of battle.) Fourth Field Army forces continued the campaign against the Nationalists through 1950, when they finally overwhelmed the defenders of Hainan Island in a costly amphibious operation during mid- to late April 1950.

As the Fourth Field Army cadre began to take up administrative and government posts in Guangdong and Guangxi Provinces, the **Korean War** broke out.

Lin Biao was the commander of the **Chinese People's Volunteers** who formed the first echelon of Chinese forces that entered North Korea.

REFERENCES

William W. Whitson with Chen-Hsia Huang, *The Chinese High Command* (New York: Praeger, 1973); ZGDBKQS.

FOURTH FRONT ARMY The Fourth Front Army was organized in the Dabie Mountains of Henan Province, where the **Eyuwan Soviet** was located. It was heavily influenced by **Xu Xiangqian**, who in 1955 was made a marshal of the **People's Liberation Army** (PLA). Xu emphasized maintaining a conventional military force that was well trained and highly disciplined, as opposed to the more egalitarian, ''guerrilla-style'' approach to military leadership and discipline taken by the leaders of the **First Front Army**. In particular, the leaders of the Fourth Front Army, especially **Zhang Guotao**, rejected the political commissar system and its internal political security mechanisms embedded in each PLA unit, which they believed divided the soldiers as well as the leaders and turned them against each other.

In fall 1932, the Fourth Front Army was driven out of the Eyuwan Soviet by the Nationalist **Encirclement Campaigns**. The army embarked on a **Long March** to Sichuan Province at that time, where, under Zhang Guotao's direction, it established a new **Soviet Base Area** in the northern part of Sichuan. By 1935, Zhang Guotao's Sichuan–Shaanxi Soviet Base Area was well established, and the Fourth Front Army had grown to a strength of about 60,000 armed soldiers. In spring 1935, about 15,000 of these troops stayed in the Sichuan–Shaanxi base area to secure the territory, while the remainder of the army, some 45,000 soldiers strong, marched to Mougong. They linked up there with forces of the First Front Army on June 16, 1935. Serious disagreements broke out almost immediately among the leaders of the two armies over matters of military style and strategy but especially over discipline issues and the First Front Army's political commissar system.

On matters of strategy, Zhang Goutao wanted to move the entire People's Liberation Army further into northwest China, closer to the Soviet Union, into the Gansu-Xinjiang Province area. There, he believed, the PLA could recover from the Long March, build its strength in relative safety and security, and be resupplied by Stalin more easily. **Mao Zedong** opposed this course of action, arguing that, strategically, it was more important to engage the Japanese. A final decision on the two opposing courses of action was postponed temporarily, with **Zhou Enlai** mediating the dispute, and both armies moved a short distance to Maoergai, where another conference was convened. The **Maoergai Conference**, held in July 1935, led to a serious split between the two armies, with Zhang Guotao and **Zhu De** eventually splitting from the First Front Army and moving the Fourth Front Army north to Gansu Province. Mao Zedong, **Lin Biao, Peng Dehuai**, and **Ye Jianying** moved the First Front Army into Shaanxi Province.

There they established the revolutionary base area at **Yan'an. Liu Bocheng** went to Gansu Province with the Fourth Front Army, splitting the senior leadership of the PLA.

By the time that the **Eighth Route Army** was established, in August 1937, the Fourth Front Army had reestablished itself as a People's Liberation Army force. The Fourth Front Army was converted into the 129th Division of the Eighth Route Army. Many senior Fourth Front Army cadres also joined the **New Fourth Army**, which moved to the area around the Dabie Mountains, when it was established in 1938. The bulk of the Fourth Front Army, after becoming the Eighth Route Army's 129th Division, evolved into the **Second Field Army** after 1949. Liu Bocheng was the most senior of the Fourth Front Army leaders associated with the Second Field Army.

REFERENCES

Xiao Zhaoran, *Zhonggong Dangshi Ming Cidian* [A Concise Dictionary of the Chinese Communist Party's History] (Beijing: Jiefangjun Chubanshe, 1986); William W. Whitson with Chen-Hsia Huang, *The Chinese High Command* (New York: Praeger, 1973); ZGDBKQS, vol. 1.

FOURTH RED ARMY. See NEW FOURTH ARMY

FRANCO–CHINESE WAR (1884–1885) (Sino–French War) During the mid-eighteenth century, as France opened treaty ports in China, French interests in Southeast Asia also increased significantly, especially in Annam and Cochin China (now Vietnam). During the **Taiping Rebellion** (1850–1864), while French forces and military leaders were assisting the Qing Court in controlling the rebellion in China, France took control of Saigon, in South Vietnam, and most of the Red River Delta, including Hanoi and the port of Haiphong in the north. The ruler of Annam, resident in Saigon, signed a treaty with France in 1874, making all of Annam a French protectorate. However, the area had traditionally been a tributary state of China, and the Qing dynasty vigorously protested the French–Annamite Treaty. China took no action against France or Annam, however, because the Qing rulers at the time were more concerned with their dispute with Japan in the **Formosa Crisis** of 1874. In 1880, the ruler of Annam sent tribute to China, an action that reinforced the Qing dynasty claims to the area. Meanwhile, troops subordinate to a former Taiping rebel, Liu Yongfu, based in South China, operated along China's southern border, harassing French shipping on the Red River and making forays into northern Cochin China. The Qing emperor sent troops to suppress these rebels, and by 1884 the troops under Liu Yongfu were incorporated into the Qing Army. The Qing Court decided to negotiate the problems along the coast and off Indochina, resulting in the Li–Fournier Agreement of 1884. However, the agreement was so vaguely worded that, although it recognized the 1874 French treaty with Annam, it also contained stipulations that China continued to exercise suzerainty over Annam.

By mid-1884, in response to acts of piracy and attacks on French ground forces, soldiers of the French and Chinese armies were fighting along the Yunnan border area at the point where the Red River flows south from China into Vietnam. France sought indemnity from China for damage in a battle in June 1884 and began a naval buildup along the Chinese coast. **Li Hongzhang**, who had negotiated the Li–Fournier Agreement, warned the Qing Court that France might attempt to seize the **Fuzhou Dockyard** to reinforce its indemnity claims, while the Chinese governor-general in Canton, who was responsible for administering southern China, including the area around Fuzhou, tried to first fortify, then blockade the Min River, where the Fuzhou Dockyard was located. This disrupted normal trade on the coast. In response to the disruption of trade, the French sent a fleet into Fuzhou Harbor, on the southern end of the Taiwan Strait, and anchored it off **Ma Wei**, on the east bank of the Min River opposite the city of Fuzhou and not far from the Fuzhou Dockyard. On August 23, 1884, China had 11 ships in the area of Ma Wei on the Min River, ranging from three 1,400-ton ships to smaller gunboats. France had eight ships at anchor there. Although war had not been declared, the French admiral, A. A. P. Courbet, attacked the Chinese navy in the Min River. In less than 30 minutes, the firing ended with the French victorious. The French then proceeded to destroy the entire Chinese dockyard complex at Fuzhou. There are varying accounts of the total damage and casualties to China from that battle. One account says that 22 Chinese warships were sunk, killing 39 Chinese officers and 2,000 sailors. By another account, 11 ships of the ''Chinese Self-Strengthening Fleet'' were sunk, and a total of 521 Chinese sailors were killed and 51 missing after the attack by Admiral Courbet (see **self-strengthening movement**).

In the wake of the battle, China declared war on France on August 26, 1884. Admiral Courbet took the French fleet to Formosa (Taiwan) on the opposite side of the strait but was unable to find a suitable place to land forces on the island. Instead, he established a blockade along the west coast of the island, where all the usable ports are located, which he maintained until April 23, 1885, when Sino–French negotiations resumed. On land, the war was prosecuted by the Chinese with some success, and the Qing Army scored a significant victory over the French at Langson. Meanwhile, Chinese naval forces sought to relieve the naval blockade on Formosa. Li Hongzhang sought to raise and train a new fleet, which moved up and down the southern Chinese coast attempting to avoid engagements with the French. On February 13, 1885, another major naval engagement was fought against the French at the Yangtze River estuary. Admiral Courbet sent two launches with quiet running engines into the estuary armed with torpedoes. The Chinese ship *Yu Yuan* was hit by torpedoes and sunk, and the *Deng Qing*, although not hit, was apparently scuttled and abandoned by its crew in their confusion and fear. Because of the defeat, the Chinese admiral, Zhang Cheng, was beheaded by the Qing Court. The war ended with the French victorious, primarily because their naval training, seamanship, and tactics were better. Despite the new ships that the Chinese had purchased from abroad in the self-strengthening movement, the Chinese navy was never unified and fought under foreign leadership, rendered ineffective at times by poor Chinese naval leadership and the political structure of the Qing Court, which tended to make military leaders indecisive.

REFERENCES

Stanley Karnow, *Vietnam: A History* (New York: Viking Press, 1983); John L. Rawlinson, *China's Struggle for Naval Development, 1839–1895* (Cambridge: Harvard University Press, 1967); Jonathan Spence, *The Search for Modern China* (New York: W. W. Norton, 1990); Bruce Swanson, *Eighth Voyage of the Dragon—A History of China's Quest for Seapower* (Annapolis, MD: Naval Institute Press, 1982).

Vance A. Morrison

FU ZUOYI (Fu Tso-Yi) (1895–1974) Fu Zuoyi was born in 1895 in Jungho County, Shanxi Province. After graduating from Baoding Military Academy, he entered the military service of **Yan Xishan**, the **warlord** in Shanxi Province. Fu joined the Guomindang together with Yan in 1927. During his military and political career Fu was seen as both a hero and eventually a traitor to the Guomindang. From October 12, 1927, to January 12, 1928, he captured and held Chochow for the Guomindang and earned respect as a courageous military leader. In 1930, with Yan Xishan, Fu joined a coalition against **Chiang Kai-shek**. The coalition failed, but Fu maintained his power and popularity in the Shanxi Province area, becoming governor of Suiyuan in 1931. For the next 17 years his efforts focused in the Suiyuan area, first fighting the Japanese and later the Communist forces. From 1932 until the end of **World War II** (1945), he fought local bandits in Suiyuan and the Japanese in the Great Wall areas of Jehol and Chahar Provinces. He was responsible for victories against pro-Japanese Inner Mongolian leader Te Wang, driving him from Suiyuan. Fu became famous in Suiyuan's resistance at the Battle of Pailingmaio in November 1936, against the Japanese and their Mongol allies. During this time he moved up in political circles, becoming a member of the Central Committee of the Guomindang while continuing through 1945 as a military leader and local hero.

After the Japanese surrender in 1945, Fu's efforts were concentrated against the Communist forces, and he won a major, though brief, victory against the Communists in North China in 1946, lifting the siege of Datong in Shanxi and taking Kalgan (Zhangjiakou) from the Communists. He stayed active in the campaign against the Communists for some time, but by mid-1947, the Guomindang offensive was failing. Fu became commander in chief of the North China Bandit Suppression Headquarters in December 1947. His armies were focused in the Beijing-Tianjin-Baoding area and Tangshan, which is south of Tianjin. Though for a time his armies held firm in east Hebei through summer 1948, by December 1948 Fu had suffered a major defeat, losing most of his 11th Army group in Kalgan as well as the 35th Army, 105th Army, 106th Army, and part of the 16th Army Group in Xinbaoan. The Communists outnumbered Fu's forces, which consisted of 50 divisions of about 10,000 troops each, by about 2 to 1. The Communist campaign strategy of keeping Fu Zuoyi's forces confined to these northern regions and preventing him from reinforcing Nationalist armies fighting in the **Huai-Hai** or the Yangtze River campaigns was successful. After these losses the remainder of Fu's troops were one division in Beijing and one in Suiyuan. Acting on his own and influenced by his daughter, who was a Communist, and other colleagues, he sent his representative to secretly sign a surrender agreement with the Communists, which resulted in the fall of Beijing, the conclusion of the **Beijing–Tianjin Campaign** in favor of the Communists, and the Guomindang's loss of North China. More seriously, since Fu had agreed to evacuate Beijing and incorporate his remaining troops in the **People's Liberation Army** (PLA), the PLA ranks were increased by over 250,000 Nationalist troops, which filed out of the city of Beijing.

Fu later became the minister of water conservancy for the People's Republic of China and was awarded the Liberation Medal by Mao Tse-tung. He died in Beijing in 1974.

REFERENCE

William W. Whitson with Chen-Hsia Huang, *The Chinese High Command* (New York: Praeger, 1973).

FUJIAN INCIDENT. See FRANCO–CHINESE WAR

FUTIAN MASSACRE (Futian Incident) (December 1930) Using the Red **First Front Army** as a base of operations, **Mao Zedong** organized an investigation into the activities of local soviets in the area of southwest Jiangxi Province, concentrating in Zhi An and Futian. Mao's goal was to link supporters of Li Lisan with the "Anti-Bolshevik" (A-B) Corps operating in the area. The A-B Corps was an organization of the Nationalists designed to purge Communists from Nationalist units. It was organized by the Guomindang in 1927. Mao charged that members of the Jiangxi Provincial Action Committee and the **Jiangxi Soviet** leadership were rightists because they opposed his own ideas and supported those of Li Lisan. Mao ordered his subordinates to arrest and imprison the Jiangxi leaders. As many as 4,000 officers and men of the Jiangxi Soviet were allegedly arrested.

Following their arrest, a battalion of the Red Army's 20th Corps (Red 20th Army) revolted on December 7, 1930, and attacked Futian, where the arrested men were held. According to Nationalist reports, as many as 10,000 Red Army soldiers and civilians of the Jiangxi Soviet were "suppressed" and liquidated by forces loyal to Mao. **People's Liberation Army** histories make no mention of this liquidation on Mao's orders but say that "because of internal contradictions the Red 20th Army was reorganized and redesignated the 64th Division."

REFERENCES

Stephen C. Averill, "The Origins of the Futian Incident," in Tony Saich, ed., *New Perspectives on the Chinese Communist Revolution* (M. E. Sharpe, 1995), pp. 79–115; JGRMJFJZS, vol. 1; Warren Kuo, *Analytical History of the Chinese Communist Party* (Taipei: Institute for International Relations, 1968); William W. Whitson with Chen-Hsia Huang, *The Chinese High Command* (New York: Praeger, 1973).

FUZHOU DOCKYARD The Fuzhou Dockyard was one of several regional weapons arsenals and shipyard facilities established in China by local officials with foreign equipment or assistance. Other arsenals and dockyards were established in Shanghai (Jiangnan), Suzhou, Nanjing, Tianjin, and Jilin. The Fuzhou Dockyard was established primarily with French assistance after two Frenchmen who had helped raise armies during the **Taiping Rebellion, Prosper Giquel** and **Paul D'Aiguebelle**, signed a contract with **Zuo Zongtang** to operate the dockyard. The contract was signed in 1866, and the dockyard was actually established under contract to build 16 ships in 1867.

In fact, the contract with Zuo Zongtang called for more than ship construction. Giquel and D'Aiguebelle were also supposed to establish a naval engineering school and a navigation school at Fuzhou, all within a five-year period. The dockyard was to be operated by Chinese natives after the expiration of the contract. The Fuzhou Dockyard launched its first ship June 1869, the *Wan Nian Qing* (Ten Thousand-Year Qing Dynasty), a steam-powered, 238-foot ship with six guns, powered by a screw propeller rather than paddle wheels. By 1873, 11 warships of various classes, from corvette to gunboat, had been produced by the dockyard. The final four ships constructed at the Fuzhou Dockyard were merchant ships built to haul passengers and cargo. Three of these, the *Chen Hang, Yong Bao*, and *Da Yu*, carried troops from **Li Hongzhang**'s Anhui Army to Taiwan during the **Formosa Crisis** of 1874 with Japan.

Between 1874 and the **Franco–Chinese War** (1884–1885), the Fuzhou Dockyard launched an additional seven ships, which were of composite, iron frame-wooden skin construction. But the dockyard was continually hampered by financial problems and, in fact, supported partially by opium taxes. After the Franco–Chinese War, the dockyard's financial situation improved, but it had to begin procurement from sources other than France, turning primarily to England and Germany.

REFERENCES

Li Hong Chang, *Memories of Li Hong Chang*, ed. William Mannix (Boston: Houghton Mifflin, 1913); Prosper Giquel, *The Foochow Arsenal and Its Results: From Commencement in 1867 to the End of the Foreign Directorate on 16 February, 1874*, trans. H. Lang (Shanghai: 1874); John L. Rawlinson, *China's Struggle for Naval Development, 1839–1895* (Cambridge: Harvard University Press, 1967).

G

GANG OF FOUR The "Gang of Four" is the term coined for the four primary leaders of the Chinese Communist Party's **Cultural Revolution** (Great Proletarian Cultural Revolution, or Wenhua Da Geming, in Chinese). Aside from **Mao Zedong**, the major Cultural Revolution leaders, who were arrested by Mao's former bodyguard and head of the Central **People's Liberation Army** Guards Bureau (8731 Unit), were Jiang Qing, Yao Wenyuan, Zhang Qunqiao, and Wang Hongwen. Jiang Qing was the wife of Mao Zedong. Zhang Qunqiao was the Shanghai Communist Party Bureau leader and, in 1969, a party Politburo member. Yao Wenyuan was an essayist and drama critique who sparked the Cultural Revolution with his critique of the play *Hai Rui Dismissed from Office* (said to have attacked Mao Zedong for dismissing **Peng Dehuai** at the **Lushan Conference** in 1959). Wang Hongwen was a young, rebellious cadre activist.

REFERENCES

Lowell Ditmer, *Liu Shao-ch'i and the Chinese Cultural Revolution* (Berkeley: University of California Press, 1974); Richard Curt Kraus, *Class Conflict in Chinese Socialism* (New York: Columbia University Press, 1981); Thomas W. Robinson, ed., *The Cultural Revolution in China* (Los Angeles: University of California Press, 1971).

GAO GANG (c.1902–1954) A member of the Chinese Communist Party's (CCP) leadership elite and a senior political commissar in the **People's Liberation Army** (PLA), Gao's fall from power is studied as the first instance of serious factionalism within the party leadership after 1949. Gao was one of the earliest members of the CCP in Shaanxi Province. It was to his redoubt near **Yan'an** that **Mao Zedong** led the **First Front Army** after the **Long March**. In June 1945, Gao Gang was elected a CCP Central Committee member, the only leader of the Shaanxi Soviet so rewarded. During the **Civil War** (1946–

1949), Gao served as political commissar to **Lin Biao** in Manchuria. After the Communist victory, Gao stayed in Manchuria and became secretary of the CCP's Northeast Bureau. Concurrently, he held the positions of commander and political commissar of the Northeast Military Region and chairman of the Northeast People's Government. Of the six regional areas into which the CCP had divided the country in the first years, only Gao held all four posts. His positions as military region commander and political commissar in the heavily industrialized northeast, a region with a tendency toward separatism and "warlordism," made him one of the most powerful party and army figures in postliberation China.

In July 1949, Gao went to Moscow, where he negotiated a barter agreement with the Soviet Union. Because in 1949 Manchuria was occupied by Soviet troops and later was a vital sanctuary for China's prosecution of the Korean conflict, there were some suggestions that Gao's alleged ties to Stalin (who died in 1953) may have emboldened Gao to engage in factionalism and to oppose Mao and the Central Committee. When the CCP decided to shift its economic and political policy from New Democracy to socialism and to recentralize administrative authority, it brought Gao and other senior officials to Beijing. In 1952, Gao became the first chairman of the State Planning Commission, a key post. Gao also became a Politburo member.

During the mid-1953 Financial and Economic Conference, Gao leveled criticism against **Liu Shaoqi** and **Zhou Enlai**. Hinting that he was a worthy successor to Liu, Gao lobbied other leaders to replace Liu during a postconference tour of the country. In accentuating cleavages within the leadership, Gao reportedly raised the questions of policy differences, fanned the resentment of PLA leaders at alleged favoritism toward cadres from "White areas" during the Civil War, and denigrated the work of Liu Shaoqi and Zhou Enlai. Gao allegedly offered positions in a future Gao-led government to his interlocutors, among whom was Rao Shushi, who was also a senior political commissar during the Civil War period. Rao served in the **Huai-Hai Campaign** under **Chen Yi**.

During a 1953 Politburo meeting, Mao called for a stop to Gao's factionalist activities. The CCP was to address Gao's challenges and criticism at the February 1954 Fourth Central Committee meeting, but Gao refused to accept blame for his alleged actions, and the meeting ended without resolution. Gao committed suicide in August 1954. The first sketchy, public accounts of Gao's challenge to the CCP emerged from the March 1955 National Policy Conference. **Deng Xiaoping** presented the charges against Gao at that conference, but Deng's report was not made public until years later.

REFERENCES

Avery Goldstein, "Trends in the Study of Political Elites in China," *China Quarterly* 139 (September 1994): 714–30; Sergei N. Goncharov, John W. Lewis, and Xue Litai, *Uncertain Partners: Stalin, Mao and the Korean War* (Stanford, CA: Stanford University Press, 1993); Donald Klein and Anne B. Clark, *Biographic Dictionary of Chinese Com-*

munism: 1921–1965 (Cambridge: Harvard University Press, 1971); Frederick C. Teiwes, *Politics and Purges in China: Rectification and the Decline of Party Norms: 1950–1965* (White Plains, NY: M. E. Sharpe, 1979); Frederick C. Teiwes, *Politics at Mao's Court: Gao Gang and Party Factionalism in the Early 1950's* (Armonk, NY: M. E. Sharpe, 1990).

David E. Reuther

GENERAL DEPARTMENTS OF THE PEOPLE'S LIBERATION ARMY

The senior military policy and operations body in China is the **Central Military Commission (CMC) of the Chinese Communist Party**. The CMC makes all policy and operational decisions for the **People's Liberation Army** (PLA), which are carried out through the General Departments: The General Staff Department (GSD), responsible for strategy and operations; the General Political Department (GPD), responsible for party indoctrination, internal security, and psychological operations against Taiwan; the General Logistics Department (GLD), responsible for supply, material, construction and medical care; and the General Armaments Department (GAD), created in summer 1998, responsible for weapons development and procurement. The GSD functions like a German or Soviet general staff, controlling the ground forces, the air and naval forces, strategic rocket (missile) forces, intelligence, training, and the combat arms of the military (armor, artillery, communication).

There is a parallel governmental structure of military command and control subordinate to the State Council, with a Ministry of National Defense and a State Central Military Commission. However, this organization exists on paper only to rubber-stamp party CMC decisions. The defense minister is a junior CMC member, coequal to the heads of the General Department of the PLA.

Military policy originates in the Central Military Commission of the Communist Party and is turned into operational directives through the General Staff Department. The GSD also serves as the operational headquarters for all of the ground forces of the PLA. The General Political Department was created in 1931 as the military "organization department" of the party. Its political commissars hold coequal status with unit commanders. The GPD was abolished in 1937 during the **United Front**, at the request of the Nationalists, who had their own GPD. The Communist Party's GPD was revived in 1946 and today is a distinct career tract for PLA officers. The GLD was created out of the General Rear Services Department to ensure a seamless system of logistics, repair, service support, and supply in the PLA. It is a powerful organization that controls a number of military industries. Many of its directors have gone on to be chiefs of the General Staff Department.

REFERENCES

Frederica M. Bunge, ed., *China, a Country Study* (Washington, DC: Government Printing Office, 1981); Bruce Gilley, "Stand-Down Order," *Far Eastern Economic Review* (September 10, 1998); James Pinkney Harrison, *The Long March to Power* (New York: Praeger, 1972); Ellis Joffe, *The Chinese Army after Mao* (Cambridge: Harvard University Press, 1987).

GENERALS—PLA OFFICERS PROMOTED TO GENERAL IN 1988

On October 1, 1988, when military ranks were restored in the **People's Liberation Army** (PLA), 17 officers were selected for promotion to full general (*shangjiang*). Ranks had been abolished in the PLA in 1965, in a wave of egalitarianism and what may be termed a ''guerrilla mentality,'' after having been conveyed in 1955. In a number of cases these promotions were more honorary than actual, since the officers were quite old (the oldest general promoted on that day was 81). However, these promotions were very important symbolically, because they restored the ranks of some officers who had been stripped of their rank, purged, and persecuted during the **Cultural Revolution**. Other promotions to general were rewards for good performance in the **Self-Defensive Counterattack against Vietnam** in 1979. The Vietnam experience of the PLA really provided the impetus for the restoration of ranks, because battlefield command and control in Vietnam were hampered by not having a clearly defined seniority system in adjoining units.

Name	Age	Year Entered PLA	Field Army Association[a]
Hong Xuezhi	81	1929	4
Liu Huaqing	78	1931	1
Qin Jiwei	80	1927	2
Chi Haotian	65	1944	3
Yang Baibing	73	1938	not available
Zhao Nanqi	63	1945	not available
Xu Xin	73	1937	North China
Guo Linxiang	77	1930	North China
You Taizhong	76	1931	2
Wang Chenghan	77	1930	2
Zhang Zhen	80	1930	3
Li Desheng	78	1930	2
Liu Zhenhua	73	1938	4
Xiang Shouzhi	77	1934	2
Wan Haifeng	74	1933	3

Name	Age	Year Entered PLA	Field Army Association[a]
Li Yaowen	76	1938	3
Wang Hai	69	1944	3

[a] Field army with which the general was associated by virtue of service. The patron–client nature of Chinese military politics is built on association in field armies. See **Field Army System**.

REFERENCE

Guo Yi, *Zhonggong Jun Tidian Jiangzhuan* (Taipei, Taiwan: Wenhua Chubanshe, 1994).

GENERALS—PLA OFFICERS PROMOTED TO GENERAL IN 1993

As Jiang Zemin sought to consolidate his power and assumed the posts of president, general secretary of the Chinese Communist Party, and chairman of the Chinese Communist Party **Central Military Commission** (CMC), he worked with the CMC to regularize and professionalize the military promotion and retirement system. In 1993, Jiang promoted 25 officers to the rank of general. Many of the generals appointed in 1988 had retired, thereby regularizing the Officer Corps of the **People's Liberation Army**.

Name	Age	Year Entered PLA	Field Army Association
Zhang Wannian	69	1944	4
Yu Yongbo	63	1947	4
Fu Quanyou	64	1946	1
Zhu Dunfa	67	1939	2
Xhang Lianzhong	63	1947	3
Cao Shuangming	65	1946	2
Song Qingwei	63	1947	4
Gu Shanqing	65	1946	3
Wang Ruilin	65	1945	3
Zhang Taiheng	63	1946	3
Xu Huizi	62	1948	4
Liu Jingsong	61	1951	not available
Li Wen	64	1948	3
Wang Ke	63	1944	3
Zhou Keyu	65	1947	3
Gu Hui	64	1947	4

Name	Age	Year Entered PLA	Field Army Association
Li Jiulong	65	1945	4
Shi Yuxiao	61	1949	1
Li Jing	64	1946	3
Cao Sheng	64	1946	4
Dai Xuejiang	64	1946	3
Yang Dezhong	71	1938	2
Li Laizhu	62	1947	2
Li Xilin	64	1945	2
Ding Henggao	63	1961	none

REFERENCE

Guo Yi, *Zhonggong Jun Tidian Jiangzhuan* (Taipei, Taiwan: Wenhua Chubanshe, 1994).

GENG BIAO (1909–1996) Geng Biao was born in 1909 in Hunan Province. He had a 56-year career that combined military service and foreign diplomacy. In 1979, he was one of the retired generals and senior military leaders reported to have signed a letter urging **Deng Xiaoping** not to use the **People's Liberation Army** (PLA) to resolve the crisis in Beijing over demonstrations for more democracy that led up to the **Tiananmen Square Massacre**. From the early days of the formation of the People's Liberation Army until 1937, when the **Anti-Japanese War** started, Geng Biao served as a senior leader in the Chinese Workers and Peasants Red Army. Geng Biao commanded the 32d Regiment of the 11th Division, Fourth Red Army, from 1932 to 1934. He later served as a regiment commander and division chief of staff in the **First Front Army**. After the **Long March**, Geng Biao was moved to the **Fourth Front Army**, where he served as an army (corps) chief of staff in 1936–1937. During the Anti-Japanese War period, Geng was a brigade chief of staff in the 129th Division, **Eighth Route Army**. In 1948, he was a deputy corps commander in the **North China Field Army** until 1950.

Geng Biao's career in the Communist Party took a turn away from military service in 1950, when he was appointed ambassador to Sweden, where he served until 1956. He was concurrently accredited as the People's Republic of China ambassador to Denmark and Finland during this time. He subsequently served China as ambassador to Pakistan (1956–1960), vice minister of foreign affairs (1960–1963), ambassador to Burma (1963–1967), and ambassador to Albania (1969–1970). His service abroad in the mid-1960s helped him to avoid the internal political clashes of the **Cultural Revolution** that killed, maimed, or jailed so many of China's revolutionary leaders.

Concurrently to Deng Xiaoping's 1978 rehabilitation and elevation to the position of chief of the PLA General Staff Department, Geng Biao was made a vice premier. He was appointed minister of national defense in 1981, when that position was revived to facilitate China's opening to the outside world. Like **Chen Yi**, who served as China's foreign minister after a distinguished military career, and **Zhou Enlai**, who started as a military leader, Geng Biao is an example of how the PLA used its experienced leaders to manage the country.

REFERENCE

Wolfgang Bartke, *Biographical Dictionary and Analysis of China's Party Leadership, 1922–1988* (New York: K. G. Saur, 1990).

GIQUEL, PROSPER MARIE (1835–1886) Prosper Marie Giquel organized the **Taiping Rebellion** era, French-led **Ever-Triumphant Army** along with Edourd le Brethon de Caligny in December 1861 after the city of Ningbo fell to the Taiping rebels. Giquel initially worked with Adrien Tardif de Mordrey in fighting the Taipings along with Tardif's **Franco–Chinese Corps** of Jiangsu and **Frederick Townsend Ward**'s **Foreign Arms Corps** fighting the Taipings around Shanghai. Giquel returned to Ningbo in June 1862 to work with Brethon de Caligny to organize the Ever-Triumphant Army. Brethon initially commanded the army while Giquel organized financial support from wealthy Ningbo Chinese merchants. Brethon was killed in winter 1862, and Tardif took command, while Giquel continued to concentrate on raising funds. Giquel did take to the field at times, however, and was seriously wounded in the attempt to attack Shangyu on November 21, 1862. He convalesced for 19 months in France and returned to China in June 1864, in time to see the Ever-Triumphant Army assist in the capture of Nanjing.

Giquel later worked with **Zuo Zongtang** on the **Fuzhou Dockyard** project between 1866 and 1877. He continued to work to assist the Chinese until his death in 1886. His dockyard project was destroyed by the French fleet in August 1884.

REFERENCES

Prosper Giquel, *A Journal of the Chinese Civil War, 1864*, trans. Stephen H. Leibo (Honolulu: University of Hawaii Press, 1985); Ssu-yu Teng, *The Taiping Rebellion and the Western Powers: A Comprehensive Survey* (Oxford: Clarendon Press, 1971).

GORDON, CHARLES GEORGE (January 28, 1833–January 26, 1885)

Charles G. "Chinese" Gordon was born in England, the youngest son in a British army family of 11. His father rose to the rank of lieutenant general, as did his older brother Henry. Charles G. Gordon was a third-generation army officer who enrolled in the Royal Military Academy at the age of 15, in 1847.

He graduated in June 1852, commissioned in the Royal Engineers, and was sent to the engineer corps depot at Chatham for his first assignment. His second posting, in early 1854, was to Pembroke Dock, Wales, where he worked building a series of forts on the coast. Gordon was ordered to Crimea in December 1954, where he took part in the siege of Sebastopol. He distinguished himself under fire in Crimea and was wounded once. In May 1856, after the Crimea War, Gordon was sent to Bessarabia, on the Russian–Romanian border, where he did survey work. He returned to England in October 1858.

In 1860, Gordon, then a captain of Royal Engineers, volunteered to go to Shanghai as part of the British military force, pressuring the Qing dynasty to ratify the **Treaty of Tianjin**. Gordon arrived in Shanghai on September 17, 1860, and joined Lord Elgin's expeditionary force in Tianjin. He was part of the joint Anglo–French force that burned the Summer Palace outside Beijing. Gordon remained in Tianjin after the withdrawal of the bulk of the force as part of a British garrison of 3,000 troops. The British commander was General Stavely, who was Gordon's brother-in-law (Henry Gordon, the older brother who Charles followed into the army, married Stavely's sister).

In May 1862, Gordon took a force of engineers to Shanghai, which traveled there with two infantry regiments. Gordon surveyed the area and passed the information on Shanghai to General Stavely. Meanwhile, the **Ever-Victorious Army**, commanded by **Frederick Townsend Ward**, was engaged in suppressing the **Taiping Rebellion**. The army's task was to clear a 30-mile zone around Shanghai of rebels. When Ward was killed in September 1862, his deputy, a fellow American, **Henry A. Burgevine**, took command of the Ever-Victorious Army. Within months, Burgevine had committed a series of atrocities, allegedly including blowing prisoners out of the mouth of cannon and robbing the bank of his financial benefactor in Shanghai to pay the army. **Li Hongzhang**, the Qing governor, dismissed Burgevine and needed a new commander. Stavely, who wanted to see increased British influence over the Ever-Victorious Army, sought to place Gordon in charge. However, he needed authority from London to second a British officer to Chinese service. On January 9, 1863, an order in council in Britain, based on the one that had earlier authorized the **Lay-Osborn Flotilla**, authorized the secondment of officers to China independent of Lay and Osborn, who had tried to raise a British-controlled naval force. As the Ever-Victorious Army began to suffer defeats because of poor leadership, the Chinese finally agreed to accept Charles Gordon as its new commander. Gordon was promoted to brevet-major on March 26, 1863, and assumed command of the Ever-Victorious Army.

Gordon first conducted a series of successful campaigns, clearing the 30-mile zone of Taipings. Following this, in 1864, Gordon moved his force, often fighting in conjunction with the French-led **Ever-Triumphant Army**, against Taiping strongholds. He captured Suzhou on December 5, 1863, and continued to pressure the Taiping capital of Nanjing when his force was disbanded by the Chinese in June 1864.

By 1865, Gordon was a lieutenant colonel, again building forts in England. He carried out antislavery missions in Central Africa and was provincial governor there from 1874 to 1876, then governor-general in Sudan in 1877. By 1881, Gordon was a major general. He returned to China once more, where he used his old personal relationship with Li Hongzhang to avert a conflict between Russia and China. In 1884, he helped to supervise the British evacuation from Sudan and was appointed governor-General in february. Gordon was killed on January 26, 1885, defending the gates of Khartoum from Sudanese rebels.

REFERENCES

Piers Compton, *The Last Days of General Gordon* (London: Robert Hale, 1974); Anthony Nutting, *Gordon, Martyr and Misfit* (London: Constable, 1966); Arthur Ormont, *Chinese Gordon: Hero of Khartoum* (New York: G. P. Putnam's Sons, 1966).

GUANGZHOU UPRISING (1927) After the **Nanchang Uprising** of August 1, 1927, the Red Army dispersed around South China to conduct the **Autumn Harvest Uprisings** (Qiu Shou Qi Yi). These ''uprisings'' were organized throughout Hunan, Hubei, Guangdong, Guanxi, and Anhui Provinces to counter the Guomindang's military pressure on the Communists. Believing the area of South and East China to be ripe for revolution, the Communist Party organized peasants, workers, and those Communists who survived the Guomindang ''White Terror'' orchestrated by **Chiang Kai-shek** in April 1927 to conduct an armed uprising in the city of Guangzhou (Canton). The uprising was planned for December 12, 1927, but because the Guomindang had already begun to react, the uprising was actually initiated on December 11. About 3,000 men of the ''Workers and Peasants Army'' seized the Public Security Bureau headquarters and military installations. In the suburbs, in all, some 20,000 people took part in the uprising, seizing about 20 artillery pieces and 1,000 rifles. The Nationalists (Guomindang) immediately counterattacked with three divisions that were garrisoned around the city. Foreign troops protecting citizens in the French, British, and Japanese concessions also fired on the Communist forces. By December 13, the Communist forces pulled out of Guangzhou, regrouping in the hills and later forming the **Jiangxi Soviet Revolutionary Base Area**, forming the Fourth Division of the Red Workers and Peasant Army.

REFERENCES

ZGDBKQS, vol. 1; ZGRMJFJZS, vol. 1.

GUTIAN CONFERENCE (December 1929) The Gutian Conference was held in December 1929 as the Ninth Party Congress of the Fourth Army (**Fourth Corps**) of the Red Army. The conference was named for the site where it was held, Gutian, Shangkang County, Fujian Province. The conference is a significant watershed in **People's Liberation Army** (PLA) history. It exemplifies the tension in the PLA among officers who sought to build a purely professional military without political commissars, those who argued for a political commissar system within a military that still produces crops and undertakes public works, and those who argued for a purely guerrilla force operating in a completely egalitarian system. This tension and the debate over the balance of these PLA missions continue to exist in the present day.

The aim of Communist Party Central Committee member **Li Lisan** at the conference was to centralize party control over the various independent Soviet bases. At the time these quasi-independent base areas were run by local soldiers' committees. In some base areas, or soviets, military forces organized themselves along the lines of **warlord** armies. The troops drilled and trained while they were supported by the local populace. In other soviet areas, the Communist Party was having a difficult time controlling military units that roamed the countryside like rebel bands. **Mao Zedong** argued for a middle line, supporting the need for the type of structure and organization called for by Li Lisan but insisting on strict party control and influence in military units through a strong system of political commissars.

Red Army field commanders like **Peng Dehuai, Chen Yi, Zhu De**, and **Lin Biao** tended to favor a structured, professional military that recruited and trained soldiers for service along the lines of old warlord units. Zhu De is known to have advocated this approach at the conference. **Zhou Enlai** and **Deng Xiaoping**, speaking at the conference, suggested that a political commissar system was important to retain. They had been trained in Moscow by the Comintern and favored the Soviet Red Army model. Mao's position at the conference retained elements of both positions.

Mao Zedong articulated his position at the Gutian Conference in his article "On Correcting Mistaken Ideas in the Party." Mao complained that the Red Army had developed "a purely military view," seeking to use the military forces only to fight like a group of mercenaries. Plus, according to Mao, like the warlord armies, the Red Army's fighting was not coordinated on a broad scale. Each **Soviet Base Area** sought to build up its own semi-independent territory and army. Like Li Lisan, therefore, Mao argued for structure and central discipline. However, Mao insisted that a strong program of political discipline and education must be maintained in the Red Army. He also insisted on a system of mutual criticism and discussion within the party to define and decide on important tasks. Mao, like the field commanders, wanted to see a set of rules and regulations drafted that would govern the actions of the army and define party–army relations. Finally, while Mao supported guerrilla tactics, he insisted in his article that there must be a cohesive military strategy applied and that units not simply roam the countryside like rebel bands.

The conference agreed on three major functions for the Red Army, which evolved into the **General Departments of the PLA**. The conference established an organization for war, which evolved into the General Staff Department; an organization for production, which evolved into the General Logistics Department; and an organization for mass political work, which evolved into the General Political Department. The PLA today still retains these three functions (the General Armaments Department, responsible for weapons and procurement, was added as a fourth department in 1998). The emphasis on the relative strengths and functions of the three departments varied and continued to be argued at all subsequent party meetings on the **Long March**.

REFERENCES

Mao Zedong, "On Correcting Mistaken Ideas in the Party," *Selected Works*, vol. 1 (Beijing: Foreign Languages Press, 1975), pp. 105–16; John E. Rue, *Mao Tse-tung in Opposition, 1927–1935* (Stanford, CA: Stanford University Press, 1966); Richard C. Thorton, *The Comintern and the Chinese Communists, 1928–1931* (Seattle: University of Washington Press, 1969).

H

HAINAN ISLAND DEFENSIVE CAMPAIGN (1950) Although the Chinese **Civil War** is said to have ended in 1949, when the People's Republic of China was established on October 1 of that year, the Chinese Communists had not yet actually gained control of Hainan Island, as well as a number of small offshore islands that remained occupied by Nationalist troops, including Jinmen (Quemoy) and Mazu (Matzu). The **People's Liberation Army** landed on Hainan Island in March 1950, using a diversionary feint near the capital, the port city of Haikou, and conducted its main landings east and west of the city. Nationalist resistance on the island was light. Most casualties during the operation were the result of motorized junks used by the landing force capsizing en route. The island was ''liberated'' from Nationalist control in April 1950.

REFERENCES

David G. Muller, Jr., *China as a Maritime Power* (Boulder, CO: Westview Press, 1983); Gerald Segal, *Defending China* (London: Oxford University Press, 1985).

Vance H. Morrison

HASEGAWA PROCLAMATION (August 25, 1937) Admiral Hasegawa, the commander of the Japanese Third Fleet, issued a proclamation on August 25, 1937, that prohibited Chinese shipping from transiting sections of China's coast. Hasegawa's proclamation prohibited all Chinese shipping, public or private, from Chinese territorial waters extending from geographical coordinates 37 degrees 40 minutes north/221 degrees 44 minutes east to 23 degrees 14 minutes north/116 degrees 48 minutes east. This effectively closed the Chinese coast between the Yangtze River and the port of Swatou, in Fujian Province. The United States, in reaction, issued a statement refusing to allow the Japanese navy to board Chinese ships, calling any such boarding an act of war.

REFERENCES

Marius Jansen, *Japan and China: From War to Peace, 1894–1972* (New York: Rand McNally, 1975); James W. Morley, ed., *The Chinese Quagmire: Japan's Expansion on the Asian Continent, 1933–1941, Selected Translations* (New York: Columbia University Press, 1983).

HE LONG (1896–1969) He Long was a legendary leader and hero of the **People's Liberation Army** (PLA) who reportedly started his military career as a bandit armed with only kitchen knives. He Long was born in Hunan Province in 1896. According to the stories circulating about him, at the age of 13, during a famine in Hunan, he killed a tax official, escaped to the hills, and joined a secret society, the Ge Lao Hui, which he eventually led. By 1916, the governor of Hunan appointed him to the Military Department of the province and gave him responsibility for collecting taxes. During the **Northern Expedition**, He Long commanded a division in the National Revolutionary Army.

He Long joined the Chinese Communist Party (CCP) in late 1926 and in April 1927 commanded a corps that formed the main Communist force in the August 1, 1927, **Nanchang Uprising**. After that failed, He Long returned to the Hunan–Hubei border to organize guerrillas. He built a force of 3,000 men by April 1928. He Long was a follower of **Li Lisan**, and he committed a force of 16,000 men, which he had personally built and trained, against Wuhan and Changsha. This attempt to execute Li's strategy of attacking and controlling urban areas cost He Long 12,000 casualties. His force rebuilt to 30,000 men by April 1932, when it was defeated in the Fourth Nationalist **Encirclement Campaign** in Hunan. With **Zhang Guotao**, He Long moved from the Dabieshan **Eyuwan Soviet** area to Sichuan but later returned to Hunan with only 10,000 survivors. After the linkup with the **Fourth Front Army** and Zhang Guotao on the **Long March**, He Long's **Second Front Army** cooperated closely with the forces of Zhang Guotao. In the **Anti-Japanese War**, He Long commanded the 120th Division of the **Eighth Route Army**, where he had the primary mission of protecting the Communist base in **Yan'an** from the Japanese **Three-All Campaign**. During the **Civil War**, He Long turned his forces over to **Peng Dehuai** and became the deputy commander of the Northwest Field Army, later redesignated the **First Field Army**. After the Civil War, He Long dominated the province of Sichuan and the **Cultural Revolution** in 1966 and 1967 made him a target of the Red Guards. He was severely attacked by them. He Long died on June 9, 1969. His son, He Pengfei, at the time of publication of this book was a vice admiral and deputy commander of the PLA Navy. One of his daughters is a major general at the People's Liberation Army Academy of Military Science.

REFERENCES

William W. Whitson with Chen-Hsia Huang, *The Chinese High Command* (New York: Praeger, 1973); ZGDBKQS, vol. 1.

HONG XUEZHI (1913–1995) Hong Xuezhi was a major figure in an effort to regularize and professionalize the **People's Liberation Army** (PLA) along Soviet lines. He is most closely associated with **Peng Dehuai**, who in 1956, in conjunction with the Eighth Congress of the Chinese Communist Party, worked against **Mao Zedong** to remove political commissars from PLA units, to move military tactics away from an emphasis on **People's War**, and to replace militia units with reserve divisions.

Hong Xuezhi was born in 1913 in Anhui Province. His early military career is obscure, but he is known to have been associated with **Huang Kecheng** before 1937, indicating that he was in the **First Front Army** during the **Long March**. By 1945, Hong Xuezhi was chief of staff of the Third Division, **New Fourth Army**, which operated in the **Dabieshan** Mountains not far from Anhui Province. During the **Civil War** period, Hong Xuezhi was deputy commander of the 15th Army and political commissar of the 16th Army, both subordinate to the Fourth Field Army. From 1951 to 1954, during the **Korean War**, Hong was the director of the Rear Services Department (Logistics Department) of the **Chinese People's Volunteers** in Korea.

Serving as the deputy to Huang Kecheng, from 1954 to 1956, Hong Xuezhi was deputy director, Rear Services Department. When Huang moved up to be chief of the General Staff Department, Hong Xuezhi replaced him as director of the General Logistics Department (GLD), as Rear Services came to be called. Tan Zheng was director of the General Political Department at that time. Hong served as GLD director from 1956 to 1959, when he was caught in the controversy surrounding the organization and structure of the PLA at the **Lushan Conference** in 1959. Like Peng Dehuai, he was purged from the ranks of the military after the Lushan Conference. Hong worked in defense industry organizations in relative obscurity for years and suffered during the **Cultural Revolution**.

By 1978, when **Deng Xiaoping** rose to power, and many of the Communist Party cadre who were purged during the Cultural Revolution were rehabilitated, Hong was appointed director of the General Office for National Defense Industries (known also as the NDIO) and director of the General Logistics Department of the PLA in 1980. In 1982, he rose to deputy secretary-general of the Central Military Commission. In 1985, along with many other older, Long March veterans, he was moved into the Central Advisory Commission of the party. Hong Xuezhi died in 1995.

REFERENCES

Wolfgang Bartke, *Biographical Dictionary and Analysis of China's Party Leadership, 1922–1988* (New York: K. G. Saur, 1990); William W. Whitson with Chen-Hsia Huang, *The Chinese High Command* (New York: Praeger, 1973).

HUAI ARMY As the Qing dynasty faced the challenge of the **Nian Rebellion** (1851–1868), it turned to locally created, recruited, and financed military forces to suppress the rebels. The Huai Army was very much like the **Xiang Army**, which **Zeng Guofan** had recruited and used to fight in the **Taiping Rebellion**. It was raised from forces around the area under threat, in this case parts of Anhui, Henan, Jiangsu, and Shandong Provinces. The Huai Army also derived its name from the major river system flowing through Anhui Province, the Huai River, just as the Xiang Army was named for a river in Hunan.

The Qing rulers appointed Zeng Guofan, who had been effective against the Taiping rebels, to suppress the Nian Rebellion. Because Zeng had disbanded his own Xiang Army after the Taipings were defeated in Nanjing in 1864, to meet the Nian threat he began to raise a new force from local gentry and peasants in northern Anhui Province, which was threatened by the Nian rebels. Zeng was dependent on **Li Hongzhang**, however, for both troops and money. Li at the time was governor of Liangjiang, an area that included Anhui Province as well as Jiangsu and Jiangxi. Li Hongzhang had been a protégé of Zeng, so there was personal trust between the two officials. Nonetheless, the Huai Army did not perform well for Zeng, probably because its leaders were motivated by personal loyalties to local officials. The Qing-appointed Zeng Guofan, from Hunan, was viewed as an ''outsider.'' The Qing Court therefore switched the positions of Li and Zeng, making Zeng the governor of Liangjiang and appointing Li to suppress the Nian Rebellion.

The Huai Army was actually a quasi-official, semiprivate regional force. But it proved effective in defeating the Nian rebels.

REFERENCES

Chiang Siang-tseh, *The Nian Rebellion* (Seattle: University of Washington Press, 1954); Elizabeth Perry, *Rebels and Revolutionaries in North China, 1845–1945* (Stanford, CA: Stanford University Press, 1980).

HUAI-HAI CAMPAIGN (November 6, 1948–January 10, 1949) The Huai-Hai Campaign was one of the three decisive military campaigns in late 1948 that decided the political fate of China (the **Manchuria**, or Northeast Campaign, and the **Ping-Jin**, or Beiping-Tianjin Campaign, were the other two). Huai was the largest battle of the Chinese **Civil War**, involving over a million combatants. This campaign destroyed Nationalist power north of the Yangtze River and triggered **Chiang Kai-shek**'s resignation as president of the Republic of China. The approximately 7,600-square-mile campaign area was located in the relatively flat North China Plain, where lines of communication were dominated by the Grand Canal (running northeast-southwest), the Tianjin-Pukou Railroad (north-south), and the Lunghai Railroad (east-west). The two railroads form a ''T'' with the city of Xuzhou at the intersection. Running northeast to southwest from Xuzhou are low hills, which restricted the movement of combat forces and helped to isolate the individual battles of the campaign. The Huai River forms the North China Plain's southern boundary. **People's Liberation Army** historians named this campaign the Huai Campaign, combining the names for the Huai River and the Lunghai Railway, two of the major geographical features important in the campaign. Nationalist historians call it the Battle of "Hsuchou-Pangfou." From the standpoint of military operational art, this campaign is a classic demonstration of tactical encirclement and the destruction in detail of enemy forces by the attacking force. By 1948, Communist forces had shifted to mobile, conventional warfare and had seized the operational and tactical initiative from the Nationalist Army. Key to the Communist success, however, was the political commissars' mobilization and direction of hundreds of thousands of peasants who provided logistical support, moved the supplies, and dug the trenches used as positions for combat. This freed front-line troops for battle. Both sides used tanks and heavy artillery during the campaign, and combat took place in urban areas and in open terrain. The Nationalists fielded the larger, better equipped force, and, although some soldiers defected, most of the Nationalist units fought tenaciously. The Guomindang leadership violated the principle of unity of effort and command throughout the campaign and remained strategically passive, garrisoning towns and guarding railroads. Generalissimo Chiang Kai-shek, the Nationalist commander, did not go to the battlefield but issued his tactical orders from Nanjing. Competition between Guomindang generals representing the Whampoa and Guangxi cliques complicated the Nationalist effort, as did personal rivalries within the Whampoa faction of the Nationalist Army. Guomindang field commander Liu Chih and his deputy Tu Yu-ming (who came to the Huai-Hai Campaign after losing in Manchuria) proved to be mediocre tacticians and did not maneuver forces on the battlefield. Rejecting suggestions to pull back to the more easily defended Huai River and to hold the cities, the Nationalist command accepted a battle at Xuzhou, the headquarters for a major Nationalist pacification command. Even as the battle started, Nationalist defenses were in flux, positions incomplete, and units were still arriving.

The Communist Party Central Committee issued orders for an attack on October 11, 1948. The bulk of the East China Field Army (ECFA) of 420,000 men, under the command of **Chen Yi** and political commissar Rao Shushi, marched from Shandong and approached the Xuzhou nexus rail line from the east. The Central China Field Army (CCFA) of 130,000 soldiers, under the command of Liu Bocheng and political commissar Deng Xiaoping, approached from the west, supported by a few ECFA units. The CCFA units struck first in the west on November 6, to fix and hold in place the Nationalist Second Army under Chiu Ching-chuan northwest of Xuzhou. Two days later the ECFA, supported by tanks and artillery, crashed into the Nationalist Seventh Army under Huang Po-tao. The Seventh Army was strung along the Lunghai Railroad as the right arm of the Xuzhou "T." Defections among Nationalist troops on the left flank of the Seventh Army allowed Chen Yi's ECFA units to maneuver between the Nationalist Seventh Army and Xuzhou, isolating and destroying the Seventh Army amid fierce street fighting in the Battle of Nienchuang (November 7–22).

After its initial feints to the west, the CCFA swung south of the hills and linked up with units of the ECFA at the railway town of Suxian on November

15, isolating the battlefield within nine days. This action also pushed the Nationalist Sixth and Eighth Armies south to Pangfou. The Nationalist 12th Army under Huang Wei, which included the last U.S.-trained units in the Nationalist Army, marched toward Suxian from the southwest but were delayed, blocked, and finally surrounded at the Battle of Shuangtuichi (November 23–December 17). During this battle the Nationalist 12th Army used massive labor gangs to build concentric trenches around fortified villages, stripping the mechanized units of the 12th Army of their mobility. Ordered to rescue the isolated 12th Army, units of the Nationalist Sixth and Eighth Armies from Pangfou were not able to break though Communist defenses. Unable to break out and maneuver on the battlefield and hobbled by defections, the 12th Army went on the defensive around November 28 and collapsed two weeks later after what Communist accounts call vicious, hand-to-hand fighting.

At the beginning of the Huai-Hai Campaign, units from the Nationalist Second and 13th Armies attempted to rescue the Nationalist Seventh Army. They made little progress. Later, the Second Army was moved south to trap the Communist forces, surrounding the Nationalist 12th Army, but it was repulsed. Within two weeks the Nationalist Seventh Army was destroyed, the 12th Army encircled, and the Sixth and Eighth Armies were blocked at Pangfou. Chiang Kai-shek then ordered the Second, 13th, and 16th Armies, as well as the divisions of the Pacification Headquarters, to break out of the Communist encirclement to the west and to relieve the 12th Army. On the evening of November 30–December 1, the Nationalist troops in Xuzhou departed, probably attempting to swing around the northeast-southwest ridge line. Command quickly broke down, allowing the Communist ECFA to isolate each of these three major Nationalist commands. Employing artillery and relying on labor gangs to dig fortifications, the ECFA encircled and reduced the Nationalist units in the Battle of Chenguangjuang (December 16–January 16, 1949), ending the successful campaign.

REFERENCES

Edmund O. Clubb, "Chiang Kai-shek's Waterloo: The Battle for Hwai-Hai," *Pacific Historical Review* vol. 25, no. 4 (November 1956); Donald S. Detweiler and Charles B. Burdick, eds., *War in Asia and the Pacific, 1937–1949*, vol. 15: *Civil War in China, 1945–50* (Garland, TX: Garland Press, 1980); Lloyd E. Eastman, *Seeds of Destruction: Nationalist China in War and Revolution, 1937–1949* (Stanford, CA: Stanford University Press, 1984); E. R. Hooten, *The Greatest Tumult: The Chinese Civil War, 1936–1949* (New York: Brassey's, 1991); Tang Tsou, *America's Failure in China, 1941–1950* (Chicago: University of Chicago Press, 1963); *United States Relations with China: With Special Reference to the Period 1944–1949* (Washington, DC: Department of State, August 1949); William W. Whitson with Chen-Hsia Huang, *The Chinese High Command* (New York: Praeger, 1973).

David E. Reuther

HUANG KECHENG (1902–1986) Huang Kecheng was one of the 10 **People's Liberation Army** (PLA) officers appointed to the rank of general when

military ranks were given to the PLA in 1955. He was born in Hunan Province in 1902 and was a graduate of the **Whampoa Military Academy**. Huang Kecheng joined the Chinese Communist Party in 1925, took part in the **Northern Expedition**, and participated in the **Autumn Harvest Uprisings**. He served in the Red Fifth Army (later renamed the Third Army Corps) as a subordinate of **Peng Dehuai** from 1929 through 1935. During that period Huang Kecheng was a political commissar. Huang took part in the **Long March**, and when the PLA regrouped at Yan'an, he was assigned to the Political Department of the **Eighth Route Army**.

In 1938 and 1939, Huang Kecheng was director of the Logistics Department for the Eighth Route Army. He also served in the 115th Division of the Eighth Route Army. He was then sent south by the Communist Party leadership to lead a unit of the **New Fourth Army**. Between 1941 and 1946, Huang Kecheng commanded the third division of the New Fourth Army and the military district in north Jiangsu Province. When New Fourth Army units were sent to northeast China to fight the **Civil War** in Manchuria, Huang commanded the third Division of the **Northeast Democratic Allied Army**. He seized the city of Changchun with the division on April 18, 1946.

In 1949, when the **Fourth Field Army**'s 12th Army was sent into Hunan Province, Huang Kecheng served as political commissar of the Hunan Military District. Under the newly imposed Communist Party structure, he concurrently served as the secretary of the Hunan Provincial Communist Party Committee as well as vice chairman of the Hunan Province Military Affairs Committee. He was deputy to **Xiao Jingguang**, who was also made a general in 1955. Huang Kecheng was moved to the party center in Beijing in 1952, where he was appointed director of the PLA General Logistics (Rear Services) Department as well as a member of the National Defense Council and the Communist Party Central Committee. In 1958, Huang Kecheng became chief of the General Staff Department of the PLA. However, when **Mao Zedong** removed Peng Dehuai and purged Peng's close associates in 1959, after the **Lushan Conference**, Huang Kecheng was also purged. Later in 1967, he was labeled a "rightist" during the **Cultural Revolution**. With the rehabilitation of **Deng Xiaoping** in 1978, many of those purged by Mao between 1959 and the Cultural Revolution were rehabilitated. Huang Kecheng was appointed to the 11th Central Committee of the Chinese Communist Party as a full member and was put into the powerful post of secretary of the Party Central Commission for Discipline Inspection (which rooted out many of Mao's leftist supporters). When a Party Central Advisory Commission was formed in 1985, Huang was a member. He died on December 28, 1986.

REFERENCES

Wolfgang Bartke, *Biographical Dictionary and Analysis of China's Party Leadership, 1922–1988* (K. G. Saur: New York, 1990); William W. Whitson with Chen-Hsia Huang, *The Chinese High Command* (New York: Praeger, 1973).

HUANGAN-MACHENG UPRISING (1927) The Huangan-Macheng Uprising was part of a series of **Autumn Harvest Uprisings** ordered by the Chinese Communist Party Central Committee in August 1927. After the **Northern Expedition**, in response to excess taxation by local landlords and Guomindang authorities in Hunan and Hubei, peasant associations and secret societies in Central China began to organize into armed groups. In Macheng, in the **Dabie** Mountains of eastern Hubei Province, the Macheng Peasant's Association organized a "Self-Defensive Force." Armed with about 50 rifles, the Peasants Association began to conduct insurgent operations against local landlord and Guomindang forces. This force later evolved into the 31st Division of the "Red 11th Corps." By 1932, these forces were incorporated into the **Eyuwan Soviet** and the **Fourth Front Army**.

In Huangan, Hubei Province, northwest of Macheng in the Dabieshan Mountains, local peasants were organized and armed by Xie Fuzhi into an armed group. These armed peasants were led by Xie to attack the county magistrate in a dispute over taxes. Xie Fuzhi later became the commander of all public security forces in China in 1963, and during the **Cultural Revolution**, he was political commissar of the Beijing Military Region.

REFERENCES

Frederick Fu (F. F.) Liu, *A Military History of Modern China: 1924–1949* (Princeton: Princeton University Press, 1956); William W. Whitson with Chen-Hsia Huang, *The Chinese High Command: A History of Military Politics, 1927–1971* (New York: Praeger, 1973).

HUANGPU MILITARY ACADEMY. See WHAMPOA MILITARY ACADEMY

HUNDRED REGIMENTS CAMPAIGN In the summer of 1940, the strength of the Nationalists was seriously degraded as a result of Japanese successes in the Yangtze River valley. Meanwhile, on the Communist side, forces of the **Eighth Route Army** and the **New Fourth Army** were placed under the command of **Zhu De**. This gave Zhu De control over civil governments and all military forces in the area of Shanxi, Hebei, eastern Inner Mongolia, and parts of Shandong and Liaoning. Zhu De then gave an order in early August 1940, directing planning for the Hundred Regiments Campaign. **People's Liberation Army** (PLA) units immediately began reconnaissance and logistics preparation for a major effort to disrupt all Japanese lines of communication and to attack Japanese forces in the North China Plain area. The effort called the Hundred Regiments Campaign had three phases.

Phase one of the campaign was conducted from August 20 to September 10, 1940. The objective of the first phase was to disrupt Japanese lines of communication, especially rail lines in the North China Plain. Forty-seven regiments under the command of **Liu Bocheng** and 46 regiments under **Nie Rongzhen** attacked along the rail lines during this phase. The major effort was against the rail line between Shijiazhuang, capital of Heibei Province, and Taiyuan, capital of Shanxi Province. The campaign was supported by some 40,000 militia forces that built fortifications and transported supplies and the wounded. A blocking action north of Taiyuan, at Datong, to prevent reinforcement, was conducted by **He Long** (whose family still has strong financial interests in coal mining and mills in the Datong/Shanxi area). Another feature of this campaign was the use of the ''dare to die regiments'' (*gansidiu*), which had defected from the forces of the **Yan Xishan** to the Communists.

Phase two of the campaign lasted another 20 days, until early October, and aimed at destroying Japanese forces in the Wu Tai and Taihang Mountains along the Hebei-Shanxi border. This phase drained the strength of Communist forces because it called for attacks on fixed Japanese garrisons and installations that, although cut off from reinforcement by phase one, fought back hard.

Phase three of the campaign lasted from October 6 to December 5, 1940, and was designed as a reconsolidation and withdrawal by Communist forces. In phase three Japanese forces counterattacked and put into effect their **"Three-All Campaign"** (destroy all, kill all, burn all) in retaliation against the local populace and Communist forces. **Peng Dehuai** directed the Communist defense against the counterattack, centered primarily in the Taihang Mountains of Shanxi Province, east of Taiyuan. This savage Japanese counterattack began in March 1941 and shook the Chinese populace, as the Japanese intent was to break peasant support for the Communists.

One result of the campaign was that the successful Japanese counteroffensive (the ''Three All Campaign'') reinforced the need for the Communist leadership of the PLA to concentrate on guerrilla warfare. Conventional action by PLA forces proved too costly for the military and the local populace. This strengthened the position of Mao Zedong over his combat commanders who had argued for conventional battles. The Hundred Regiments Campaign may also have helped the Allied war effort by preventing a Japanese attack north into the Soviet Far East in coordination with a German offensive against Russia in the West.

REFERENCES

Zhongguo Renmin Jiefangjun Zhanshi Jianbian [An encyclopedia of the wartime Chinese people's liberation army] (Beijing: Jiefangjun Chubanshe, 1992); William W. Whitson with Chen-Hsia Huang, *The Chinese High Command* (New York: Praeger, 1973).

HURLEY, PATRICK J. Major General Patrick J. Hurley, a U.S. Army officer, served as the personal representative of President Roosevelt to Generalissimo **Chiang Kai-shek** from August 18, 1944, until January 8, 1945. He was appointed ambassador to China on January 8, 1945. His tenure in this position saw the recall of General **Stilwell** from China, the appointed of Lieutenant General **Wedemeyer** as successor to Stilwell, and the collapse of any Nationalist–Communist cooperation into **civil war**.

Patrick J. Hurley was a Republican lawyer who had served in the army in World War I. He was later secretary of war under President Herbert Hoover. In November 1942, Hurley had been sent to Moscow, where he met with Stalin and Molotov. He returned from that mission with the distinction of having been the first American to whom Stalin and the Soviet General Staff had disclosed their strategy. In the Soviet Union, Hurley had also been permitted to visit combat areas. Hurley's earlier success on the mission to the Soviet Union made him an appealing representative for President Roosevelt during the period that Washington was trying to increase cooperation between the Nationalists and the Communists against the Japanese.

Hurley was sent to represent President Roosevelt to Chiang Kai-shek with instructions to coordinate the military picture in China with Chiang as commander in chief, to attempt to resolve the problems between Stilwell and Chiang, and to assist in getting China to continue supporting the Burma Campaign with military forces. By Hurley's own congressional testimony in 1949, Hurley said that he inferred three general missions from the instructions he got from the president: keep China in the war, keep the Chinese (Nationalist) Army in the field fighting the Japanese, and attempt to unify the Chinese Communists and Nationalists as a fighting force against Japan. Major General Hurley was accompanied on his mission to China by Donald Nelson, a former head of the U.S. War Production Board and a U.S. corporate official. While Hurley was to attend to political and military affairs, Nelson was supposed to help the Nationalists plan their own defense production and to advise them on administering and using Lend-Lease aid.

Patrick Hurley held his first discussions with Chiang Kai-shek on September 7, 1944, in Chongqing. Hurley insisted on the unification of command of all forces in China under Stilwell, while Chiang objected that without limits on Stilwell's authority, it would exceed that of Generalissimo Chiang. Chiang also fought for control over the distribution of Lend-Lease supplies coming into China. In the middle of these talks, the military situation in South China so deteriorated that Stilwell had to fly to Guilin, in southeast China, where he ordered the demolition and abandonment of the U.S. air base there in the face of advancing Japanese forces. By September 19, after Chiang had essentially agreed to all of the proposals put forth by Hurley, Stilwell returned and delivered a stern message from President Roosevelt, with Hurley present and over Hurley's objection. The United States insisted that Chinese armies continue the Salween offensive in Burma to open the land supply route and that Stilwell be placed in unrestricted command of all forces. This démarche firmed Chiang Kai-shek's resistance to Stilwell. Chiang agreed to place all Chinese forces under an American but insisted that Stilwell be relieved. Stilwell was recalled from China in October 1944, and General Wedemeyer was assigned to replace him with the concurrence of Chiang Kai-shek.

Hurley remained in Chongqing, despite his failure to have achieved his objectives with regard to Stilwell. However, the U.S. ambassador to China,

Clarence D. Gauss, resigned on November 1, 1944. Hurley flew to **Yan'an** on November 7, still working to bring the Communist forces into the war under Wedemeyer's command. He was present in Yan'an on November 10, when **Mao Zedong** signed an "Agreement between the National Government of China, the Guomindang of China and the Communist Party of China," the "Five Point Draft Agreement." This plan called for the Nationalists and Communists to work together against the Japanese, for the organization of a coalition national government in China, for the formation of a united national military council, and for the legal recognition of all anti-Japanese political parties in China. The Nationalists countered this with a "Three Point Plan" agreeing to reorganize Communist forces and incorporate them into a National Army and to recognize the Communist Party as legal. However, the Nationalist counterproposal insisted that the Communists give up control of all their troops to the Nationalists and that the Communists subscribe to the National ideology, Sun Yat-sen's "Three Principles of the People."

The Nationalist counterproposal was presented to **Zhou Enlai**, who represented the Communist Party in Chongqing on November 22, 1944. After returning to Yan'an for consultations, Zhou told Hurley that cooperation between the Nationalists and the Communists seemed impossible. Nonetheless, Hurley continued to aggressively pursue his mission, as charged by President Roosevelt, albeit sometimes blindly, if not naively. Hurley, at this stage, apparently believed what he had been told by Stalin and Molotov in Moscow—that the Chinese Communist Party was not controlled or affiliated with the Moscow-directed Communist International. By December 24, Mao Zedong told Hurley that the Nationalists must "release all political prisoners, withdraw all forces surrounding Communist areas, abolish all oppressive regulations which restricted the people's freedom, and end all secret service activity." Hurley eventually came to believe that his combined efforts to develop a political rapprochement between the Nationalists and the Communists, combined with having broached two different projects for arming the Communists with American weapons, might not succeed (see **Cooperation Plans, U.S. and Yan'an Communist Forces**).

During the time that all these negotiations were taking place, after Gauss resigned on November 1, Hurley was nominated to replace Gauss as U.S. ambassador to China. After consultation among the White House, State Department, and the War Department, Hurley's nomination was forwarded to the Congress on November 30, 1944. He presented his credentials to China on January 8, 1945. Hurley was convinced that the "Nationalist Government had the firm support of the people, and that the Communists could never control China." However, within his mission, several officers were not so confident of this. Raymond P. Ludden, a second secretary at the embassy detailed to work at Headquarters, U.S. Forces, China Theater, in support of Wedemeyer, sent two dispatches to the contrary. Ludden believed that the Communists had widespread popular support in the areas under their control. Moreover, Ludden reported, each time Nationalist forces withdrew in the face of Japanese military advances, the Communists filled the political void, gaining more converts. Between the time that he became ambassador on January 8 and the time Ludden issued the two reports mentioned herein (February 12 and February 16), Hurley grew disenchanted with Wedemeyer and his headquarters. The third event that drew Hurley's pique was the OSS-advanced cooperation plan between American forces and Yan'an-based Chinese Communist forces. In Hurley's view, this plan was advanced without his authority. Then Ludden's two reports contained assessments that directly contradicted Hurley's own views. Hurley was also aware that the Communists were trying to split the U.S. team, seeking a $20 million loan from Wedemeyer, which they asked be kept secret from Hurley. Wedemeyer refused and informed Hurley.

On February 19, 1945, Hurley returned to Washington for consultations. At that time he was certain that some military officers in Wedemeyer's headquarters, members of his own embassy mission, and officers in the Far Eastern Division of the State Department were critical of his handling of matters in China and were seeking to undermine him. Meanwhile, also in mid-February, Roosevelt had acceded to many of Stalin's demands. In the **Yalta Agreement**, Roosevelt ceded parts of China to Soviet influence. Hurley was sent to Moscow to discuss the Yalta Agreement with Stalin and Molotov, but on Roosevelt's (and, subsequently, Truman's) orders, the agreement was kept secret from Chiang Kai-shek. Although he stayed on for another nine months, Hurley eventually resigned his post on November 26, 1945, during another trip to Washington to consult on China. On November 27, 1945, in a speech at the National Press Club, Hurley attacked the administration for not having a clear policy toward China. However, Secretary of State Byrnes refused to pass on the resignation to the President. Hurley had tried to resign twice before in November and had accused a group of Foreign Service officers of disloyalty. After agreeing to go back to China, a speech by Congressman DeLacy and two press articles, one in the *Chicago Sun* and one in the Communist-controlled *Daily Worker*, convinced Hurley that his own secret reports and confidential conversations at the State Department were being leaked to the press and to the Communists. Instead of returning to China, Hurley went to New Mexico, having already tendered a written resignation, and General Marshall was sent on the **Marshall Mission** to China.

REFERENCES

Russell D. Buhite, *Patrick J. Hurley and American Foreign Policy* (Ithaca, NY: Cornell University Press, 1973); James Chace, *Acheson, the Secretary of State Who Created the American World* (New York: Simon and Schuster, 1948); Herbert Feis, *The China Tangle: The American Effort in China from Pearl Harbor to the Marshall Mission* (Princeton: Princeton University Press, 1953); Donald Lohbeck, *Patrick Hurley* (Chicago: Henry Regnery, 1956); U.S. Department of State (Far Eastern Series 30), *United States Relations with China: With Special Reference to the Period 1944–49* (Washington, DC: Government Printing Office, 1949).

I

ICHIGO, OPERATION (April 1944–February 1945) The goal of Operation Ichigo, whose name was taken from the Japanese "Ichigo Plan," was to conduct an offensive operation throughout East China to open a major strategic line of communication between Saigon, Vietnam, and Pusan, Korea. Japan needed to relieve the strain on its maritime supply routes along the China coast and in the South and East China Seas because of the heavy toll taken by submarine attacks on the Japanese supply lines. To accomplish this task, Japanese forces had to secure the Beijing to Changsha rail lines, in the northern part of China; the routes in South China from Changsha to Nanning and the Vietnamese border; as well as the Guangzhou to Hengyang rail line. In preparation for Operation Ichigo, Japan first moved some units to concentrate forces from their Mongolian Garrison Army and the Kwantung Army, located in Manchuria, into eastern China. This brought Japanese forces in China to a total of about 820,000 men. About 15 divisions of this force, 150,000–200,000 men, participated in Operation Ichigo.

The initial phase of Operation Ichigo was launched on April 19, 1944. This offensive focused on eliminating Chinese armed resistance in the Henan Province area. The Japanese 11th Corps, which controlled the area around Wuhan (Wuchang and Hankow), drove north into Henan, while the Japanese 12th Corps drove south from Kaifeng. A supporting offensive by the Japanese First Corps, which occupied the area west of the Beijing-Kaifeng railway, neutralized Chinese forces in the area of Luoyang, west of Kaifeng. Although Chinese Nationalist forces at that time had about 34 divisions in the Henan Province area, the units were disorganized and fell apart in the face of the Japanese offensive, turning into roving bands of soldiers who only worsened the situation in Central China by looting Chinese peasant houses and villages. This first phase of Operation Ichigo was accomplished successfully by May 1944.

By June, the second phase of the operation was initiated. This called for a drive south from the Wuhan area, securing the Hankow-Changsha-Hengyang rail link. This was a particularly important area since Changsha was the center of what was then the largest rice-producing area in China. It was also in an area that was essentially under the control of Chinese forces, both Nationalist and Communist, making it an important objective for Japan. After securing the rail line to Hengyang by July, Japanese forces continued the drive south and southeast along axes of attack. The main attack from Hengyang resulted in the capture of Guilin, Liuzhou, and Nanning, opening the rail line from Hanoi to China for use by the Japanese. This axis also eliminated Allied (Nationalist–U.S.) control over four airfields that had been used for offensive operations against the Japanese by the 14th Air Force: Hengyang, Lingling, Guilin, and Liuzhou. By September 1944, the Japanese had captured Liuzhou, and the final phases of Operation Ichigo, securing the rail line to Nanning and south to the Vietnam border, was completed in January 1945. The southeastern thrust, designed to open the Hengyang-Guangzhou rail link, was completed by October 1944, with the help of a supporting attack by the Japanese 23d Corps, based in Hong Kong and Guangzhou.

Although the U.S. 14th Air Force was able to slow the offensive by continuous bombing, Operation Ichigo succeeded in opening the Japanese supply line. It also precipitated a major dispute among American leaders, with **General Chennault** of the 14th Air Force arguing for more supplies to be flown over the "Hump" while the air force bombed the Japanese. **General Stilwell** argued that only the buildup of a strong offensive ground capability, such as what he envisioned in the U.S.-trained, jointly manned **Y-Force** and **Z-Force**, would succeed in winning the war and driving the Japanese out of China. This was a common, almost classic argument between the ground forces and the air forces over the utility of airpower versus land power and the efficiency of strategic bombing that contributed to a bitter dispute between Stilwell and Chennault. The success of the Japanese Operation Ichigo and the rift it opened within the ranks of the American leadership and between the two American generals also precipitated President Roosevelt's recall of Stilwell on October 18, 1944, and Stilwell's replacement by **Lieutenant General Wedemeyer**.

REFERENCES

Akira Iriye, *Across the Pacific* (New York: Harcourt and Brace, 1967); Barbara W. Tuchman, *Stilwell and the American Experience in China, 1911–45* (New York: Macmillan, 1971); U.S. Army Center of Military History, *China Defensive* (Washington, DC: Government Printing Office, 1996).

INDUSTRIAL SECTOR The People's Republic of China, at the time of its establishment in 1949, inherited an industrial base that was ravaged by almost continuous warfare from the time of the **Northern Expedition** (1926, the campaign to put an end to warlordism) and the end of the **Civil War**. Prior to 1949, much of China's industry was concentrated in the major coastal cities, along the Yangtze River, and in the northeast. The largest industrial center in China was Shanghai, followed by such coal mining and steel production cities as Anshan, Fushun, and Shenyang. Much of this industry, concentrating on steel, coal, refining, and concrete production, had been introduced by foreign companies. In the interior, Wuhan, Chongqing, and Taiyuan developed modern industrial enterprises, including armaments and ship construction in the cases of Wuhan and Chongqing. The industrial sector in Taiyuan, Shanxi Province, owed much of its development to rich coal deposits and to the foresight of the **warlord Yan Xishan**. Tianjin, 90 miles east of Beijing, which was a gateway for major foreign intrusion aimed at the capital, was an important industrial city, and Qingdao, on the Shandong Peninsula, was an industrial center that owed its productive capacity primarily to German and Japanese influence. Today such interior cities as Lanzhou, Gansu Province, and Baotou, in Inner Mongolia, have developed into centers of heavy industry.

China has a good base of natural resources on which to build its industry, but the distribution of those resources is not balanced. The distances between resources and energy users for instance are great, stressing the transportation infrastructure. Energy resources are heavily concentrated in the north of the country, whereas industrial output remains concentrated along the east coast and in Guangdong Province. There are adequate supplies of iron and coking coal on which to base a steel industry, but China has still imported steel, pig iron, copper, and aluminum because of domestic demands and because its transportation infrastructure could not meet distributed demand in the mid-1980s and the 1990s. China's proven coal reserves are more than 700 billion tons, and its estimated reserves are over 3,000 billion tons. However, much of it is low-quality bituminous coal. The 1995 output of coal was 1.36 billion metric tons, but China consumed almost as much coal as it produced. Onshore and offshore oil reserves are estimated at around 5.3 billion tons, most of which are not tapped, but China consumed even more than the 150 million metric tons of oil it produced in 1995. In fact, China has committed as much as $12.5 billion to the development of oil fields in neighboring Kazakhstan, to the west, and the construction of a pipeline stretching across the northern rim of the Tarim Basin in Xinjiang to Lanzhou. China also has natural gas reserves, but the exact size of these resources is not known; it is estimated at between 129 billion and 24 trillion cubic meters.

The major limitation of China's energy industry has been extraction and distribution, with few major pipelines built up to the 1990s. In nonmetallic materials, China has adequate supplies of barite, fluorite, salt, talc, sulfur, graphite, magnetite, and phosphates to meet domestic needs. Although deficient in chromium, platinum, and gold, there are large reserves of rare earth materials such as beryllium, tungsten, molybdenum, barium, manganese, mercury, zirconium, and titanium. China also has abundant uranium deposits. Despite these resources, industrial development has been spotty, and many of the industries still rely on technologies of the 1950s. Central bureaucratic planning is partially responsible for the uneven development of industry and the concentration of development in a few sectors such as nuclear weapons and their delivery systems and naval shipbuilding. China's gross domestic product (GDP) in 1997 was 7.47 trillion *yuan* (one dollar equals 8.28 *yuan*).

The balance of the industrial structure in China has also shifted among sectors in the last 20 years. Primary industry, centering around agriculture, accounted for 31.1 percent of GDP in 1979 and 18.7 percent of GDP in 1997. The shifts in the percentage of the secondary sector of the economy (industry) in total GDP rose from 47.3 percent in 1979 to 49.2 percent in 1997, while in the tertiary sector (services) the percentage of GDP rose from 21.4 to 32.1 percent in the same period. The Communist Party has managed its economy in Five-Year Plans, which set broad targets in industrial and agricultural capacity. The First Five-Year Plan, beginning in 1949, succeeded in doubling China's post–Civil War industrial capacity by 1952. It concentrated on the secondary sector of the economy, constructing plants and equipment for heavy industry, and depended primarily on assistance from the Soviet Union.

Two factors in the late 1950s and early 1960s resulted in the establishment of new plants beyond China's operating capacity and local, inefficient plants around the country that could not be supported by the resource and transportation infrastructure. The first factor was the overambitious and eventually disastrous Great Leap Forward (1959–1960; see **Peng Dehuai**). This plan was conceived by **Mao Zedong** and adopted in 1958 as a means of achieving national self-sufficiency through accelerated collectivization, especially in agriculture, and the use of labor-intensive methods in a widely distributed new industrial sector. The result was a major famine in 1960–1961 and a monumental waste of resources. The second factor that damaged the growth of China's industrial capacity was the withdrawal of Soviet assistance in the early 1960s and the establishment of the **Third Line** (or Third Front) of industries as a means of distributing production capacity around the interior of China in case of attack by either the United States or the Soviet Union. As in the Great Leap Forward, the redistribution of industry was not always rational given the transportation and resource infrastructure of China.

Broadly speaking, China's defense industrial base is inadequate to meet its demands for new weapons and weapon platforms. Much of Chinese military production is still based on designs introduced by the Soviet Union in the 1960s, and most of the improvements since that time have been with foreign assistance, especially from Israel, France, Italy, Pakistan, and Iran. China has excelled in the manufacture of a few types of systems, such as short-range surface-to-surface missiles and nuclear delivery systems, but its defense industries have been unable to effectively produce major new aircraft or armor systems. Even efforts at reverse-engineering imported or pirated foreign technologies have had only limited success. Like the industry that developed in China during the **self-strengthening movement**, China's defense industries today can produce a few things well, such as short- and medium-range missile systems and nuclear weapons. They are limited, however, by an inability to innovate and systems-engineer. This is particularly true of the aircraft industry. Moreover, a great deal of China's production methods concentrate on "batch-building" large orders according to a state plan. However, China is slowly mastering both the machine processes and the use of digital technologies in manufacturing and in its weapons systems. The **Commission of Science, Technology and Industry for National Defense** (COSTIND) and its antecedents in the defense industrial sector were transformed in 1997 to speed military industrial development. With the newly established **General Armaments Department** as a new staff element of the **People's Liberation Army**, China's military industrial capacity is expected to improve. But there are limits to the extent of this improvement since state-owned factories in the industrial sector are operating at losses and are the least productive of Chinese enterprises.

REFERENCES

Gary Bennett, ed., *China: Facts and Figures Annual-Handbook*, vol. 20 (Gulf Breeze, FL: Academic International Press, 1996); *China News Analysis*, No. 1624, December 15, 1998; *China 2020: Development Challenges in the New Century* (Washington, DC: The World Bank, 1997); Jonathan D. Pollack, *The R&D Process, and Technological Innovation in the Chinese Industrial System* (Santa Monica, CA: Rand, 1985); Jonathan D. Pollack and Richard H. Yang, *In China's Shadow* (Santa Monica, CA: Rand, 1998); Jan S. Prybyla, *The Political Economy of Communist China* (Scranton, PA: International Textbook, 1970); Larry M. Wortzel, *China's Military Potential* (Carlisle, PA: Strategic Studies Institute, 1998).

J

JAPANESE REPATRIATION FROM CHINA (1945–1946) At the end of **World War II** there were 3 to 4 million Japanese soldiers and civilians in China. Civil war was imminent, and both Chinese factions, Nationalist and Communist, sought to take advantage of the Japanese presence by using the surrendered Japanese soldiers to reinforce their own armies. The American position was that the war had been fought to remove Japanese influence from China and that the specter of civil war made the removal of Japanese troops even more important. Organizing repatriation, however, was a major effort requiring significant American logistical and political resources. The **Marshall Mission** to China was important in moving repatriation forward, because General Marshall organized two crucial cease-fires in North China and Manchuria.

Under the direction of the Supreme Command Allied Powers (SCAP) in Tokyo, the repatriation program began with the Japanese government's establishing disembarkation and demobilization facilities in Japan. SCAP placed whatever Japanese shipping remained available for use under SCAP's shipping control administration for Japanese merchant shipping (SCAJAP) established on October 12. Using this motley fleet and its own resources, the Americans began clearing Korea, the Philippines, and the Pacific Islands of Japanese troops in mid-September 1945. At the time, SCAP estimated that, given shipping limitations, it could take five years to clear all theaters.

Representatives from SCAP Tokyo, General **Wedemeyer**'s command in China, and the Nanjing government met October 25–27, 1945, in Shanghai to review American experience in Korea and plan for the repatriation of Japanese from China. Under General Order Number One, Japanese troops from Manchuria to the 16th parallel in Vietnam were to surrender to Generalissimo **Chiang Kai-shek**. Chinese authorities were responsible under the repatriation guidelines for transporting the Japanese to embarkation ports, enforcing health and contraband regulations, and screening for war criminals. Because North China ports were mine-free, repatriation started there first, in October. In fact, because of a paucity of shipping and lack of priority with the Nanjing government, no significant movement took place through December.

In the meantime, the growing **Civil War** in China swirled around the Japanese troops and the U.S. Marines sent to assist in their repatriation (see **U.S. Marines in North China**). As an example of just how bad the situation was, U.S. sources indicated that, from the end of the war to December, Japanese troops in one North China sector suffered 2,653 killed, 3,507 wounded, and 2,194 missing/defected.

On December 6, General Wedemeyer traveled to Tokyo to discuss the China program with SCAP planners. The meeting supported a Joint Chiefs of Staff proposal to significantly augment SCAJAP with American shipping. On January 15, 1946, SCAP and China Theater representatives met again in Tokyo to finalize plans to give priority to the China Theater. Repatriation of Japanese forces from the Philippines and the Ryukyus was postponed until midyear. The meeting assigned 58 percent of all SCAJAP shipping to China, including an augmentation (of SCAJAP resources) by 100 Liberty ships, 85 Landing Ships (Tank) (LSTs), and seven hospital ships from American stocks. Chinese repatriation ports of embarkation were prioritized: first priority went to Shanghai, where there were 741,070 persons; North China with 505,280 persons; and Taiwan with 490,384 persons. Second priority went to South China, where there were 135,090 Japanese; third priority went to French Indochina, which had 30,958 Japanese; and fourth priority went to Manchuria with 1,603,000 Japanese.

The logistics problems were enormous. Just as the program gained speed, a cholera epidemic erupted among the evacuees. Those ships with cholera cases were quarantined. At one point 22 ships with a total of 76,000 repatriates were in quarantine in the port of Uraga, Japan. The program suffered one accident. The former Japanese vessel *Enoshima Maru* struck a floating mine 50 miles outside Shanghai on January 22, 1946. The vessel carried approximately 4,300 civilians, of whom all but 77 were rescued.

By March the new shipping allocated to the task came on line. Guangzhou was cleared by April 6; Taiwan was cleared by mid-April, one week ahead of schedule; North China was cleared by May. After General Marshall engineered a cease-fire between the Nationalist and Communist Chinese factions, 800,000 Japanese were shipped out of Manchuria between the end of July and the end of October. Japanese soldiers and civilians held in the Soviet occupation area were not released until after SCAP declared the program over on December 31, 1946. Excluding Manchuria, 97.8 percent of all Japanese in China (some 3,015,133 persons) had been repatriated.

Approximately 67,000 Japanese technicians and dependents, however, were retained throughout China at the request of the Nationalist government, usually under U.S. protest. Most of these civilians were located in Manchuria and Taiwan (over 23,000 persons). The Nationalist government retained the services of other Japanese personnel by giving them Chinese documentation. One way or another, both factions in the Chinese Civil War still gained access to some Japanese military personnel.

REFERENCES

Foreign Relations of the United States, 1946, Vol. 10: *The Far East: China* (Washington, DC: Government Printing Office, 1972); Donald G. Gillin with Charles Etter, "Staying On: Japanese Soldiers and Civilians in China, 1945–1949," *Journal of Asian Studies* 42, no. 3 (May 1983): 497–518; National Archives, *Record Group 331* (Supreme Command Allied Powers G-3—SCAP G-3); National Archives, *Record Group 334* (Army Military Advisory Group in Nanking: 1942–1949); National Archives, *Record Group 353*, Records of the State War Navy Coordinating Committee (SWNCC); Reports of General MacArthur, *MacArthur in Japan: The Occupation: Military Phase*, vol. 1: *Supplement*, prepared by his General Staff (San Francisco: Presidio, 1966).

David E. Reuther

JIANG JIESHI. See CHIANG KAI-SHEK

JIANGNAN ARSENAL (Kiangnan Arsenal) As the foreign-led and regional forces in coastal China coalesced to victory in the **Taiping Rebellion**, local military and government leaders in South China established arsenals to manufacture weapons and ammunition. One of the most prominent of these was the Jiangnan Arsenal in Shanghai, which was established in spring 1865 by **Li Hongzhang**, using some machinery from the arsenal established in Suzhou by **Zuo Zongtang** and other machinery bought in the United States by Yong Wing, an agent operating on behalf of **Zeng Guofan**. The arsenal was established as part of the **self-strengthening movement**.

By the time it had been operating for only six months, in fall 1865, the Jiangnan Arsenal had expanded to become a shipyard. By 1868, the arsenal had produced a first ship, which was a steam-driven paddle wheeler. Although the ship, the *Tian Zhi*, utilized a foreign engine, the hull and the armor were made at the arsenal. By 1872, the Jiangnan Arsenal had built six ships, all of about 200 feet in length and all armed with 24-pound howitzers made at the arsenal. The Jiangnan Arsenal slowed its ship production program in 1875 and concentrated on guns and rifles.

By 1876, Jiangnan was one of four arsenals operating in China, the others being the **Fuzhou Dockyard** and arsenal, the Tianjin Arsenal, and the Nanjing Arsenal. Between 1865 and 1904, the Jiangnan Arsenal produced 25 ships for Li Hongzhang's fleet. It also produced cannon and breech-loading rifled guns for the army. The Chinese navy still operates the same shipyard in Shanghai at part of the navy base there and has named a new class of frigate, Jiangnan, after the shipyard. Like its sister shipyards and arsenals, the Jiangnan Arsenal was established and run by a regional governor. This regional orientation contributed to the period in the early twentieth century, when China was dominated by regional armies and **warlords**.

REFERENCES

Charles G. Gordon, *General Gordon's Private Diary of His Exploits in China* (London: Simpson Low, 1885; New York: Kraus Reprint, 1971); Li Hong Chang, *Memoirs of Li Hong Chang*, ed. William F. Mannix (Boston: Houghton Mifflin, 1913); Rhoads Murphy, *The Treaty Ports and China's Modernization: What Went Wrong?* (Ann Arbor: University of Michigan Press, 1970); John L. Rawlinson, *China's Struggle for Naval Development, 1839–1895* (Cambridge: Harvard University Press, 1967).

JIANGXI SOVIET. See JINGGANGSHAN REVOLUTIONARY BASE AREA

JINAN INCIDENT (1928) As the **Northern Expedition** began to close in on its goal of placing China's **warlords** under control as means to create a unified nation, **Chiang Kai-shek** was made commander in chief of the Guomindang Army and a member of the standing committee of the Nationalist Party Central Executive Committee in January 1928. In order to solidify support for his drive north from Nanjing, Chiang attempted to ally with the warlords **Feng Yuxiang**, based in Henan at the time, and **Yan Xishan**, the Shanxi military governor and leader. By March 1928, Chiang was trying to secure the North China Plain, including the area around Beijing, which was controlled then by **Zhang Zuolin**. The Guomindang Army sent troops into Jinan, Shandong Province, on April 30, 1928, with the objective of securing the capital and hub of road and rail lines of communication on the Shandong Peninsula. The issues of Chinese warlord forces in Shandong and control of the railroads in the province were major topics of discussion at the **Washington Conference** of 1921–1922. From a strategic standpoint, however, the most important issue may have been the Jinan-Qingdao railway, which linked the port of Qingdao with the Yellow Sea. In April 1928, Chiang's troops seemed on the brink of securing the city, but there were about 2,000 Japanese nationals living in Jinan. Because the Japanese government had already seen its nationals living in China threatened by fighting among Chinese military forces in Hankou and Nanjing during the early part of the Northern Expedition, the city of Jinan was garrisoned with about 500 Japanese soldiers. Chiang sought the withdrawal of Japanese forces. Instead, however, the small garrison was reinforced quickly when Japan decided to move 5,000 troops into the area to protect its citizens. On May 3, Nationalist and Japanese troops clashed in a fight that escalated and ended up with the military forces of each side committing a number of atrocities against each other. Reportedly, some prisoners from each side were blinded or castrated by the other side. By May 11, the Japanese drove the Nationalist forces out of Jinan, and Chiang instead turned his forces against Beijing and Zhang Zuolin instead of continuing to fight the Japanese. Chiang appealed to the League of Nations, protesting the Japanese presence and intervention.

REFERENCES

Donald G. Gillin, *Warlord: Yen Hsi-shan in Shanxi Province, 1911–1949* (Princeton: Princeton University Press, 1967); James E. Sheridan, *Chinese Warlord: The Career of Feng Yu-Hsiang* (Stanford, CA: Stanford University Press, 1966); Jonathan R. Spence, *The Search for Modern China* (New York: W. W. Norton, 1990).

JINGGANGSHAN REVOLUTIONARY BASE AREA In winter 1927, in the Hunan-Jiangxi border area, the Red Workers and Peasants Revolutionary Army was engaged in the **Autumn Harvest Uprisings** under the general supervision of Mao Zedong. The elements of the army under Mao joined forces with Workers and Peasants Revolutionary Army elements under the command of **Zhu De** and **Chen Yi**, who had taken part in the **Nanchang Uprising**. These forces subsequently moved to the Jiangxi-Guangdong border area to continue the uprising. The forces established contact and began coordinating their operations in January 1928. The units under Zhu and Chen moved into Hunan to participate in the ''Hunan Uprising.'' By March 1928, the Nationalist Army had assembled a force of seven divisions, which pursued the Communist forces along two major axes of advance. Zhu and Chen withdrew to preserve their forces. By April 1928, forces under Zhu and Chen joined with the units under Mao's command to form the Fourth Army Corps of the Workers and Peasants Revolutionary Army. Three divisions were established, the 10th under Zhu De, the 11th under Mao's command, and the 12th commanded by Chen Yi. Zhu De was concurrently the fourth Army commander, and Mao was the party representative (political commissar) and secretary of the Military Commission. The forces then totaled about 10,000 men when they established a secure, protected base area in the Jinggangshan (*shan* means ''mountains'' in Chinese). This base area was the headquarters for the **First Front Army** until October 1934, when the **Long March** took place.

REFERENCES

Jonathan D. Spence, *The Search for Modern China* (New York: W. W. Norton, 1990); ZGDBKQS.

JING-JIN CAMPAIGN. See BEIPING-TIANJIN CAMPAIGN

JINMEN (Quemoy) Jinmen is a small island five miles off the coast of Xiamen (Amoy), to which Nationalist forces fled and which Chinese Communist forces failed to take after an attempted amphibious assault in 1949. These islands were still occupied by a sizable and well-entrenched Nationalist Army force in 1999.

When **People's Liberation Army** (PLA) assault forces were unable to defeat a small Nationalist garrison on Jinmen on October 25, 1949, it gave the first hint of just how deficient the PLA was in naval power. Jinmen at that time was defended by approximately 15,000 Nationalist troops when an assault force of about 30,000 soldiers from the PLA **Third Field Army** organized as an amphibious assault force from Xiamen to take the island. The PLA forces loaded onto motorized junks in an attempt to cross the five miles from Xiamen to the island of Jinmen. The island's Nationalist defenders kept a steady stream of fire on the overloaded PLA junks, capsizing and sinking many of them before the attackers reached the island. According to Nationalist accounts, the bodies of Communist forces littered the beaches of Jinmen for days after the attempted assault. One effective tactic used by the Nationalists against the Communist troops was to dump thousands of gallons of fuel from aircraft onto the water near the beach and to set it aflame to thwart the landings.

A successful Communist landing on one beach established a small beachhead, but their advance was stopped by two Nationalist tanks, which the PLA could not fight their way past. By October 27, the PLA had to pull back from the island after failing to reinforce its tenuous beachhead. This defeat blunted further attempts to achieve a final victory over the Nationalists, who had withdrawn to Taiwan. Nationalist forces still occupied most of the offshore islands in 1950. In December 1952, the People's Republic of China (PRC) conducted a number of small raids against the offshore islands, including Jinmen and **Mazu**, but did not attempt to seize them.

In one particularly notable incident, after 1964, General **Luo Ruiqing**, then chief of the **General Staff Department** of the PLA, ordered a more aggressive show of force in the Taiwan Strait. On November 13, 1965, allegedly on Luo's orders, eight PRC Navy patrol craft engaged two Nationalist ships at sea off Jinmen. The Nationalists lost a minesweeper in the battle, while the PLA Navy lost seven of its own patrol craft. Luo was purged two weeks after the incident.

REFERENCES

David G. Muller, Jr., *China as a Maritime Power* (Boulder, CO: Westview Press, 1983); Bruce Swanson, *Eighth Voyage of the Dragon—A History of China's Quest for Seapower* (Annapolis, MD: Naval Institute Press, 1982).

Vance H. Morrison

K

KIANGNAN ARSENAL (1865). See JIANGNAN ARSENAL

KONG KA PASS INCIDENT (1959) Despite the August 25, 1959, **Longju Incident** between Chinese and Indian border guards along the **McMahon Line**, Indian army units in the western sector of the Sino–Indian border, at the directions of the Indian government, began to reconnoiter and establish outposts. On October 20, 1959, a patrol of about 70 Indian Special Border Police ran into a Chinese-established border post at the Kong Ka Pass, which the Chinese regarded as the geographic boundary between the two countries. The Chinese border guard forces detained three Indians on the twentieth. On October 21, there was a skirmish in which nine Indians were killed, and seven more were taken prisoner. The prisoners were later released to the Indian government, but the Indian press portrayed the incident as an unprovoked attack by the Chinese, as it had portrayed the Longju Incident.

Following these two incidents (Longju and Kong Ka) the Indian and Chinese governments began a round of diplomatic negotiations over demarcation and agreed, in the interim, to pull back from the disputed border and not to establish more outposts. This policy broke down in 1962, when the **Sino–Indian Border War** broke out after India began to push the disputed border further, allegedly encroaching on Chinese territory.

REFERENCES

Deng Liqun et al., eds., *Zhongguo Renmin Jiefangjun* (Beijing: Dangdai Chubanshe, 1994); Neville Maxwell, *India's China War* (Bombay: JAICO Press, 1971).

KOREAN WAR (1950–1953) The 38th parallel, which cuts the Korean Peninsula almost in half, was designated as the temporary line of demarcation between the Soviet and American areas of responsibility for the surrender of

Japanese forces on the peninsula by the protocol for the surrender of Japan in 1945, at the end of **World War II**. After no progress in the United Nations on unifying the peninsula, South Korea held elections on May 10, 1948. The Soviets protested and established the Democratic People's Republic of Korea (DPRK, or North Korea) in the zone north of the 38th parallel in September 1948. The Republic of Korea (ROK, or South Korea) was recognized as a state on January 1, 1949, by the United States. As a global political background to this, U.S. policy in Asia after World War II was to control air and sea lines of communication but to avoid ground hostilities on the Asian mainland. On January 12, 1950, Secretary of State Dean Acheson, in an address to the National Press Club in Washington, drew a security cordon for the United States in the Far East that did not include Korea or Taiwan. He was supported in this view by the Joint Chiefs of Staff.

The Korean War began on June 25, 1950, when North Korea invaded the Republic of South Korea. On the same day, the United Nations (UN) Security Council condemned the aggression. The UN Security Council convened in special session at 2:00 P.M. New York time on the twenty-fifth and, taking advantage of the absence of the representative of the Soviet Union, Andrei Gromyko, who could have exercised a veto, called the assault into the ROK a breach of the peace. The UN resolution at that session called for (1) an immediate cessation of hostilities and (2) the authorities of North Korea to withdraw their forces back to the 38th parallel. On June 27, 1950, the UN asked its members to assist South Korea. The ROK forces were rapidly overcome by the North Korean People's Army (NKPA). Seoul fell on June 27, 1950, to the NKPA, and the South Korean army was defeated.

Responding to the crisis, President Truman ordered U.S. ground forces into Korea on June 30, 1950. The newly formed Task Force Smith, established out of units from U.S. occupation forces in Japan, confronted NKPA forces in the vicinity of Osan on July 5, 1950. The U.S. force was easily overrun. It had been part of an occupation army, and the leaders had not concentrated on unit tactics or combat preparations; its outdated equipment and antitank weapons were no match for the NKPA's Soviet-made T-34 tanks. U.S. forces fell back and established the Pusan Perimeter on August 4, 1950. The U.S. commander, Lieutenant General Walton H. (Johnny) Walker, firmly told his staff there would be no more retreating. Walker's Eighth Army held the perimeter stubbornly against determined attacks from the NKPA forces August 5–19, 1950, during the first battle of Naktong Bulge (a bend in the Naktong River, west of Obong-ni); during general perimeter fighting from August 25 to September 15; and during the Naktong Offensive on September 1–5, 1950. Outnumbered along the entire defensive line, General Walker employed a strategy of active defense, in which he accepted risk at certain points along the line, thinning his forces, while repositioning forces where and when they were most needed. It was an effective strategy born of necessity.

At the outbreak of the Korean War, China and the United States had no diplomatic relations because the United States recognized the Republic of China as the legitimate government of China. The lack of direct diplomatic communication proved to be a critical factor in each side's calculations and contributed more than once to a failure to heed warnings about the war that came from China. On September 1, 1950, referring to North Korea, **Mao Zedong** publicly stated that China could not tolerate the invasion of a neighbor. **Zhou Enlai** formally passed a warning to the United States and the UN through Indian ambassador Sardar K. M. Panikkar on September 3, 1950, that the Chinese would intervene if U.S. forces entered North Korea. These warnings, however, were dismissed by the United States as mere bluff or unreliable.

The momentum of the war dramatically shifted against the NKPA after the successful Inchon landing by U.S. forces on September 15, 1950. On September 18, 1950, UN forces broke out from the Pusan Perimeter and shifted from defense to pursuit and exploitation. The NKPA, caught between MacArthur's "anvil" at Inchon and Walker's Eighth Army "hammer" in pursuit, rapidly collapsed. Seoul was liberated on September 26, 1950.

Zhou Enlai publicly reiterated on September 30, 1950, the warning that the Chinese would not tolerate the invasion of one of its neighbors. Continuing to discount these warnings, however, UN forces pushed north, crossing the 38th parallel into North Korea. General MacArthur, the commanding general of the U.S. Far Eastern Command, called for the surrender of the North Korean capital, Pyongyang, which was captured on October 19.

China began to react to the U.S. move into the north. The first campaign of the **Chinese People's Volunteers** (CPV) was secretly launched between October 25 and November 5, 1950. Chinese forces, primarily consisting of light infantry, began crossing the Yalu River October 18–19, under the command of **Peng Dehuai**. The CPV moved at night and hid during the day to conceal its movements.

On November 1, the CPV ambushed the U.S. First Cavalry Division at Unsan on the western side of the peninsula. Although intelligence reports indicated the presence of Chinese forces, they were believed to be no more than about 70,000 strong. In reality, there were over 200,000 Chinese soldiers already in Korea. The U.S. X Corps, which had landed on the east coast of North Korea at Wonsan on October 26, advanced toward the Yalu River, which forms the border between North Korea and China on the east, November 10–26, while the U.S. Eighth Army advanced in the west. General MacArthur launched his final, "Home by Christmas" offensive on November 24, 1950.

The second CPV campaign, from November 25 to December 24, 1950, not only succeeded in stopping the UN drive to the Yalu River but also succeeded in driving UN forces completely out of North Korea. On November 25, the CPV attacked the Eighth Army a Ch'ongch'on River in the east. Two days later, CPV forces in the west attacked the U.S. first Marine Division and the Army's Seventh Infantry Division at Changjin Reservoir (Chosin Reservoir). Between November 26 and December 1, the U.S. Second and 25th Infantry Divisions were defeated along the Ch'ongch'on River and forced to retreat. In the west, between November 27 and December 10, the X Corps fought through CPV forces to Hungnam port, while the First Marine Division was forced to retreat from Kot'o-ri. By December 24, when the X Corps sailed from Hungnam, UN forces had been completely evacuated from North Korea.

The third CPV campaign was launched on December 31, 1950, against Peng Dehuai's advice. Peng attempted to convince Mao that the UN forces had consolidated into an in-depth defensive position, and the CPV forces lacked experience with fighting fortified positions. The CPV logistics lines were also overextended, Peng argued, and Chinese troops lacked sufficient food, winter clothing, and other essential supplies. Further, after pushing from northeast China through the length of North Korea, CPV soldiers were exhausted and in need of rest and reorganization. A desire to end the war quickly, combined with the sweeping victories over the UN forces, however, encouraged Mao to order the CPV to continue to push southward, despite Peng's concerns. On January 4, 1951, Seoul fell, and by January 14, the UN line was pushed back to the 37th parallel.

On January 25, 1951, UN/U.S. forces resumed the offensive. Within two days, the CPV began its Fourth Campaign, which lasted until April 21. On February 14, the U.S. 23d Infantry Regiment, with help from the French Infantry Battalion, turned back a CPV counteroffensive at Chipyong-ni. The UN seized the initiative between February 17 and March 17, 1951, and moved north. Seoul was liberated for a second time on March 18. During this fighting, on April 11, General MacArthur was relieved of command, and General Matthew B. Ridgway assumed command of UN and U.S. forces.

The fifth CPV campaign was launched between April 22 and May 21, 1951. On April 22, the CPV attacked the British Brigade northwest of Seoul near the Imjin River with a force of 50,000 men. UN lines held, and the CPV broke contact on April 30. U.S. forces halted the CPV Soyan Offensive May 16–22, 1951. Between May 23 and June 1, UN forces pushed north and reached the 38th parallel on June 13. On June 23, 1951, the Soviet Union delegate Malik proposed a truce in the UN. Truce talks began at Kaesong on July 10. The most bloody fighting of the war (and an intensive Communist propaganda campaign) occurred during this time, until the armistice was signed on July 23, 1953.

From August 1 to October 31, 1951, in an effort to straighten its lines, the UN launched a series of limited battles, known as Bloody Ridge and Heartbreak Ridge. In late October, peace talks resumed at Panmunjon, and a cease-fire was agreed to at the line of contact. Between November 1951 and April 1952 there was a stalemate along the battlefront while the Panmunjon talks continued.

On January 28, 1952, the CPV headquarters reported that U.S. and United Nations planes had spread smallpox in areas southeast of Inchon. On February 18, Radio Moscow accused the United States of using bacteriological warfare against North Korea. By March, an antigerm warfare campaign was launched in China. Although the Chinese have never recanted their charge that the Americans used germ warfare in Korea, their claim has never been supported by scientific evidence.

Between June and October 1952, truce talks were deadlocked over how to handle the repatriation of prisoners of war (POWs). A number of hill battles were fought, including Baldy and Whitehorse. On October 8, 1952, truce talks recessed and fighting resumed. The ROK sector was faced with particularly heavy battles in the Kumwha sector until November 1952, during which the South Koreans distinguished themselves as tough and courageous soldiers.

In November 1952, India offered a proposal on the POW issue to the UN, and President Truman lost the U.S. national election to Eisenhower.

On March 30, 1953, Zhou Enlai indicated the Chinese would accept India's proposal on the POW issue. The talks at Panmunjom resumed. As the negotiations continued, one of the bloodiest battles of the war, the Battle of Pork Chop Hill, occurred April 16–18, 1953. During April 20–26, both sides exchanged sick and wounded POWs at Panmunjom.

On June 4, 1953, the Chinese and North Koreans agreed to accept UN truce proposals, and the fighting ceased. On September 4, screening and repatriation of POWs began at Freedom Village, Panmunjom. CPV forces remained in North Korea until 1958.

REFERENCES

Roy E. Appleman, *U.S. Army in the Korean War: South to the Naktong, North to the Yalu (June–November 1950)* (Washington, DC: Center of Military History, U.S. Army, 1961); T. R. Fehrenbach, *This Kind of War* (New York: Macmillan, 1963); Alexander L. George, *The Chinese Communist Army in Action: The Korean War and Its Aftermath* (New York: Columbia University Press, 1967); Walter G. Hermes, *United States Army in the Korean War: Truce, Tent and Fighting Front* (Washington, DC: Center of Military History, U.S. Army, 1965); Callum A. MacDonald, *Korea: The War before Vietnam* (New York: Free Press, 1986); Billy C. Mossman, *United States Army in the Korean War: Ebb and Flow, November 1950–July 1951* (Washington, DC: Center of Military History, U.S. Army, 1966); Nie Rongzhen, *Inside the Red Star: The Memoirs of Marshal Nie Rongzhen*, trans. Zhong Renyi (Beijing: New World Press, 1988); James F. Schnabel, *United States Army in the Korean War: Policy and Direction, the First Year* (Washington, DC: Center for Military History, U.S. Army, 1988); Shuguang Zhang, *Mao's Military Romanticism: China and the Korean War, 1950–1953* (Lawrence: University Press of Kansas, 1995); John Toland, *In Mortal Combat: Korea, 1950–1953* (New York: William Morrow, 1991).

Susan M. Puska

KOREAN WAR ARMISTICE AGREEMENT (July 27, 1953) The United Nations (UN) sought a way to end the war soon after the entry of the **Chinese People's Volunteers** (CPV) and a general retreat of UN forces in late 1950. The UN established a cease-fire group on December 14, 1950; however, China rejected its draft 13-point cease-fire resolution on December 22, 1950. On January 11, 1951, the UN cease-fire group proposed five principles to resolve the

Korean War, which the American negotiators were instructed to support. These were rejected by the Chinese on January 17, 1951. As a result, the UN officially declared China an aggressor in the Korean War on February 1, 1951.

On March 24, 1951, General Douglas MacArthur, commander of UN/U.S. forces, issued an unauthorized statement suggesting talking with the Chinese on a cease-fire. He was subsequently relieved by President Harry S Truman on April 11, 1951, and replaced by General Matthew B. Ridgway.

George Kennan met with Jacob Malik, Soviet ambassador to the UN, on May 31, 1951, to discuss ways to end the war. On June 23, 1951, in a radio address, Malik called for a cease-fire and withdrawal of forces to positions along the 38th parallel. Truce negotiations began at Kaesong on July 10, 1951. The agenda was agreed on by July 26. However, the Chinese and North Korean delegates suspended talks on August 23, after accusing UN forces of violating the neutrality of Kaesong. Talks resumed on October 25, 1951, at Panmunjom. Between November 1951 and April 1952 there was a general stalemate along the battlefront while the Panmunjom talks continued.

Between June and October 1952, truce talks were deadlocked over prisoner of war (POW) repatriation. Several hill battles were fought, including Baldy and Whitehorse. On October 8, 1952, truce talks recessed, and fighting resumed. The South Koreans on the Kumwha sector were hardest hit in the new offensive.

India presented a proposal in the UN on the POW issue in November 1952. Meanwhile, President Truman was defeated that month by Dwight D. Eisenhower. Realizing that the new president was a former wartime commander, Mao suspected President Eisenhower would escalate the war. Within a month after Joseph Stalin's death on March 5, 1953, **Zhou Enlai** indicated the Chinese would accept India's proposal on the POW issue. The talks resumed at Panmunjom, and, in the following month, April 20–26, the sick and wounded POWs were exchanged at Panmunjom.

On June 4, 1953, the Chinese and North Koreans agreed to accept the UN truce proposal, and the fighting ceased. On September 4, 1953, screening and repatriation of POWs began at Freedom Village, Panmunjom. The Korean War Armistice Agreement was signed on July 27, 1953, after almost two years of protracted negotiations. This two-year period marked the bloodiest fighting of the war, as the belligerents attempted to extract concessions from each other. Although the armistice marked the end of fighting and resolution of the prisoner of war problem, it did not end the war. Officially, a state of war still exists on the Korean Peninsula in the absence of a peace agreement.

REFERENCES

Rosemary Foot, *A Substitute for Victory* (Ithaca, NY: Cornell University Press, 1990); Walter G. Hermes, *United States Army in the Korean War: Truce, Tent and Fighting Front* (Washington, DC: Center of Military History, U.S. Army, 1965); Shuguang Zhang, *Deterrence and Strategic Culture: Chinese-American Confrontations, 1949–1958* (Ithaca, NY: Cornell University Press, 1992); Shuguang Zhang, *Mao's Military Roman-*

ticism: China and the Korean War, 1950–1953 (Lawrence: University Press of Kansas, 1995).

Susan M. Puska

KOREAN WAR, DECISION BY THE CENTRAL MILITARY COMMISSION (CMC) TO MOBILIZE FOR Although preparations for possible war in Korea began as early as July 1950, the final Chinese decision to intervene was not reached until September, after much debate within the Beijing leadership. The formal decision to intervene in the **Korean War** was published in a Central Military Commission (CMC) document dated September 20, 1950. **Mao Zedong** was a strong advocate of intervention and believed Soviet support, in the form of aircraft and equipment, would be forthcoming. Once the decision was reached, however, Joseph Stalin balked on support. Mao hesitated in implementing the order, but after a short delay and more discussion, he decided to proceed, unaided if necessary. Perhaps impressed by the Chinese determination to assist North Korea, Stalin provided some military aid after he was informed of Beijing's final decision.

At the time of the Chinese intervention, the Truman administration believed the Chinese would never risk war with the United States, despite any gains that China might achieve through intervention. The warning to the United States, which was passed from **Zhou Enlai** through the Indian ambassador on October 3, 1950, was unheeded, as Washington interpreted it as either a mere bluff or an unauthorized threat.

Many Western scholars have interpreted China's decision to intervene in the Korean War as the direct consequence of the U.S. decision to cross the 38th parallel and push to the Yalu River in the November 1950 "Home by Christmas" drive. According to this argument, U.S. actions directly threatened Beijing and compelled the Chinese to intervene. Some have further suggested that if the United States had adopted a different strategy, China's entry into the war could have been avoided.

Telegrams that were sent at the time between Mao and Stalin, as well as other Chinese sources on the Korean War that were published in the 1980s, however, have provided additional insight into the decision. In the early days after the founding of the People's Republic of China, Mao was convinced the United States would invade and/or bomb China to restore the Nationalist government to power. He believed war with the United States was inevitable and concluded that it was better for China to pick the time and place than to allow the United States to seize the initiative. Once the United States crossed the 38th parallel, Mao was convinced that the United States could not go unchallenged because he thought U.S. intervention in Korea ultimately threatened China and all of northeast Asia. As the UN forces crossed the 38th parallel, and the North Korean People's Army (NKPA) rapidly collapsed, Mao concluded it was in China's self-interest to intervene in North Korea rather than wait to fight the United States on Chinese territory in the future. Although China was still devastated from years of war, Mao believed it was better to fight the United States before China was rebuilt and had even more to lose.

REFERENCES

Chen Jian, *China's Road to the Korean War: The Making of the Sino-American Confrontation* (New York: Columbia University Press, 1994); Sergei N. Goncharov, John W. Lewis, and Xue Litai, *Uncertain Partners: Stalin, Mao, and the Korean War* (Stanford, CA: Stanford University Press, 1993); Melvin Gurtov and Byong-Moo Hwang, *China under Threat: The Politics of Strategy and Diplomacy* (Baltimore: Johns Hopkins University Press, 1980); Hao Yufan and Zhai Zhihai, ''China's Decision to Enter the Korean War: China History Revisited,'' *The China Quarterly* 121 (March 1990): 94–115; Russell Spurr, *Enter the Dragon: China's Undeclared War against the U.S. in Korea, 1950–51* (New York: Newmarket Press, 1988); Allen S. Whiting, *China Crosses the Yalu: The Decision to Enter the Korean War* (Stanford, CA: Stanford University Press, 1960); Shuguang Zhang, *Mao's Military Romanticism: China and the Korean War, 1950–1953* (Lawrence: University Press of Kansas, 1995).

Susan M. Puska

L

LAY–OSBORN FLOTILLA On August 30, 1862, an Order in Council by the British government authorized the head of the Imperial Customs Service, Horatio Nelson Lay, to create a modern navy for China. The concept, suggested by Lay, was to assist the Manchu dynasty with a fleet of British gunboats, commanded by British officers. The fleet was to respond to the orders of the Qing leaders, but Lay was to screen and approve them. A British Royal Navy officer, Captain Sherard Osborn, was chosen to command the fleet, really only a flotilla, with a mission to suppress piracy in Chinese waters and to assist the customs service in trade oversight. Lay argued, in suggesting the concept, that if the flotilla was left under the command of local Chinese authorities, it would be misused. The Order of Council also authorized British officers to serve in the armed forces of the emperor of China, in the Lay–Osborn fleet.

Prince Gong, the Qing negotiator, and **Zeng Guofan** sought to train Chinese sailors to be gunners and officer cadets on the ships. Some believe that Zeng actually sought to control the entire flotilla himself, as a means to enhance his regional power base. However, Lay signed a secret, personal agreement with Osborn on January 16, 1863, stipulating the fleet would obey only Lay's orders, not the orders of the Chinese. Osborn resigned after arriving in Shanghai, when Prince Gong learned of the secret agreement. The fleet was dissolved on November 10, 1863, and the Chinese-purchased ships were sold by the British at a financial loss to the Chinese.

REFERENCES

J. S. Gregory, *Great Britain and the Taipings* (New York: Praeger, 1969); Teng Ssu-yu, *The Taiping Rebellion and the Western Powers* (Oxford: Clarendon Press, 1971).

LI DESHENG (1916–) Li Desheng joined the **Fourth Front Army** in the Eyuwan **Revolutionary Base Area** at the age of 16, in 1932. Li was born in

Hubei Province in 1916. He participated in the **Long March** and later commanded a battalion in the Fourth Front Army. By 1937, Li was operating in combat as a unit commander in the 129th Division of the **Eighth Route Army**, which was then led by **Liu Bocheng**, with **Xu Xiangqian** as Liu's deputy. **Deng Xiaoping** was traditionally also an instrumental actor in the patron–client network that evolved into the **Field Army System**. When the **Second Field Army** formed, Li Desheng served as both a regimental and divisional commander in the Second Field Army. Liu Bocheng was field army commander during that time (1949–1953), and Deng Xiaoping was political commissar. Li Desheng fought in the **Korean War** as commander of the 35th Division of the 12th Army. He also served as chief of staff of the 12th Army toward the end of the war (1953). From 1958 to 1967, Li was commander of the 12th Army in the Anhui Province area. As the **Cultural Revolution** heated up, Li moved into the leftist camp and was the commander of the Anhui Military District from 1967 to 1970. He was elected to the Ninth Central Committee (strongly leftist) in 1969.

From 1970 to 1974, Li Desheng was the director of the **General Political Department**, which firmly supported **Mao Zedong**'s policies during the Cultural Revolution. Li was moved to command the Shenyang Military Region in Northeast China (Manchuria) in 1974. The region remained virtually his private fiefdom for 11 years, until 1985, when Deng Xiaoping managed to have him moved into the position of political commissar of the **People's Liberation Army**'s National Defense University. Li Desheng was a Politburo member on the 12th Central Committee. He retired in 1992 from his political commissar position.

REFERENCES

Wolfgang Bartke, *Biographical Dictionary and Analysis of China's Party Leadership, 1922–1988* (New York: K. G. Saur, 1990); *Who's Who in China: Current Leaders* (Beijing: Foreign Languages Press, 1991).

LI HONGZHANG (Li Hung-Chang) (1823–1901) Although primarily a statesman and a diplomat, rather than a military figure, Li Hongzhang was one of the most instrumental actors on the Chinese domestic political and international scene in the late nineteenth century. Li was involved in all of the most critical negotiations with foreign powers on behalf of the Qing dynasty, including the resolution of the **Taiping Rebellion**, the development of the **self-strengthening movement**, and the resolution of the **Sino–Japanese War** (1894–1895). Li Hongzhang was born in Hefei, Anwei Province, into a family of officials. His father was an imperial degree holder who was a classmate of **Zeng Guofan** (see **Taiping Rebellion**), and Zeng took a personal interest in Li Hongzhang and his career.

Li attended the prestigious imperial Hanlin Academy in 1847, after studying under Zeng Guofan in Beijing. When the Taiping rebels moved into Anwei

Province in 1853, Li and his father, Li Wenan, returned to Hefei from Beijing, where they organized a local militia to defend their properties and county against the Taipings. Li Hongzhang actually led his own recruits in battle and received an imperial promotion to a higher official degree for a victory he won. Li moved on to the staff of the Anhui provincial governor based on the recommendation of Zeng Guofan in 1854 and became a county official. His father died in 1855, requiring Li to return to Hefei, but he continued to lead militia in defense against the Taiping rebels. In 1858, Li went to Nanchang and took part in the campaign by Qing dynasty forces to recapture Jiangxi Province from the Taiping rebels. When Zeng Guofan was appointed governor-general of Jiangxi and Jiangnan in 1860, Li left his service for a short period but returned in 1861, after recruiting a militia force that assisted in the capture of Anjing from the rebels. Li was made acting governor of Jiangsu Province and took part in the defense of Shanghai from the Taipings.

By the age of 39, Li Hongzhang was head of a province that included the wealthy treaty port and city of Shanghai. Li found that the city was defended by a Chinese–foreign force, the **Ever-Victorious Army**, founded by **Frederick Townsend Ward**. Li got along well with Ward but fell out with **Henry Burgevine**, who replaced Ward as head of the army after Ward's death in 1862. Burgevine was replaced by **Charles George Gordon** (''Chinese Gordon''), with whom Li got along well.

By 1865, Li Hongzhang took command of Nanjing, where he established an arsenal for the manufacture of weapons under the supervision of Halliday McCartney. However, as the **Nian Rebellion** festered, Li was again pressed into military service by the Qing dynasty and made imperial commissioner and governor-general of Hunan and Hubei. After the suppression of the Nian Rebellion in 1868, Li was made governor-general at Wuchang, assuming the post on March 1, 1869. He continued to move around China, primarily in ''hot spots'' of domestic unrest, until 1879, when he became superintendent of northern trade. From this position he established the **Beiyang Army**, which was one of the contributing factors to the problem of **warlordism** in China in the early twentieth century. Li divided his time between Baoding and Tianjin and established the Baoding Military Academy as well as the **Beiyang Fleet**.

In 1871, Li Hongzhang was called on by the Qing dynasty to negotiate a treaty with Japan. The treaty eventually precipitated the **Liuzhou Island Incident**, in which Japan occupied parts of Taiwan. On the mainland, Li helped develop railroads and a linked mining–rail–port industrial system in North China. (See the discussion of the importance of roads and railways in the Introduction.) He was also instrumental in establishing the **Navy Board**, or Admiralty Board, designed to develop a unified, modern navy for China. He established a military academy in Tianjin and facilitated the development of a fleet of 28 ships for the northern fleet. Li was also called on to help resolve problems with the French government over Chinese suzerainty over Annam (northern Vietnam) in 1883. Li's refusal to renegotiate the agreement with France over Annam led to the **Franco–Chinese War**. Li was later made responsible for settling the Sino–Japanese War (1894–1895) and was shot by a Japanese fanatic while negotiating the Treaty of Shimonoseki. In the treaty, China ceded the Liaodong Peninsula, Taiwan (Formosa), and the Pescadore Islands to Japan. Li's next call to negotiate a treaty with foreigners came after the **Boxer Rebellion**. He concluded the Boxer Protocols in 1901, to the advantage of the foreign powers. Li Hongzhang died on November 7, 1901.

REFERENCES

Arthur Hummel, *Eminent Chinese of the Ch'ing Period (1644–1912)* (Washington, DC: Government Printing Office, 1943); John L. Rawlinson, *China's Struggle for Naval Development, 1839–1895* (Cambridge: Harvard University Press, 1967).

LI LISAN LINE (The "left opportunism" of Li Lisan) This term refers to the ideas advanced by Li Lisan when he was a member of the Politburo of the Chinese Communist Party (CCP). The basic principles advanced by Li Lisan called for an early regularization of military forces and the abandonment of a rural, agrarian-based guerrilla strategy in favor of a general insurrection and attacks on the cities. Li Lisan called for an attack on Wuhan and Changsha in the summer of 1930. Li Lisan was in fundamental disagreement with **Mao Zedong** over the prosecution of the CCP political program and the fight in China's revolution. Mao believed that peasants from the rural areas should be recruited and that cities should be encircled by strong peasant movements and guerrilla bases. After a meeting on June 11, 1930, the Politburo supported Li Lisan's position and passed a resolution calling for the organization of the armed forces into numbered units and armies to conduct an armed insurrection by Red Army groups against key cities and in several provinces around China. Li Lisan believed that by doing so, the CCP could speed the revolution in China but also concurrently spur a greater number of countries in the world on to revolution, supporting the Moscow-based Comintern (Communist International). After attacks on Changsha, Nanchang, and Guangxi through September 1930 failed, resulting in severe losses for the Red Army, the CCP reexamined Li Lisan's strategy. Mao Zedong at the time was one of three directors of the military committee of the Red Army. Mao continued to have faith in his strategy of relying on guerrilla base areas and rural insurrection, labeling Li Lisan's tactics as "leftist opportunism and adventurism." The Communist International and the CCP Central Committee blamed all of the failures on Li Lisan, who made a self-criticism at the Third Plenary Session, Sixth Central Committee (September 1930), relinquished his position in the Politburo, and was put on party probation. Li was born in Hunan Province in 1890 and died as a result of persecution at the beginning of the **Cultural Revolution** on August 24, 1966.

REFERENCES

Lau Yee-fui et al., *Glossary of Chinese Political Phrases* (Hong Kong: Union Press, 1977); Tong Bai, ed., *Junren Shouce* (Beijing: Changzheng Chubanshe, 1987); William

W. Whitson with Chen-Hsia Huang, *The Chinese High Command* (New York: Praeger, 1973); Xiao Zhaoran, *Zhonggong Dangshi Jianming Cidian* [A Concise Dictionary of the History of the Chinese Communist Party] (Beijing: Jiefangjun Chubanshe, 1987); ZGRMJFJZS, vol. 1.

LI XIANNIAN (1909–1993) Li Xiannian was born in 1909 in Huangan, Hubei Province, in the **Dabie Mountains** (Dabieshan). His father was a poor peasant, and he was a carpenter's apprentice when, in 1927, he joined a self-defensive peasant association that was involved in the **Huangan-Macheng Uprising** and later turned into the **Fourth Front Army**. From 1983 to 1988 Li Xiannian was the president of the People's Republic of China.

After the Huangan-Macheng Uprising in late 1927, Li Xiannian joined the Communist Party. He continued to organize peasant resistance and self-defense forces to fight landlords and Guomindang (KMT) government officials in the Huangan area through 1932, commanding his own detachment of guerrillas. When the **Eyuwan Soviet** formed in 1932, Li Xiannian became the political commissar of the 30th Army (Corps) of the **Fourth Front Army**. He took part in the Fourth Front Army's **Long March** and the linkup between the Fourth Front Army and the First Front Army. Li's formal military education took place after the Long March in 1927, at the **Anti-Japanese Military and Political College** at Yan'an. He was then sent back to the Eyuwan Soviet area in the Dabieshan Mountains and commanded a column of the **New Fourth Army**. From 1941 to 1945, Li Xiannian was commander and concurrently political commissar of the Fifth Division, New Fourth Army.

During the **Civil War**, Li Xiannian continued to lead **People's Liberation Army** (PLA) forces in the Hubei-Central China area as commander of the Central Plains Liberation Army. After 1949, Li was deputy commander of both the **Second Field Army** and the Fourth Field Army. He stayed in Hubei as commander and political commissar of the Hubei Military District from 1950 to 1952 and later as the vice chairman of the Central–South China Military and Administrative Council from 1952 to 1954.

Continuing to serve China in a civilian capacity, Li Xiannian was minister of finance from 1954 to 1975 and a vice premier of the State Council from 1954 to 1980. He was president of China from 1983 to 1988. Li Xiannian died in 1993.

REFERENCE

Wolfgang Bartke, *Biographical Dictionary and Analysis of China's Party Leadership, 1922–1988* (New York: K. G. Saur, 1990).

LIANGHEKOU CONFERENCE (June 25–26, 1935) During the **Long March**, the Red **First** and **Fourth Front Armies** managed to link up in western Sichuan. However, the leaders of the two armies continued to disagree over strategy and tactics in the wars against the Japanese and the Guomindang. Red

Fourth Front Army leaders **Zhang Guotao**, Chen Changhao, and **Xu Xiangqian** wanted to march south and west to establish a base area in southern Sichuan. They also raised again the decisions of the **Zunyi Conference** and sought to reverse some of the results of that conference. Fourth Front Army members also wanted stronger representation on the Chinese Communist Party (CCP) Central Committee and wanted Zhang Guotao to act as secretary-general of the CCP. **Mao Zedong** argued strongly for the continued march to the north to establish a guerrilla base in the Shaanxi area (**Yan'an**), where the fight could be carried on against the Japanese. In a compromise solution to the competing demands, brokered by **Zhu De**, the decisions of the Zunyi Conference remained intact but were to be revisited after consultation with the Soviet-based Comintern; no changes were made in the leadership body of the CCP; and Zhang Guotao was made political commissar of the Red Army. In addition, eight members of the Red Fourth Front Army were elected to the Central Committee. However, Mao Zedong's strategy of moving north into the Shaanxi area to establish a guerrilla base prevailed. In the end, the meeting produced a serious split between the Fourth and First Front Army members over such issues as ranks and titles, military discipline, and regular versus guerrilla operations, with Mao and **Zhou Enlai** (arguing for the guerrilla base in Shaanxi) allied against Zhang Guotao. The dispute continued at the **Maoergai Conference** two months later, in August 1935.

REFERENCES

Warren Kuo, *A Comprehensive Glossary of Chinese Communist Terminology* (Taipei: Institute of International Relations, 1978); William W. Whitson with Chen-Hsia Huang, *The Chinese High Command* (New York: Praeger, 1973); ZGDSJMCD; ZGRMJFJZS, vol. 2.

LIAONING-SHENYANG CAMPAIGN (Liao-Shen Campaign) (September 12–November 2, 1948) Considered one of the three most important **Civil War** campaigns (Liao-Shen, Ping-Jin, and Huai-Hai), the Liaoning-Shenyang Campaign was conducted by the **People's Liberation Army** (PLA) between September 12 and November 2, 1948. Capitalizing on the battlefield successes of the **Sungari River** and **Siping Offensive Campaigns**, PLA forces waged a 52-day strategic campaign to secure the geographic corridor from Shanhaiguan, where Hebei Province meets the Bohai Gulf, northeast to Shenyang, Siping, and Changchun. The PLA also calls this the Liao–Shen Campaign, using the first characters of the name of Liaoning Province and the city of Shenyang. The campaign is also significant because it marked a turn from guerrilla warfare-based, hit-and-run tactics, to maneuver operations. During the campaign, the PLA captured a great deal of Nationalist and former Japanese military equipment, which helped the army build up for the next phase of the Civil War, the **Beiping-Tianjin** (Ping-Jin) **Campaign**.

After the seizure of Changchun and Siping, securing the major rail and road corridors north and east from Shenyang, the relative strengths of Nationalist and Communist forces changed. Nationalist troop strength in China dropped from 4.3 million troops to about 3.6 million, of which about 2 million were deployed in field armies (combat corps) in strategic point defensive missions around major cities such as Shenyang, Beijing, Xi'an, Hankou (Wuhan), and Xuzhou. Communist strength, meanwhile, had increased from a total of 1.2 million troops to 2.8 million, of which 1.5 million were deployed in field armies capable of maneuver warfare. In northeast China, by 1948, the Nationalists had 480,000 soldiers deployed in 14 corps (armies) with a strength of about 44 total divisions. These Nationalist troops were employed in combat as four army groups. The Nationalists also had about 55,000 troops involved in garrison duties in major northeastern cities. PLA forces at the same time were employed as a Field Army Group comprising 12 infantry columns (each roughly equivalent to a corps), an artillery column, and 17 independent divisions. Total Communist troop strength was about 1.1 million soldiers, 700,000 of whom were employed in 53 combat divisions. Given their superiority in forces in the northeast, by early 1948, the Communist Party **Central Military Commission** decided to conduct a major strategic offensive to seize control of northeast China and to destroy the Nationalist Army there. The PLA forces were commanded by **Lin Biao**, and the political commissar in the northeast was **Luo Ronghuan**, who had replaced **Peng Zhen** after the failed summer 1947 **Siping Offensive Campaign**.

At a planning conference in Harbin during April and May 1948, Lin Biao decided to conduct the campaign in three phases. Phase 1 began on September 12, 1948, and was designed to gain control of the main rail and road supply route from Jinzhou, north of Shanhaiguan, to Shenyang and Jilin, thus isolating Nationalist forces and cutting them off from resupply. Phase 2, October 18 through 28, 1948, consolidated control of Jinzhou and continued the PLA drive northwest, seizing control of Heishan, between Jinzhou and Shenyang. The third phase of the campaign took place between October 30 and November 2, concluding with the Communist occupation of Shenyang and the evacuation of the remaining Nationalists from the northeast through the port of Huhidao.

During the second phase of the campaign, which ended on October 28, 1948, the Communist forces succeeded in blocking Nationalist movement west of Heishan. One of the major battles of the second phase of the campaign took place around a gap between mountains in the Heishan area (Hill 101). Commanding the Communist 10th Column (about two divisions strong), reinforced by one division from the First Column, Liang Xingchu was directed to block Nationalist forces from impeding Lin Biao's redeployment of forces from Jinzhou to the northeast. Liang Xingchu's force of three divisions fought a pitched defensive battle from October 23 to 25 against five Nationalist divisions with air and artillery support. At one time, on October 25, Liang's forces lost Hill 101 to the Nationalists but later recaptured that key terrain feature in a counterattack. The defensive at Heishan cost the 10th Column 10,000 casualties, about the strength of a division. Of these, 6,000 casualties were taken in the first day of the battle. The 10th Column earned the honorific title of a ''Hero Unit.'' This successor in unit lineage to the column, the PLA 38th Group Army, still carries the honorific title. At the end of Phase 2 of the campaign, Lin Biao's forces captured over 38,000 prisoners, 150 artillery pieces, 22 tanks, 6,000 horses, and 600 vehicles.

The third phase of the Liaoning–Shenyang Campaign was the assault on the city of Shenyang (Mukden), which housed the Nationalist ''Northeast China Bandit Suppression Headquarters.'' The Nationalist commander, General Wei Lihuang (Wei Li-Huang), left the city before the final phase, leaving General Zhou Fucheng (Chou Fu-ch'eng) in command. The Communist 12th Column began the approach to attack on the city on October 29, 1948. By October 31, the First and Second Columns were also positioned for the assault on the city. The city fell to a strong, coordinated attack by these columns and several independent divisions, with Nationalist forces surrendering on the afternoon of November 1. The Communist Seventh Column captured Liaoyang the same day, while the Eighth Column captured Anshan, an industrial city. By November 2, the Nationalists abandoned the city of Yingkou, which was occupied by the Ninth Column. General Du Yuming (Tu Yu-ming), the Nationalist commander at the port of Huludao, to the southwest, ordered the evacuation of that city. This displaced 140,000 Nationalist troops, along with several thousand civilians plus equipment.

REFERENCES

Chen Chao, *Liao-Shen Zhanyi* (Shanghai: Renmin Chubanshe, 1959); Zhong Yufei, *Liao-Shen Zhanyi* (Beijing: Qingnian Chubanshe, 1964); F. F. Liu, *A Military History of Modern China, 1924–1949* (Princeton: Princeton University Press, 1956); William W. Whitson with Chen-Hsia Huang, *The Chinese High Command* (New York: Praeger, 1973); ZGRMJFJZS, vol. 3.

LIMITED WAR THEORY. See LOCAL WAR DOCTRINE; PEOPLE'S WAR UNDER MODERN CONDITIONS

LIN BIAO (1907–1971) Lin Biao was one of the premier soldiers, tacticians, and strategists of the **People's Liberation Army** (PLA). He was designated **Mao Zedong**'s successor by Mao himself. However, by 1971, after being accused of plotting a coup d'état against Mao, Lin died in an air crash over Mongolia, allegedly in an attempt to flee China in a military aircraft. Lin Biao was born in 1907 in Huihongshan, Hubei Province. At the age of 19 he enrolled at the **Whampoa Military Academy**. After graduating he took part in the **Northern Expedition**, during which he served as adjutant to General Zhang Fakui, commander of the Fourth Nationalist Army. At the age of 20, Lin Biao was promoted to the rank of colonel. Following the **Nanchang Uprising**, in August 1927, Lin took his regiment to join the 20th Army under **He Long** and **Ye Ting**. Later, during the **Autumn Harvest Uprisings**, he took part in the organization of peasant uprisings in Hunan Province under the leadership of **Zhu De**. When the Red Army Academy was established in the **Jinggang Mountains**, he became commandant. In 1932 Lin Biao was made commander of the First Red Army Corps. During the **Long March**, he commanded the advanced guard under the leadership of **Peng Dehuai**. After arrival in Yan'an, in Shaanxi, Lin immediately returned to his duties as head of the military academy. At the beginning of the **Anti-Japanese War** he was commander of the 115th Division of the **Eighth Route Army**, where his units won the first victory against the Japanese in the Battle of Pingxingguan in September 1937. He was wounded shortly afterward and sent to the USSR for medical treatment. During the final phase of the Anti-Japanese War, Lin Biao was a member of the Communist delegation negotiating with the Nationalist government in Chongqing. After 1945, Lin was sent to Manchuria, where his forces cooperated closely with the Soviet occupation forces. As the Communist revolution against the Guomindang (KMT) government developed, Lin's army was well enough established to limit Nationalist occupation to Manchuria's urban areas. Eventually, Lin's troops were reorganized to form the influential **Fourth Field Army**. In 1954, Lin was appointed vice premier and vice chairman of the National Defense Council and a marshal in 1955.

From 1959, following the purge of Peng Dehuai, until 1971, Lin served as minister of national defense. Lin was accused of attempting a coup d'état against Mao in 1971, with the aim of succeeding Mao Zedong as leader of the CCP. This was called the **571 Plan** (a homonym for the words "*wu qi yi*," which in Mandarin mean "armed uprising"). After the failure of the coup, the aircraft in which Lin Biao allegedly attempted to flee to the USSR crashed in the People's Republic of Mongolia. There are serious doubts about the attempted coup, and it is not known if the plane was shot down or blown up, or if an assassin was on board. It is more likely that Mao Zedong orchestrated the purge of Lin Biao under pressure from Mao's wife, **Jiang Qing**, to secure Jiang's succession to the party leadership (see **Gang of Four**).

REFERENCES

Wolfgang Bartke, *Who's Who in the People's Republic of China* (Armonk, NY: M. E. Sharpe, 1981); Michael Y. M. Kau, ed., *The Lin Piao Affair: Power Politics and Military Coup* (White Plains, NY: International Arts and Sciences Press, 1975); William W. Whitson with Chen-Hsia Huang, *The Chinese High Command* (New York: Praeger, 1973); Yao Mingle, *The Conspiracy and Death of Lin Biao* (New York: Alfred A. Knopf, 1983).

LIU BOCHENG (1892–1986) Liu Bocheng was born in 1892 in Sichuan Province. His father was a failed scholar who never attained an official position and earned a living as a traveling musician. Despite having received an early classical education, Liu was not permitted to take the Qing dynasty civil service examinations. He enrolled at the Chengdu Military Academy and joined the **warlord**-led New Army in 1911. He fought under a Sichuan warlord (Xiong

Kewu) and lost his left eye in combat. His military nickname was "Duyanlong" (one-eyed dragon). Ironically, Liu fought against the forces of **Zhu De**, later to be his colleague but then fighting for a Yunnan warlord, in the 1920s. Liu Bocheng participated in the **Northern Expedition** in 1926 and took part in the August 1, 1927, **Nanchang Uprising** (he had joined the Communist Party in 1926). Liu Bocheng later attended the Soviet Red Army College (Frunze Military Academy) in Moscow from 1928 to 1930. He returned to China and joined the **Jiangxi Soviet** in 1931, where he headed the Red Army Academy for three years. During the Jiangxi Soviet period, in the first counterencirclement campaigns, when Red Army forces successfully used the maneuver to ambush and defeat Soviet forces, Liu Bocheng was one of the **People's Liberation Army** (PLA) leaders who often attacked Mao Zedong. By the time of the Nationalist Fifth **Encirclement Campaign**, Liu Bocheng was chief of staff of the Red Army. After starting the **Long March**, Liu continued to disagree with Mao over tactics and military style, rejecting peasant-based guerrilla warfare in favor of regular units operating along Soviet tactical lines. Liu's ideas were closest to those of **Zhang Guotao** and **Zhu De**.

By 1938, Liu Bocheng was commander of the 129th Division, **Eighth Route Army**. His political commissar was **Deng Xiaoping**. Before that, however, at the close of the Long March, Liu commanded the column forming the advanced guard of the Red Army. Liu's ties to Sichuan and its minorities gave him a strong advantage in his command. His 129th Division was essentially formed from elements of the old **Fourth Front Army**. During the **Civil War**, Liu Bocheng operated along the rail lines from Beijing to Wuhan. Liu commanded forces that made a large swing from Nanjing south and west to Yunnan and Sichuan. He was the main leader associated with the **Second Field Army**. Liu Bocheng survived the **Cultural Revolution** despite his lifelong opposition to many of Mao's theories. He was one of the 10 generals appointed to be marshals of the PLA in 1955 and was a Politburo member through the Eighth, Ninth, 10th, and 11th Central Committees. Liu eventually lost most of the vision in his remaining eye, becoming less politically active after 1978 but remaining influential.

REFERENCES

Wolfgang Bartke, *Who's Who in the People's Republic of China* (Armonk, NY: M. E Sharpe, 1981); Tong Bai, ed., *Junren Shouce* (Beijing: Changzheng Chubanshe, 1987); William W. Whitson with Chen-Hsia Huang, *The Chinese High Command* (New York: Praeger, 1973).

LIU SHAOQI (1898–1973) Liu Shaoqi was a Communist Party activist who helped to form the Chinese Communist Party (CCP), organized trade unions, mass movements, and underground organizations in Nationalist-controlled areas, and was the designated successor to **Mao Zedong** and president of the People's Republic of China. During the period between the **Nanchang Uprising** on August 1, 1927, and the end of the **Anti-Japanese War** in 1945, Liu helped to organize guerrilla bases. Later, during the **Cultural Revolution**, he was vilified by Mao and the **Gang of Four** and purged from his positions.

Liu Shaoqi was born in 1898 in Ningxiang County, Hunan Province, not far from the birthplace of Mao Zedong. In 1917, while attending the Hunan First Normal School in Changsha, Liu joined the "New People's Study Society," established by Mao. He later attended a "work-study" education program in France, where he met **Deng Xiaoping** and other central figures in the Chinese Communist Party. Liu also studied at the University of the Toilers of the East along with **Zhang Guotao**, one of the founding leaders of the **People's Liberation Army** (PLA).

Although Liu was a Communist Party official throughout his career and not a soldier, he devoted much of the Anti-Japanese War to organizing guerrilla units. He operated behind Japanese lines in Anhui and Jiangsu Provinces with the **New Fourth Army** and, after **Xiang Ying**'s death in the **New Fourth Army Incident** (1941), was appointed the political commissar of the New Fourth Army. At the **Zunyi Conference**, in 1935, he supported Mao's takeover of the Communist Party's Military Affairs Commission from Zhang Guotao. Liu was attacked by **Lin Biao** and the leftists of the Cultural Revolution Group as a "rightist and capitalist roader." He was purged from his positions and died in obscurity under house arrest and "rehabilitation through labor" in 1973.

REFERENCE

Lowell Dittmer, *Liu Shao-ch'i and the Chinese Cultural Revolution: The Politics of Mass Criticism* (Berkeley: University of California Press, 1974).

LOCAL WAR DOCTRINE As Chinese military strategists assessed the strategic basis for warfare, the traditional Maoist concept of fighting a protracted **People's War** shifted to accommodate new technology and changes in the nature of warfare in what was called **People's War under Modern Conditions**. By the mid-1980s, based on their assessment of the war China had fought with Vietnam (the **Self-Defensive Counterattack against Vietnam**), the state of Sino–Soviet relations, and China's relations with the United States, Chinese strategists began once more to revise their assessment not only of the likelihood of war but of the nature, scope, and scale of future warfare. This new strategic assessment produced the doctrine of "local war" (*jubu zhanzheng*) theory, which held that China was more likely to fight a limited, regional, or local war on its periphery for the foreseeable future than a major or nuclear war. Local wars were assessed by the **People's Liberation Army** (PLA) to be a potential threat along China's periphery, at the outer limits of China's territorial claims or along contested boundaries. Thus, the space in which a local war would be fought was expected to be contained, and the time of the war, or its duration, was expected by PLA thinkers to be limited. These doctrinal shifts required a new emphasis on speed, agility, and surprise, as well as what is called "battlefield knowledge.'' The local war was expected to be resolved by negotiation and to be constrained in its scope by both domestic politics and international opinion. These factors led the PLA to develop Rapid Reaction Forces that could move swiftly to China's periphery supported by air and, where needed, naval assets.

By the end of the 1991 Gulf War between the U.S.-led United Nations coalition and Iraq, PLA strategists began a new assessment of local war doctrine. The Gulf War reinforced their beliefs about the changed nature of warfare, especially the role of high-technology weapons and intelligence or sensor systems to support combat. This led to a further revision in doctrine to account for ''local (limited, or *jubu*) war under high-technology conditions'' (*zai gao jishu xia de jubu zhanzheng*). In preparation for such a war, the PLA believed it needed better command, control, communications, and intelligence; improved electronic warfare equipment; improved logistics and maintenance for deployed forces; and the capability to conduct ''information age warfare'' (*xinxi zhan*). While this was a considerable departure from many aspects of ''people's war'' theory, aspects of that doctrine, such as the reliance on the support of the populace and the use of reserves or militia when required, were retained by the PLA. By 1998, some 12–15 divisions, distributed among the 24 group armies in the PLA, were converted to Rapid Reaction Forces intended to respond to ''local wars under high-technology conditions.''

REFERENCES

Mi Zhenyu, *Jubu Zhanzheng yu Jundui Jianshe Yanjiu* [Local War and the Study of Army Building] (Beijing: Junshi Wenyi Chubanshe, 1988); Nan Li, ''The PLA's Evolving Warfighting Doctrine, Strategy and Tactics, 1985–1995: A Chinese Perspective,'' in *The China Quarterly*, no. 146 (June 1996): 443–63; Pan Xiangting, ed., *Gao Jishu Tiaojiar Xia Meijun Jubu Zhanzheng* [American High-Technology Limited War under Modern Conditions] (Beijing: Jiefangjun Chubanshe, 1994); ZGRMJFJDSD.

LONGJU INCIDENT (1959) Indian reconnaissance of the **McMahon Line**, which established the Sino–Indian Border on March 24, 1914, sought to establish geographical points upon which to base the border, as a replacement for the line drawn by Henry McMahon after negotiations in 1913–1914. McMahon drew a map on which the borderline he drew did not follow the highest ridge near a Tibetan village called Mingyitun. The Indian reconnaissance apparently attempted to make the border clearer and to adjust the line to a boundary along a river, which put a hamlet called Longju within Indian territory. The Indians then unilaterally established a border post where their own maps showed Chinese territory without first negotiating the boundary change.

Beijing protested the moves by India and was especially concerned about the establishment of the border outpost. The Chinese government complained that, on August 25, 1959, Indian troops south of Mingyitun opened fire on Chinese border troops, which returned fire. New Delhi immediately issued its own protest, accusing China of "deliberate aggression" because Chinese troops had used force to move Indian soldiers out of Longju. The Chinese claimed the Indians had fired first and had then withdrawn from Longju on their own after the exchange of fire. This was the first clash of the boundary dispute on the Sino–Indian border.

REFERENCES

Deng Liqun et al., eds., *Zhongguo Renmin Jiefangjun* (Beijing: Dangdai Chubanshe, 1994); Neville Maxwell, *India's China War* (Bombay: JAICO Press, 1971).

LONG MARCH (1934–1935) From October 1934 through November 1935, in two echelons along two axes, the forces of the Red Army executed a strategic withdrawal and marched west and north through southern and Central China to a base area in **Yan'an**, Shaanxi Province. (see **Fourth Front Army** on the earlier Long March by that force).

In the midst of the Fifth **Encirclement Campaign**, which lasted from October 1933 to October 1934, the Communist Party began to consider completely evacuating the **Jiangxi Soviet** area. The initial conferences to develop an evacuation plan took place at Ruiqin, in the heart of the **Jinggang Mountains** (Jinggangshan). After a series of conferences in spring and summer 1934, plans were made to abandon the Jiangxi Revolutionary Base Area and to move south and west into a new Soviet base area in Central China. The main force of Communist troops was withdrawn from contact with Nationalist forces through the summer of 1934 and replaced with local troops. The **People's Liberation Army** (PLA) forces designated to conduct the withdrawal and evacuation from Jiangxi were divided into five corps and a main body or central column.

The army moved out in October 1934. The central column for the march was about 25,000 strong, comprising 15,000 party cadres and administrators, a cadre regiment and a national security regiment, and 5,000 laborers. The First Corps was composed of 18,000 soldiers under the command of **Lin Biao**. It served as the advanced guard for the march. The Third Corps, led by **Peng Dehuai**, also containing about 18,000 soldiers, was the security and screening force for the Long March, while the Fifth, Eighth, and Ninth Corps formed the rear guard of the army. Although the Fifth Corps was at full strength (18,000 soldiers), the Eighth and Ninth Corps were one division under strength and had only about 11,500 soldiers each. Chen Yi stayed behind with the local forces.

By December 1934, the difficulty of the march and a series of combat engagements against the Nationalists took a heavy toll on the Long Marchers. By the time the column passed through northern Guangdong and arrived at Yichang to cross the Xiang River, the Red Army had lost 25,000 of the original 102,000 that started the Long March. These losses were from desertion brought on by illness and disease as much as because of combat casualties.

After the Xiang River Crossing in November 1934, the **Second Front Army** under **He Long** linked up with the **First Front Army**, led by **Zhu De**. The column swung into Guizhou Province and defeated a force of local Nationalist troops at Liping, on the Guizhou Hunan border, where a planning conference was held. The Long March units then drove on Zunyi, in Guizhou, which was captured on January 5, 1935. By the time of the capture of Zunyi, only about 30,000 of the original 100,000 personnel that started the Long March were left in the Red Army. The **Fourth Front Army**, under **Zhang Guotao**, which had conducted its own march earlier, after the Nationalist fourth **Encirclement Campaign**, meanwhile continued to march west, hoping to link up with the First Front Army in Sichuan.

After a successful battle against the Nationalists in Zunyi (site of the **Zunyi Conference**) in February 1935, Mao Zedong, who had assumed political leadership of the army, tried to swing the forces northwest to link up with Zhang Guotao's Fourth Front Army in Sichuan. This effort ran into heavy Nationalist resistance and failed during March 1935. The First Front Army redirected its advance to the southwest, where a major river crossing at the Wu River near Guiyang was conducted by Xiao Hua's Third Corps on March 31, 1935. Xiao also led the eventual linkup with the Fourth Front Army at Mougong in the western Sichuan Province on June 12, 1935. By that time the strength of the First Front Army had dwindled to 10,000 men.

An internal political crisis ensued at this juncture. The First Front Army had organized itself along strong party lines with political commissars and a political security system involved in units at all echelons. The Fourth Front Army, in contrast, organized itself along classic military lines, vesting authority in unit commanders. Exacerbating the division over the political leadership of army units by Communist Party political commissars was resentment in the First Front Army that the Fourth had not fought hard enough to effect an early linkup, allowing the soldiers of the First Front Army to be killed in fighting against the Nationalists.

After decisive political meetings at **Lianghekou** and **Maoergai**, the two Front Armies split, with the First moving north through Gansu Province in July 1935. The Fourth Front Army attempted to move further west, seeking more secure base areas, but later turned north again. By October 1935, the First Front Army finally reached Shaanxi Province, where it set up base areas around Yan'an and established the **Anti-Japanese University**.

REFERENCES

Chen Chang-feng, *On the Long March with Chairman Mao Beijing* (Beijing: Foreign Languages Press, 1959); Agnes Smedley, *The Great Road* (New York: Monthly Review Press, 1956); Edgar Snow, *Red Star over China* (New York: Grove Press, 1956); William W. Whitson with Chen-Hsia Huang, *The Chinese High Command* (New York: Praeger, 1973); Dick Wilson, *The Long March: The Epic of Chinese Communism's Survival, 1935* (New York: Viking Press, 1971); ZGRMJFJZS.

LU GOU QIAO INCIDENT. See MARCO POLO BRIDGE INCIDENT

LUOCHUAN CONFERENCE (August 1937) This meeting, held in Luochuan, Shaanxi Province, on August 27, 1937, approved the Chinese Communist Party (CCP) Organization Department plan calling for the conduct of a single, focused military campaign against the Japanese. The Red Armies were reorganized into the **Eighth Route Army** for the conduct of the war. At the conference, **Mao Zedong** called for a two-front conflict against both the Japanese and the Nationalists. All of the overt activities of the party and the **People's Liberation Army** (PLA) were to fight the Japanese, in coordination with the Nationalists, but PLA leaders were secretly to concentrate on conserving strength. Covertly, Communist forces were to continue to build their base areas in preparation for a future fight with the Nationalists. Mao was opposed in this strategy by **Zhang Guotao, Zhou Enlai**, and **Peng Dehuai**, who insisted that the main fight was against the Japanese. Following the conference, Mao revised his views, eventually sending all three divisions of the Eighth Route Army into neighboring Shanxi Province to fight the Japanese. Other cadres were sent throughout China to organize underground activities against the Japanese while building Communist strength.

REFERENCES

Mao Zedong, *Selected Works*, vol. 2, ''For the Mobilization of All the Nation's Forces for Victory in the War of Resistance,'' pp. 23–28 (Beijing: Foreign Languages Press, 1965); William W. Whitson with Chen-Hsia Huang, *The Chinese High Command* (New York: Praeger, 1973).

LUO RONGHUAN (1902–1963) Luo Ronghuan was appointed one of the 10 marshals of the **People's Liberation Army** (PLA) when military ranks were instated in 1955. Luo was born in 1902 in Hengshan County, Hunan Province, and is a graduate of Sun Yatsen University in Guangzhou. He was a participant in the August 1, 1927, **Nanchang Uprising**, which is commemorated as the founding of the PLA.

From 1928 to 1933, Luo Ronghuan was part of the Fourth Red Army, which expanded into the First Red Army and eventually the **First Front Army**. After commanding a battalion, he was a political commissar and a Political Department director. He was the Communist Party representative in **Lin Biao**'s guerrilla column that operated in Jiangxi and Fujian in 1929. At the **Gutian Conference**, he backed **Mao Zedong**, supporting an irregular, guerrilla-like military force with strong Communist Party leadership over the type of well-drilled professional army advocated by leaders like **Zhu De**.

Luo Ronghuan took part in the **Long March** as part of the First Front Army. He continued his strong support for Mao Zedong at the Long March Conferences (**Zunyi, Lianghekou,** and **Maoerhgai**). In 1937, when the Long Marchers reached **Yan'an**, Luo became a political instructor at the **Anti-Japanese Military and Political College**. Later he served as the political commissar of the **Eighth Route Army**'s 115th Division and in 1944–1945 was the acting commander of the division. During the period that the division operated in the Shandong area, 1939–1945, Luo Ronghuan concurrently served as the commander and political commissar of the Shandong Military District. He was reportedly hospitalized in the Soviet Union for two years. From 1947 to 1948, Luo was a deputy political commissar and political commissar of the **Northeast Democratic Allied Army**, which conducted the **Sungari River** and **Liaoning-Shenyang Campaigns**. During this period he was, for a while, **Peng Zhen**'s deputy. When the **Fourth Field Army** was formed, Luo Ronghuan was its political commissar. He then served as director of the General Political Department of the People's Revolutionary Military Council from 1950 to 1954 and director of the PLA General Political Department from 1954 to 1956. From 1956 to 1963, Luo Ronghuan was a member of the Politburo of the Chinese Communist Party, the last two years of which he was a Standing Committee member. He was a strong supporter of Mao throughout his career and was very closely associated with Lin Biao. In the **Beiping-Tianjin Campaign** (1948–1949), Luo was Lin Biao's political commissar. In addition to being a strong and influential military leader, Luo Ronghuan helped to build a strong civil Communist Party infrastructure in Communist-controlled areas and after the formation of the People's Republic of China in 1949. He died on December 12, 1963, after a long illness.

REFERENCES

Wolfgang Bartke, *Biographical Dictionary and Analysis of China's Party Leadership, 1922–1988* (New York: K. G. Saur, 1990); William W. Whitson with Chen-Hsia Huang, *The Chinese High Command* (New York: Praeger, 1973).

LUO RUIQING (1906–1978) Luo Ruiqing was one of the victims of the **Cultural Revolution**, a major leader of the **People's Liberation Army** (PLA) and the Communist Party who saw the Chinese Communist Party (CCP) and the **Red Guards** turn on him and literally drive him to attempt suicide. He was a strong supporter of **Mao Zedong** and at one time director of all public security forces in China. Luo stood for the rationalization of the allocation of economic and military resources to support China's development. Luo Ruiqing was born in 1906 in Nanchong, Sichuan Province. He attended the fifth training course at **Whampoa Military Academy** in 1926, took part in the August 1, 1927, **Nanchang Uprising**, and joined the Red Fourth Army in the Jinggangshan Mountains. In 1930 he was political commissar of the 11th division in the Fourth Front Army commanded by Lin Biao. In spring 1932, Luo was severely wounded in a battle with Nationalist units and was sent to the Soviet Union, where he received training in security and intelligence work. During the **Long March**, Luo served initially as director of the Army's Security Bureau and later with the **First Front Army**. After the Long March, Luo attended the Red Army University and in 1936 was appointed to direct the university's Education Department. In 1940 he was made head of the **Anti-Japanese Military and Po-**

litical College. Shortly afterward he was appointed director of the Political Department of the **Eighth Route Army** as well as member of the North China Bureau of the CCP Central Committee. In 1944, Luo headed the Political Department of the 18th Army Corps. During the **Civil War** he served in various posts in the PLA, primarily in North China. In September 1949, Luo Ruiqing was elected a member of the Chinese Communist Party Central Committee. He was also appointed minister of public security, a post he held until 1959, when he was appointed vice premier, vice minister of national defense, and chief of the PLA General Staff, replacing Peng Dehuai. He held all three positions until the Cultural Revolution.

In December 1965, as the Cultural Revolution began, Luo Ruiqing was required to appear at a criticism conference in Shanghai. He rejected all charges against him. In March 1966 Luo again had to answer leftist criticism under the supervision of the CCP Central Committee. On March 18, he attempted suicide by jumping from the top floor of the building where the investigation was being conducted. Prior to his attempted suicide, he had prepared a document in which he allegedly revoked all his previous confessions of guilt. The charges against Luo were hostility, opposition to Mao Zedong thought, bourgeois military thinking, undermining of centralized party democracy, adherence to the point of view of the exploiting class, and urgent appeals to Lin Biao to step down in favor of himself. Luo made no public appearances after this incident until after the Cultural Revolution in 1975. In 1977 he was elected member of the CCP Central Committee by the 11th Party Congress and identified as member of the Central Military Commission. Luo died on August 3, 1978.

REFERENCES

Wolfgang Bartke, *Who's Who in the People's Republic of China* (Armonk, NY: M. E. Sharpe, 1981); JFJJLZ, vol. 2.

LUSHAN CONFERENCE (Lushan Plenum) (August 2–16, 1959) The Eighth Plenary Session of the Eighth Chinese Communist Party (CCP) Central Committee was held at a meeting site in Lushan, Jiangxi Province, August 2–August 16, 1959. At this meeting, **Mao Zedong**'s plans for a "Great Leap Forward" in the Second Five-Year Plan (1958–1962), which had been in progress from the beginning of the year, were severely criticized by the chief of the General Staff Department of the People's Liberation Army, **Peng Dehuai**. Peng had long opposed Mao on matters of military strategy and tactics, preferring large-scale combat and maneuver warfare along Soviet lines over Mao's guerrilla warfare tactics. Peng also opposed Mao on the strength of the political commissar system and wanted a regular army employing ranks.

In the "Great Leap Forward," Mao sought to reverse the priorities among heavy industry, light industry, and agriculture, placing greater emphasis on communal living and ownership; creating local, indigenous blast furnaces for backyard industry (backyard furnaces), which were not supported by local and natural resources; and emphasizing political affilition over performance. After an extended fact-finding trip to his own home and to poorer parts of Central China, Peng Dehuai criticized the Great Leap Forward in a letter to Mao, which was supported by many party members. Earlier, when Peng had objected to the insertion of Mao Zedong's philosophy and thought into the 1956 Communist Party Constitution, he had angered Mao. Then, at the Lushan Conference, Peng attacked Mao's attempts to establish a system of communes and to accelerate the transition to Communism in China. Mao mounted a strong attack against Peng, criticizing him as a rightist who was giving in to the pressure of the bourgeoisie. Peng was dismissed at the conference, and many of his direct subordinates, like, **Ye Jianying, Hong Xuezhi**, and **Huang Kechang**, also suffered. The clash between Mao and Peng at Lushan was characterized by the Communist Party as a "class struggle between the bourgeoisie and the proletariat." Peng and his subordinates were accused by Mao of having a "simple military mentality." Mao said that Peng was forming a "rightist opportunist clique" and organizing factions to oppose Mao. The dismissal of Peng Dehuai and the concomitant dismissal of critiques of the economic program of the Great Leap Forward resulted in a massive famine that claimed as many as 20 million lives. Many of the dead, half in 1963, were under 10 years old. The rank system of the **People's Liberation Army** (PLA) was later abolished, and the role of the military commissar in the PLA was strengthened. Lin Biao supported Mao in the criticism of Peng Dehuai and replaced Peng as the defense minister after the Lushan Conference. Lin Biao came to the fore as the major military leader until he, too, was dismissed as a threat to Mao in 1971.

REFERENCES

Hong Qi, no. 13 (1967): 21–24; Roderick MacFarquhar, *The Origins of the Cultural Revolution*, vol. 2: *The Great Leap Forward* (New York: Columbia University Press, 1983); William W. Whitson with Chen-Hsia Huang, *The Chinese High Command* (New York: Praeger, 1973).

LUSHAN CONFERENCE OF 1970 (August 23–September 6, 1970) Despite **Mao Zedong**'s having declared and confirmed at the Ninth Party Congress, in April 1969, that his official successor was **Lin Biao**, a general reassessment of international and domestic policy ensued after the Sino–Soviet clashes on and near **Damansky Island**. Meanwhile, Mao and Lin Biao disagreed over domestic political matters, such as the continuation of the position of state president and a military buildup in preparation for an expected war against the Soviets. At the conference, Lin Biao presented a speech entitled "Discourse on Genius," which was critical of Mao. Moreover, a number of internal discussion groups at the congress began to consider the speech. Mao retaliated by calling together the Chinese Communist Party Politburo Standing Committee, claiming that Liu's speech was a veiled way to engage in domestic political struggle against Mao. Mao also published his own article, which was very critical of Liu's speech.

Mao managed to institute an inner-party investigation of Lin's associate, Chen Boda, at the conference, for attacking Mao and supporting Liu. A final party report on the conference continued an investigation of Chen and his political position. Mao continued to pursue Lin, leading to the alleged coup attempt of the "571 Incident" (see the entry on **Lin Biao**).

REFERENCES

Martin Ebon, *Lin Biao: The Life and Writings of China's New Ruler* (New York: Stein and Day, 1970); Michael Y. M. Kau, ed., *The Lin Biao Affair* (White Plains, NY: International Arts and Sciences Press, 1975).

M

MA WEI, BATTLE OF (August 23, 1884) The naval battle off Ma Wei, a small town at the mouth of the Min River on the east bank opposite the city of Fuzhou, was the opening of hostilities in the **Franco–Chinese War**. At the time of the battle, China had more than 50 ships in its navy, including German- and American-built gunboats and cruisers. About half of the ships were constructed in China, some at the **Fuzhou Dockyard**, near where the battle was fought. However, China had not organized its ships into a national fleet. Instead, they were controlled by regional governors-general appointed by the Qing dynasty.

The governor-general of Canton (Guangzhou) had constructed a series of fortifications along the Fujian Province coast, including along the Min River. The Fuzhou Dockyard superintendent was He Ruzhang, who had overall responsibility for the Fuzhou fleet. However, the tactical control of the fleet's ships was the responsibility of Zhang Cheng, the captain of the fleet's flagship, the *Yang Wu*.

The French fleet in Asia was dispersed off the South China and Indochina coast under the control of Admiral Amende A. P. Courbet. Although no formal state of war was declared, there were serious disputes between the French and the Chinese over control of the northern part of Vietnam (Cochin China) and the surrounding waters. The French fleet in the area in July 1884 was led by Courbet's flagship, the *Volga*, and consisted of four other warships. By August 22, 1884, the French naval presence off the Min River had grown to eight warships, all anchored in the approaches to Fuzhou in the Ma Wei roads. The ships all had armor-clad hulls and were considered modern for the time. The Chinese had a fleet of 11 wood-hulled ships with modern armament at Fuzhou. In addition there were seven steam-driven launches and 12 armed junks used for troop transport. After a dispute over unimpeded access to the river for the purpose of trade, which had gone on since early July, the French initiated action against the Chinese fleet on the afternoon of August 23, 1884.

There are different accounts of whether there was any declaration of war on the part of the French before hostilities commenced. As a minimum, it seems likely that some Chinese official, if not the governor-general in Canton himself, was notified that the French would attack if a blockade of the Min River by the Chinese was not lifted. Perhaps an official of the Fuzhou Dockyard was given an ultimatum by Courbet. In any case, within about 12 minutes of the commencement of action by the French at about 2:00 P.M. on August 23, they had sunk almost the entire Chinese fleet. Varying accounts of the battle say that all 11 Chinese warships were sunk; others say that as many as 22 Chinese ships of different classes were sunk. Only five Chinese ships were reported to have gotten under way from the Fuzhou Dockyard, and only two of these escaped upriver unscathed. Command and control signals for the Chinese fleet were poor, whereas only two French ships suffered minor damage from fire.

China declared war on France on August 26, 1884.

REFERENCES

John L. Rawlinson, *China's Struggle for Naval Development, 1839–1895* (Cambridge: Harvard University Press, 1967); Bruce Swanson, *Eighth Voyage of the Dragon—A History of China's Quest for Seapower* (Annapolis, MD: Naval Institute Press, 1982).

MANCHUKUO The puppet state of Manchukuo was established by the Japanese in Manchuria and ruled by the last emperor of the Manchu (Qing) dynasty, Pu Yi. After the **918 Incident** (September 18, 1931, or the Mukden Incident), where Japanese officers contrived and took over the region with the Kwantung Army, the Japanese established a separate state in Manchuria. The capital of Manchukuo was established at Changqun, and the city was renamed Xinjing, or "New Capital." Pu Yi was made chief executive of the state and later, on March 1, 1934, was elevated to emperor with the imperial name of Kangde. As it was established by the Japanese, the state of Manchukuo included the provinces of Liaoning, Jilin, Heilongjiang, and Jehol (which no longer exists but encompassed parts of what is now Hebei and Inner Mongolia). Manchukuo was populated primarily by Han Chinese (about 20 million people, or two-thirds of the population of Manchukuo). Only about 12 percent of the population was actually Manchu, with the remainder of the population being ethnic Korean, Japanese, and White Russian.

Manchuria had long been coveted by China's neighbors. It has abundant natural resources, including lumber, coal, iron, ore, and petroleum. It is also well located strategically, with an important peninsula, the Liaodong Peninsula, that juts into the Yellow Sea. The **Sino–Japanese War** was fought over control of the area in 1894–1895, after which the Chinese were forced to cede the Liaodong Peninsula to Japan. The Russo–Japanese War was fought in 1904–1905, in part over access to a strategic rail line connecting Port Arthur (Dalian) to

Changqun and Russian territory. The Japanese gained the rights to the leased Russian railroad after the war and in 1906 created the Kwantung Army to guard the Liaodong Peninsula, which they renamed the Japanese province of Kwantung.

The creation of Manchukuo was allegedly the idea of two senior staff officers of the Japanese army, Colonel Seishiro Itagaki and Lieutenant Colonel Kanji Ishiwara, who are also credited with having staged the 918 (Mukden) Incident. These two officers were concerned that the growing power of the Guomindang government would lead to the complete unification of China and the loss of Manchuria for Japan. Their broader strategic intent was allegedly to create a buffer area between Japan and the Soviet Union in the event of war between the two countries. After the establishment of the state of Manchukuo, Japan and Manchukuo signed the Manchukuo–Japan Protocol, recognizing the independence of the state but giving the commander of the Kwantung Army control of economic, political, and military matters related to the state. The combined Russian, Chinese, and Mongolian attacks into Manchukuo in August 1945 restored Manchuria to Chinese control, but the area was quickly the focus of the civil war between communists and nationalists.

REFERENCES

An Tai-sung, *The Sino–Soviet Territorial Dispute* (Philadelphia: Westminister Press, 1973); Liang Chin-tun, *The Sinister Face of the Mukden Incident* (New York: St. John's University Press, 1969); Lucian W. Pye, *Warlord Politics: Conflict and Coalition in the Modernization of Republican China* (New York: Praeger, 1971); Jonathan D. Spence, *The Search for Modern China* (New York: W. W. Norton, 1990).

MANCHURIA, CIVIL WAR CAMPAIGN, NATIONALIST AND COMMUNIST APPROACH MARCH (August 1945–March 1946) As the end of **World War II** drew near, both the Communists and the Nationalists prepared to occupy Manchuria, which had been under Japanese control as the Republic of **Manchukuo** since 1935, in anticipation of a Japanese collapse. When the atomic bomb was dropped on Hiroshima, the Soviets declared war on Japan and invaded Manchuria, sweeping east from Mongolia and west from the area of Khabarosk and Vladivostok. On the same day as the Soviet invasion, August 9, 1945, **Mao Zedong** ordered a general offensive by Communist forces in North China, including Manchuria. **Lin Biao** rallied Communist forces in **Yan'an** and set out for Manchuria from the Shaanxi-Shanxi border base area with a column of about 10,000 troops. Lin's forces first marched east, toward Hebei Province and the Yellow Sea, and then north, into Jehol Province (north of Beijing, since incorporated into Hebei and Liaoning Provinces). As he reached the pass between the Great Wall and the Yellow Sea at Shanhaiguan, on the Jehol-Hebei border, Lin collected further strength, bringing along captured Japanese equipment and more Communist troops. At the same time, more Communist forces, some of which were transported by sea, were ordered to move north from Shandong and Jiangsu Provinces. By the time that Lin Biao's forces crossed through Shanhaiguan, in the October–November 1945 time frame, their strength had expanded to about 100,000 troops.

Nationalist forces, meanwhile, were ordered by **Chiang Kai-shek** to move quickly to occupy Manchuria. With the assistance of U.S. Navy vessels of the Seventh Fleet, by the beginning of November 1945, while Lin Biao's forces were still marching through the pass between the Great Wall and the sea at Shanhaiguan, six Nationalist Divisions, comprising the 13th and 52d Corps, were brought to the Manchurian ports of Huludao, Dalian, and Yingkou. However, occupying Soviet forces refused permission for the Nationalist troops to disembark. Finally, on November 5, 1945, the Nationalist troops disembarked from the U.S. Seventh Fleet vessels at the port of Qinhuangdao, near Tianjin, south of Shanhaiguan. On November 16, 1945, after fighting and breaking through Communist defenses at Shanhaiguan, the Nationalist 13th and 52d Corps began their approach march up the coast into Manchuria, passing through Huludao and Shenyang (Mukden) toward the city of Siping (see **Siping, Defensive Campaign of** and **Siping, Offensive Campaign Against**) which had been the center of two decisive episodes of the Manchurian campaign of the **Civil War**.

The Japanese-trained Manchukuo Army, with a strength of about 300,000 troops, was disbanded by the Soviets. However, many of the Manchukuo provincial units still resisted Communist advances in Manchuria. Still, some of the former soldiers of the Manchukuo Army were recruited by Lin Biao's forces, swelling Communist strength to about 200,000 troops by early 1946. At the time that Soviet forces began their own withdrawal from Manchuria, which took place during February and March 1946, Nationalist troop strength in the area was much lower, about 150,000 troops. By early spring 1946, the stage was set for the decisive Northeast Campaign of the Chinese Civil War.

REFERENCES

Trevor N. Dupuy, *The Military History of the Chinese Civil War* (New York: Franklin Watts, 1969); ZGDBKQS; ZGRMJFJZS, vol. 3; ZGRMJFJZSJB.

MAO ZEDONG (1893–1976) China's most influential and mercurial twentieth-century leader, Mao Zedong developed a strategy for guerrilla warfare that eventually defeated the Nationalist Army in China. The relentless attacks against Communist forces by the Nationalists and the crucible of **World War II** combat against Japan forged Mao's politicomilitary system into an ideology that consumed China. Taken to its most radical ends, Mao's ideological system turned Communist Party leaders against each other and consumed some of the very leaders that carried the party to power in 1949.

Mao Zedong was born in 1893 in the village of Shaoshan, Hunan Province, into a moderately wealthy peasant family. Mao was tutored in reading and writing between the ages of 8 and 13, after which he continued to read Chinese novels while assisting in managing his father's affairs. In 1907, a marriage was arranged for him by his parents, but Mao refused to acknowledge the arrangement. In 1911 Mao passed the examination for Changsha Middle School, but the turmoil at the end of the Qing dynasty sidetracked his education for a short while. He continued his education at Fourth Teachers Training College, where he focused on history, geography, and philosophy. While studying at the college, Mao was active in the Student Society, where he served as secretary and later as chairman. In September 1918, Mao went to Beijing, where Li Dazhao, an early organizer of the Chinese Communist Party and a Marxist proponent, helped him find work at the library of Beijing University. In spring 1919, Mao returned to Changsha to teach at an elementary school. He was active in the Xinmin Xuehui (New People's Study Society), which he helped to found one year earlier. After the May 4 Movement, Mao initiated a weekly journal, *Xiang Jiang Pinglun*, which was later banned by local authorities. Despite the ban, Mao published articles in another journal, *Xin Hunan* (New Hunan), and organized the Wenti Yanjiu Hui (Problem Discussion Group) in Changsha. In 1920, Mao studied Communist writings, having been acquainted with Communism only through his contacts with Li Dazhao and Chen Duxiu.

In July 1920, Mao headed the Elementary Education Department at the First Teachers Training College in Changshen. He also founded a bookstore that sold Communist literature and that he expanded to seven branches in other cities of Hunan. Mao used this enterprise for fund-raising to finance political activities. Mao was one of the 12 founding members of the Chinese Communist Party (CCP) on July 1, 1921, in Shanghai. He was party secretary during the CCP First Congress. After the Congress he returned to Changsha to set up the CCP Hunan Branch, of which he became secretary.

Mao began efforts to organize workers in coal mines in Anyuan, West Jiangxi, where he met **Liu Shaoqi**. In the winter of 1921, he married the daughter of his former teacher, Yang Kaihui, with whom he had one son, Mao Anying, and a daughter. Yang Kaihui was executed in 1930, with Mao's sister, Mao Zehong, by order of the Nationalist (KMT) governor of Hunan Province, He Jian.

An arrest warrant was issued against Mao in early 1923, causing him to flee to Shanghai, where he served in the party headquarters. Mao was one of the advocates of a **United Front** policy with the KMT in accordance with the directives of the Comintern (the Soviet-controlled Communist International, also known as the ECCI, or Executive Committee of the Communist International). However, he gradually grew more critical of Comintern directives from Moscow. Dissatisfied with the United Front policies, Mao returned in late 1924 to Hunan, where he began to turn seriously toward the problems of peasant organizations. A new arrest warrant for him, issued by the governor of Hunan, persuaded Mao to leave his native province for Guangzhou, where he lectured at a training school for cadres in the peasant movement. In late 1925, Mao moved to Shanghai to organize resistance against the Nationalist governor of Hunan, General Zhao Hengti, who had brutally suppressed a strike by the workers in Anyuan.

After **Chiang Kai-shek**'s ascension to power in the Guomindang in spring 1926, Chiang moved to purge Communist cadres from the labor movement, leaving Mao without a position. Mao returned in summer 1926 to Hunan, where he worked mobilizing peasants until the collapse of the Guomindang–CCP United Front. After the failure of the **Nanchang Uprising** on August 1, 1927, Mao led the **Autumn Harvest Uprisings** in Hunan and Jiangxi. Executing a hasty, strategic retreat when the uprising failed, Mao set up the armed Communist base in the **Jinggang Mountains**.

In early 1928, Mao was joined by **Zhu De, Chen Yi**, and **Lin Biao**, who combined their forces at this base to form the Red Army. Zhu De was commander, and Mao was political commissar. In 1929, the army moved its base to Ruijin in Jiangxi. At about roughly this same time, Mao Zedong married He Zizhen, a schoolteacher. From 1930 until the beginning of the **Long March** in 1935, Mao consolidated his position in the CCP and the Ruijin Soviet. However he was not particularly influential within the CCP, which at that time was dominated by a group of people trained in Moscow. During this period, **Zhou Enlai** was one of the strongest opponents of Mao. After the successive **Encirclement Campaigns** conducted against the Communists by Nationalist forces, the Communists retreated once more from their opponents. This marked the beginning of the Long March. In January 1935, the CCP Politburo Conference in **Zunyi** gave the party leadership to Mao. He was elected Politburo chairman. Most analysts assume that this change in leadership validated Mao's concept of mobile guerrilla warfare by the Politburo, but really it marked a departure from orthodox, Soviet Comintern policy, which called for leadership of a revolution by the urban proletariat.

The Long March ended in October 1935, leaving the CCP and the **People's Liberation Army** (PLA) headquartered in **Yan'an**, Shaanxi Province. While in Yan'an, Mao shaped the ideological course of the party through a number of essays including "Strategic Problems of the Chinese Revolutionary War" (December 1936); "On Delaying Action" (March 1938); "Strategic Problems of the Anti-Japanese War and of Delaying Actions" (May 1938); "Problems of War Strategy" (November 1938); "On the New Democracy" (January 1940).

In 1939, Mao separated from his second wife, He Zizhen, and married Jiang Qing (an actress whose stage name was Lan Ping, or "Blue Apple") that same year. Mao Zedong's leadership of the CCP by that time was unchallenged. After the defeat of the Japanese in August 1945, Mao initiated a prolonged revolutionary war of liberation to defeat the Nationalists.

In 1949, after the collapse of the Nationalist government, Mao was elected chairman of the Central People's Government (head of state) and chairman of the Revolutionary People's Military Council. Later that year, Mao took his first trip abroad, conducting negotiations in Moscow. During the **Korean War**,

Mao's eldest son, Mao Anying, was killed while serving as a division commander. Mao stepped down as head of state in favor of Liu Shaoqi in December 1958.

Mao's influence in the party began to diminish after 1958. The failure of his ill-advised ''Great Leap Forward Campaign,'' touted as a means to propel China's economy to enormous achievements within a short period of time, was disastrous for China. In 1961, Mao purged the PLA's senior leader, **Peng Dehuai**, after Peng criticized the ''Great Leap.'' For a short time, Mao nearly disappeared from the scene in eclipse, under the political shadow of **Liu Shaoqi** and **Deng Xiaoping**. Seeking to shore up his own power, Mao secured the support of the military by establishing a close relationship with its leader, Lin Biao. With military support, Mao initiated the **Cultural Revolution** in late 1965, with the aim of regaining a powerful role within the party. Mao succeeded by giving young **Red Guards** the chance to question the authority of elders, parents, and teachers. Mao's wife, Jiang Qing, emerged from the shadows to encourage the fanatical Red Guards in cooperation with the **Gang of Four** during the Cultural Revolution. Two-thirds of the leading cadres of the Communist Party were attacked, humiliated, and sent to labor camps, including prominent military leaders of the revolutionary period. Eventually, order was restored only after the intervention of the PLA to suppress the Red Guards. In 1969, the Ninth CCP Central Committee was elected, and Mao had regained absolute control over the party. When Lin Biao died in 1971, following an alleged attempt to stage a coup d'état, rumors also circulated that Mao allowed his wife to purge his most serious rival for succession. Mao Zedong died on September 9, 1976. One month later his widow and her three major supporters, who have since become known as the ''Gang of Four,'' were arrested by Mao's former security guard, **Wang Dongxing**. Deng Xiaoping returned to power for the second time, giving China an opportunity to gradually eliminate the Jiang Qing faction. This process was largely completed by the end of 1978. Although the PLA continued to legitimate itself by citing the will of Mao Zedong, the post-Mao leadership whittled Mao down to a more historically accurate scale.

REFERENCES

Wolfgang Bartke, *Who's Who in the People's Republic of China* (Armonk, NY: M. E. Sharpe, 1981); Maurice Meisner, *Mao's China: A History of the People's Republic* (New York: Free Press, 1977); C. Martin Wilbur and Julie Lien-ying How, *Documents on Communism, Nationalism and Soviet Advisers in China, 1918–1927: Documents Seized in the 1927 Peking Raid* (New York: Columbia University Press, 1956); Dick Wilson, ed., *Mao Zedong in the Scales of History* (London: Cambridge University Press, 1977).

MAOERGAI CONFERENCE (Maoerhkai Conference, August 5, 1935) On August 5, 1935, the Politburo of the Chinese Communist Party (CCP) held a meeting at Maoergai, north of Chengdu in Sichuan Province, to discuss future military strategy and leadership. This was one of a series of meetings along the course of the **Long March**, during which there were serious factional splits within the CCP over the direction of military leadership and deployments of the **People's Liberation Army** (PLA). The Maoergai meeting took a step toward resolving a dispute over how to break the **First** and **Fourth Front Armies** out of a nearly complete encirclement by Guomindang forces in western Sichuan Province. **Zhang Guotao** and **Mao Zedong** argued over how to respond to Nationalist military successes against the CCP. There were continuing disputes among the military and political leadership of the CCP over how to lead the army. Some advocated leadership through a General Staff, along Soviet or German lines, while others wanted to maintain a flexible, guerrilla-like force that responded to developing situations. Mao Zedong (along with **Peng Dehuai, Lin Biao**, and **Ye Jianying**) supported a more open management style and a guerrilla-based army. Zhang Guotao, **Zhu De**, and **Liu Bocheng**, among other officers, believed that the army should develop along the lines of the Soviet or German General Staff System, with a strong staff, a rank structure, and mutually supporting arms and services. The Maoergai meeting really continued a line of argument that developed at the **Zunyi Conference** (January 1935) and the **Lianghekou Conference** (June 1935). At the tactical and operational level, the dispute between Zhang Guotao and Mao Zedong involved where and how to reconstitute and restage the army after the defeats suffered in the successive Nationalist **Encirclement Campaign** and during the Long March. Zhang argued for a retreat farther west into the Qinghai-Gansu area, which was inaccessible and would permit the army to train and rebuild its strength. Mao argued that this was a retreat from the task of defeating the Japanese. Mao further argued that the sort of restructuring of the army sought by Zhang was really an attempt at right-wing opportunism on Zhang's part and was an attempt to control the Red Army. Finally, Mao also argued that movement west was a retreat and not a reconsolidation. He advocated a march north to the **Yan'an** area of Shaanxi, the establishment of guerrilla base areas there, and continued attacks on the Japanese using guerrilla tactics while still building military strength. At that point, Zhang ignored the directions of the Politburo and moved south, hoping to eventually occupy the Sichuan plain. Not until December 1936 did Zhang Guotao, with Zhu De, Liu Bocheng, and his Fourth Front Army, move north to Shaanxi to rejoin Mao and the remainder of the Red Army at Yan'an.

REFERENCES

Lincoln Li, *The Japanese Army in North China, 1937–1941* (London: Oxford University Press, 1975); Jonathan D. Spence, *The Search for Modern China* (New York: W. W. Norton, 1990); William W. Whitson with Chen-Hsia Huang, *The Chinese High Command* (New York: Praeger, 1973); ZGDSJMCD; ZGJFJZS, vol. 2.

MARCO POLO BRIDGE INCIDENT (Lugouqiao, July 7, 1937) The incident at the Marco Polo Bridge, in the Fengtai District about 10 miles southwest of Beijing, marked the start of the **Anti–Japanese War**. It is considered by some (Spence) to be the first battle of **World War II**. Based on the Boxer Protocol of 1901, Japanese troops had the authority to conduct military maneuvers in China. On July 7, in the vicinity of Wanping, a company of soldiers from Eighth Company, Third Battalion, of the Japanese garrison in Beijing went to the area for training. The Marco Polo Bridge, a stone structure over 700 years old, spans the Yongding River at what was in 1937 (and remains today, as the Changxindian railroad yard) a critical rail junction and railroad switching yard. Less than a year earlier, at the Fengtai rail junction just east of the bridge, a major incident had occurred between Japanese soldiers and Chinese army stable hands. The Fengtai area was strategically important as a transportation hub and rail shipment point for the main rail trunk lines to Taiyuan, Tianjin, and south from Hebei.

On July 7, 1937, the Japanese troops made the bridge the base of operations for company-level training exercises, which included the firing of blank cartridges. During night operations by the Japanese, soldiers of the Nationalist Chinese (Guomindang) 110th Brigade, 37th Division, 29th Army, fired shells into the Japanese assembly area. Although there were no Japanese casualties, one Japanese soldier was alleged to have been missing. The Japanese commander ordered an attack on Wanping, where the Chinese forces were quartered. The Japanese were repulsed by the Nationalist troops on the night of the seventh. On July 8, the Chinese counterattacked but failed to dislodge the Japanese infantry company from its position near the bridge. **Chiang Kai-shek** ordered four Chinese divisions to Baoding, about 40 miles to the south, and negotiations began between China and Japan. The Japanese, meanwhile, embarrassed by their failure to take the bridge and fearing a severe loss of face as an occupying force, moved over five divisions into the area. This increased total Japanese strength in the vicinity of Beijing from approximately 10,000 soldiers to about 100,000. Japanese reinforcements arrived by July 19, and the Japanese started full-scale operations in the Tianjin-Beijing sector on July 25, 1937. The sector fell to the Japanese within two weeks.

REFERENCES

Lincoln Li, *The Japanese Army in North China, 1937–1941* (London: Oxford University Press, 1975); Jonathan D. Spence, *The Search for Modern China* (New York: W. W. Norton, 1990); William W. Whitson with Chen-Hsia Huang, *The Chinese High Command* (New York: Praeger, 1973); ZGRMJFJZS, vol. 2.

MARINES. See NAVAL INFANTRY

MARSHALL MISSION TO CHINA (1945–1947) At the end of **World War II**, China's Nationalist government was in an excellent financial situation with an estimated $900 million in gold reserves and U.S. dollars. In addition to the financial reserves of the Nationalist government, private Chinese citizens also held even more foreign exchange assets. The political situation in China, however, was not as stable as the financial situation. Chinese Communist forces largely ignored the instructions of the Nationalist government regarding accepting the surrender of Japanese military forces in China and the forces of the puppet **Wang Jingwei** government, which was controlled by Japan. Forces of the Chinese Communist Party seized all the enemy arms, ammunition, and equipment they could, occupied territory controlled by the Japanese, and sometimes incorporated the Wang Jingwei's puppet government forces into their own. As a result, throughout China, there were frequent clashes between the Nationalist and Communist armed forces.

The commanding general of U.S. forces in China and of the China Theater, Lieutenant General **Albert C. Wedemeyer**, who had replaced General **Stilwell**, reported back to Washington on November 14, 1945, that the Nationalist government was not capable of occupying Manchuria, especially against Communist opposition. The Communists had rushed in forces from **Yan'an**, creating the **Northeast Democratic United Army**. Wedemeyer had met with **Chiang Kai-shek** to assess the situation and recommended to Chiang that, instead of immediately trying to occupy Manchuria, Chiang's army should consolidate Nationalist control in Central China, north of the Yangtze River and south of the Great Wall. This, in Wedemeyer's estimation, was needed because it would secure the necessary lines of communication to support a subsequent entry into Manchuria. If Chiang failed to follow this course of action, Wedemeyer told Washington on November 20, 1945, logistical resupply for Nationalist forces would be impossible and subject to harassment by Communist guerrillas and saboteurs. Wedemeyer further told Washington that unless Chiang accepted foreign assistance and instituted an honest, competent administration that engaged in political economic and social reform, it was unlikely that Chiang could even stabilize the situation in South China. In North China, it could take years to stabilize the political situation and to control the area.

The key to creating the climate needed to establish military control of Manchuria, in the view of Wedemeyer, was to reach an agreement with the Russians and the Chinese Communists. However, since the Russians were helping the Communists to create conditions favorable to Communist control of Manchuria, which violated the recently established Sino–Soviet Treaty of Friendship and Alliance of August 14, 1945, Wedemeyer concluded that there was little hope of a satisfactory agreement between the Nationalist government and the Communists. As the situation continued to deteriorate, **Patrick J. Hurley** resigned as the U.S. ambassador to China on November 27, 1945, and general of the army George C. Marshall was appointed as the president's special representative to China with the personal rank of ambassador.

General Marshall's instructions from President Truman on December 15, 1945, were to use U.S. influence to achieve the "unification of China by peaceful, democratic methods" as quickly as possible and to work toward a cessation of hostilities between the rival Chinese political parties, especially in North China, where it was obvious that **Civil War** was about to break out. In discussion with Chiang Kai-shek, Marshall was authorized by the president to tell Chiang that "a China disunited and torn by strife" would not receive military, economic, or technical assistance from the United States.

In October 1945, the Nationalist government of China agreed with the Chinese Communist Party leaders to convene a "Political Consultative Conference" to discuss with the Chinese Communist Party a way to establish a constitutional government. The provisional list of delegates to the conference, in fact, had been drawn up on the day that Hurley had resigned as ambassador to China, November 29, 1945. In what seemed to be a positive step in view of the goals of Marshall's mission, Chiang announced on December 31 that the Political Consultative Conference would convene in Chongqing on January 10, 1946. Prior to the conference's convening, the Nationalists proposed to the Communists the establishment of a committee composed of three people, General Marshall, a Communist representative, and a Nationalist representative, to discuss the cessation of hostilities and related military matters. The Communists sent **Zhou Enlai**, with the protocol rank of general, as their representative, and General Zhang Jun (Chang Chun) was the Nationalist representative. This working group, called the "Committee of Three" because of its composition, held its first meeting on January 7, 1946. Prior to that date, on January 4, General Marshall had informed Zhou Enlai that the United States was committed to moving Nationalist troops into Manchuria. Zhou responded that such a movement was consistent with the Sino–Soviet Treaty of August 1945, and since the troop movement was a matter of U.S. policy, the Communist side agreed to the movement of Nationalist troops as an exception to the agreement on the cessation of hostilities.

On January 10, both **Mao Zedong** and Chiang Kai-shek issued orders to cease hostilities in China. All troop movements were also to be halted, with the exception of the movement of Nationalist troops into Manchuria, as agreed in the discussions between Zhou and Marshall. This cease-fire order was to be effective at midnight, January 13, 1946. On January 14, as provided for in the agreement, Executive Headquarters was to be established in Beijing, headed by three commissioners and staffed by officers drawn from the respective sides (American, Nationalist, and Communist). The Political Consultative Conference, meanwhile, convened in Chongqing from January 10 through 31, 1946, composed of representatives from all political parties in China, the Communists, the Guomindang (KMT), the Democratic League, the Youth Party, and other nonparty representatives. The Chinese People's Political Consultative Congress (CPPCC, or PCC) still meets and functions in Beijing as an organization to let non-Communist parties and nonparty-affiliated people speak out on matters related to the government.

General Marshall did not participate in the Political Consultative Conference meetings but separately made some suggestions for the establishment of a unified government, which was agreed to by the PCC. The resolutions from the PCC called for the establishment of a government organization, a program for national reconstruction, a resolution of military problems, agreement among the parties on the establishment of a national assembly, and implementing the 1936 draft constitution. The Chinese armies, Nationalist and Communist, were also to be reorganized, and a "Three-Man Military Commission" was to be established to integrate the two armies into a national, nonpolitical military force of about 50–60 divisions, with each division having about 14,000 soldiers. General Marshall, on March 11, 1946, returned to the United States to brief President Truman on the situation, which seemed to have stabilized.

However, as the Nationalist troops were moving to Yingkou, in south Manchuria, fighting broke out again. The cease-fire gradually eroded, and the Nationalist and Communist armies began their approach marches into Manchuria for the **Manchurian Civil War Campaign**.

REFERENCES

John Robinson Beal, *Marshall in China* (New York: Doubleday, 1970); Herbert Feis, *The China Tangle: The American Effort in China from Pearl Harbor to the Marshall Mission* (Princeton: Princeton University Press, 1953); U.S. Department of State (Far Eastern Series 30), *United States Relations with China: with Special Reference to the Period 1944–1949* (Washington, DC.: Government Printing Office, 1949).

MAZU (Madzu, Matsu) Mazu was one of two major offshore islands off the Fujian coast that remained occupied by Nationalist (Republic of China) troops after the establishment of the People's Republic of China (PRC) on October 1, 1949. In December 1952, the PRC conducted a number of small raids against the island of Mazu as well as the other Nationalist coastal stronghold, Quemoy (**Jinmen**). However, no attempt was made to seize Mazu. Tensions and threats of invasion continued along the Fujian coast through 1955, particularly in the area of Mazu, but then decreased significantly as China moved forward, concentrating on its economic development. In 1958, as internal tension increased in Beijing, another round of military pressure in the Taiwan Strait began, subsiding after the United States put its Seventh Fleet into the Taiwan Strait. In 1965, however, tensions again increased significantly. The Nationalists began a major buildup of forces on Jinmen and Mazu, with Jinmen eventually building to a force of 66,000 Nationalist troops and Mazu building to 19,000 defenders. On April 30, 1965, a 450-ton Republic of China (ROC) Navy patrol craft operating in Chinese coastal waters north of Mazu was attacked by as many as six PRC gunboats. The ROC naval craft escaped with heavy damage from gunfire.

REFERENCE

David G. Muller, Jr., *China as a Maritime Power* (Boulder, CO: Westview Press, 1983).

Vance H. Morrison

MCMAHON LINE. See LONGJU INCIDENT; SINO–INDIAN BORDER WAR

MILES, MILTON E. (1900–1961) Milton E. Miles was born in Jerome, Arizona, on April 6, 1900. He enlisted in the navy on April 6, 1917, at the age of 17. He graduated from the U.S. Naval Academy in 1922. After graduation Miles was assigned to the China Station, where he remained for five years patrolling China's main rivers and ports. He then earned a master's degree from Columbia University in 1929. He returned to China again in 1936, remaining there through the **Marco Polo Bridge Incident** and returning to the United States with his wife in 1939.

In July 1939, Miles reported for duty to the control board of the Navy Department in Washington, D.C., where he worked in Fleet Training. During an informal conversation over coffee in the Navy Department with his superior, Captain (later, Admiral) Willis ''Ching'' Lee, and the assistant Chinese military attaché to the United States, Major Hsiao Sin-ju, Lee and Miles raised the issue of putting a U.S. naval observer into China with the object of collecting weather data for U.S. forces and order of battle on Japanese ships operating on the coast. For about six months after this, Miles and Hsiao (who had been promoted to lieutenant colonel) began to cultivate each other. Their discussions focused on what the United States could do in China, since the United States was not at war with Japan, but China had been fighting the Japanese since July 7, 1937.

As December 1941 approached, Miles began to pursue the idea of putting naval observers into England to collect data on the Germans. After Pearl Harbor was attacked by the Japanese (December 7, 1941), Miles and Lee took more seriously the concept of putting people into China. Lee was promoted to rear admiral, and Miles, a lieutenant commander, was promoted to commander. Lee seized upon the earlier conversations and began serious plans to put into effect Miles' idea of naval observers in China. Miles finally obtained orders to China before Hong Kong collapsed on Christmas Day 1941 and attempted to approach the senior Chinese military attaché in Washington, Major General Chu Shih-ming. Miles was stopped by Hsiao, however, who revealed that he worked for Major General **Dai Li** (Tai Li), **Chiang Kai-shek**'s chief of intelligence, and suggested that Miles work only with Dai Li's people. On April 5, 1942, Miles set off on a fact-finding mission to China by way of Brazil and India. Coincidentally, he was accompanied on the flight to China by the journalist Edgar Snow, who had chronicled the experiences at Chinese Communist forces in **Yan'an**. In Calcutta, India, Miles met the U.S. ambassador to China, Charles E. Gauss, and the U.S. naval attaché at Chongking, Colonel McHugh, a Marine Corps officer. Miles took a flight into China with Gauss and McHugh aboard a China National Airways Corporation aircraft without first revealing his actual mission to Gauss. On arrival in Chongqing, a Chinese customs inspector questioned Miles about his association with Hsiao Sin-ju, the military attaché in Washington, and about Dai Li. However, no more was said at that time. Gauss and McHugh, meanwhile, were met by Al Lusey, a representative of the U.S. Office of Strategic Services in China. Miles and McHugh were summoned to see Dai Li within a short time after arrival in China, after which Miles and Dai Li sketched out a plan to conduct a covert reconnaissance of the Fujian coast. Miles was escorted into the Fujian area and was met there by Dai Li. Miles left Chongqing on May 26 and arrived in the Fujian area on about June 9, 1942. Miles continued his reconnaissance for about two months. On return to Chongqing, Miles, Lusey, and Dai Li sketched out a plan to train Chinese guerrillas to operate against Japanese forces and to collect intelligence on the Japanese.

Miles, with General **Stilwell**'s support, sketched out an agreement for a combined Office of Strategic Services (OSS)–Chinese organization, the **Sino-American Cooperative Organization** (SACO) to function in China under the direct control of the Joint Staff in Washington. Dai Li was the commander (or director), and Miles the deputy. Returning to Washington, Miles secured the concurrence of navy secretary Frank Knox, General **Marshall**, and China's representative to Washington, **T. V. Soong**. The SACO agreement was initialed in Washington on April 1, 1943, and signed in Chongqing by Dai Li on July 4, 1943.

While the agreement was being signed, without formal approval, Miles had gone ahead and established the organization, training people in the United States and organizing the first SACO unit near Hangzhou, on April 1, 1943. SACO expanded into an organization of 14 attachments and trained 16,000 Chinese guerrillas to operate against the Japanese.

Miles went on to reach the rank of vice admiral and retired from the navy in 1958. He died in 1961 of cancer before completing work on the manuscript of a book describing SACO.

REFERENCES

Milton E. Miles, *A Different Kind of War: The Little-Known Story of the Combined Guerrilla Forces Created in China by the U.S. Navy and the Chinese during World War II* (New York: Doubleday, 1967); Milton E. Miles, "U.S. Naval Group China," *U.S. Naval Institute Proceedings* 72, July 7, 1946; Ray Olin Stratton, *SACO—The Rice Paddy Navy* (Pleasantville, NY: C. S. Palmer, 1950).

MILES MISSION. See MILES, MILTON E.; SINO–AMERICAN COOPERATIVE ORGANIZATION

MILITARY OBSERVER'S MISSION. See DIXIE MISSION

MILITARY RANKS, PEOPLE'S LIBERATION ARMY Because the **People's Liberation Army** (PLA) got its start as an essentially peasant-based force that depended on close bonding and interpersonal relationships, there were always tension and conflict in the army over the importance of military rank. The PLA tradition was essentially to emphasize position in the Communist Party (e.g., cadre, party member) and military position (e.g., squad leader, division commander) over military rank. The Nationalist (Guomindang) Army, in contrast, modeled itself on German General Staff and Russian General Staff lines and used ranks. The emphasis on the egalitarian nature of the PLA reinforced the Communist Party's ideology of a classless society. There was always a tension, however, between those who wanted a military structure along more professional, Western lines, like Marshals **Zhu De** and **Peng Dehuai**, and those who pushed for the continuation of the peasant-based, egalitarian army without ranks, like **Mao Zedong**. Even some officers who would have been expected to support a strong rank structure in the military, like **Ye Jianying**, who was an instructor in the first class to graduate from the **Whampoa Military Academy** in 1924, strongly embraced the rankless, peasant tradition of the PLA. In 1981, in a speech commemorating the 60th anniversary of the Chinese Communist Party, Ye Jianying said that he hoped that the egalitarian tradition of the PLA would be maintained throughout his lifetime.

Military ranks were introduced into the PLA in 1955. In that year, the PLA appointed 10 officers whose military careers had been particularly distinguished to the rank of marshal: **Chen Yi, He Long, Lin Biao, Liu Bocheng, Luo Ronghuan**, Peng Dehuai, **Xu Xiangqian**, Ye Jianying, **Nie Rongzhen**, and Zhu De. Ten other officers were appointed to the rank of army general: **Chen Geng**, **Huang Kechang**, **Luo Ruiqing**, **Su Yu**, **Tan Zheng**, **Wang Shusheng**, **Xiao Jingguang**, Xu Guangda, **Xu Huaidong**, and **Zhang Yunyi**. The general officer ranks were modeled on Soviet lines and included colonel-general, lieutenant general, and major general.

By 1965, as Sino–Soviet relations began to sour, and ideology became more important in the PLA, the status of military rank in the PLA was examined again. On the eve of the **Cultural Revolution**, ranks were abolished. Military assessments of the performance of Chinese troops in the attack on Vietnam in 1979, attribute the poor command and control and poor lateral coordination among the PLA units to the fact that there were no visible indicators of rank or position among the officers and leaders (see **Vietnam, Self-Defensive Counterattack Against**). Despite the post-1979 debate over the importance of military rank, ranks were not restored until October 1, 1988. No marshals were appointed when ranks were restored. The new officer ranks were general, lieutenant general, major general, senior colonel, colonel, lieutenant colonel, major, captain, first lieutenant, and second lieutenant.

REFERENCES

Harlan W. Jencks, ''China's Punitive War on Vietnam, A Military Assessment,'' *Asian Survey* 19 (August 1979); Ellis Joffe, *Party and Army: Professional and Political Control in the Chinese Officer Corps, 1949–1964* (Cambridge: Harvard University Press, 1965); Ellis Joffe, *The Chinese Army after Mao* (Cambridge: Harvard University Press, 1987); William W. Whitson with Chen-Hsia Huang, *The Chinese High Command* (New York: Praeger, 1973).

MILITIA The militia system in the People's Republic of China is dated by the Chinese Communist Party to the January 1, 1922, meeting of Communist production workers in Hong Kong, when a workers' protective organization was formed. Then in August 1924, in Guangdong Province, the Peasant Self-Protection Army (*Nongmin Ziwei Jun*) was formed. The Communist Party decided at the Fourth Party Congress in January 1925 to further strengthen this Peasant Self-Protection Army, by assigning such strong Communist Party cadre as **Lin Biao, Mao Zedong**, and **Zhou Enlai** to work in Guangdong and Hubei Provinces. Lin, Mao and Zhou formed more of these local Peasant Associations and Peasant Self-Protection Armies. In March 1927, under Zhou Enlai's leadership, the Shanghai workers' association organized an armed uprising. During the period of the "Land Reform Movement," when the Communist Party retreated from the Nationalist forces into **revolutionary base areas** and created "soviets," the party mobilized the local populace into formal militia organizations. The soviet areas each had a self-protective army or organization, with participation mandatory for all males between 23 and 40 years of age. Younger people, between the ages of 16 and 23, were also organized into militia self-protection groups, but they were primarily responsible for production activities. The older militia members, however, participated in direct combat or logistical work and are credited with supporting the Counter-**Encirclement Campaigns** run by the Communist forces against the nationalists with a total force of 2.5 million militia troops. During the **Anti-Japanese War**, the **Eighth Route Army** and the **New Fourth Army** worked hard to organize the local populace into "People's Armed Departments" and militia organizations, which not only took part in direct combat with the Japanese but served as the transportation, supply, and maintenance infrastructure for the regular forces of the Communist Army. **People's Liberation Army** (PLA) histories estimate that about 10 million militia troops took part in the fighting against Japan. In the **Civil War** period (1946–1949), the PLA estimates that about 5.5 million militia members helped to drive the Nationalist forces off the mainland of China.

By 1950, the People's Republic of China had organized a formal militia organization and published regulations on the organization of the militia by local People's Armed Departments. Militia organizations were subordinated to the direction of the Military Region, but the militia organization and training requirements were organized in such a way as to avoid taking workers or peasants away from their production duties. Militia duty was voluntary for male workers and peasants, who needed the approval of the local armed department to participate in the militia group. The Communist Party counted on militia to assist People's Liberation Army and public security forces in maintaining public order, to exterminate "bandits and enemies," to catch spies, to guard construction areas, and to assist in the protection of the transportation and economic infrastructure at such places as bridges, electric stations, storehouses, and factories.

The militia today does not vary much from the original form. Its members receive rudimentary small arms and military training at the level of the small unit (squad to platoon). At the village level, the militia is organized into a company, and at some very large factories today, such as the Capital Iron Steel Company in Beijing, an entire reserve division exists. These reserve divisions are functionally like the militia. Militia organizations supported People's Liberation Army main force group armies in the **Self-Defensive Counterattack against Vietnam** in 1979 and are an integral part of campaign plans for any level of **People's War**. Although the terms "militia" and "reserve forces" are sometimes used interchangeably by the PLA, both forms exist. The PLA gives its militia strength at between 9 and 27 million at the present time.

REFERENCES

Dick Wilson, *The Long March: The Epic of Chinese Communism's Survival* (New York: Viking Press, 1971); Xin Ming, ed., *Zhongguo Renmin Jiefangjun Junguan Shouce* [A Handbook for People's Liberation Army Officers] (Qingdao: Qingdao Chubanshe, 1991); ZGDBKQS, vol. 2.

MUKDEN INCIDENT (September 18, 1931). See NINE ONE EIGHT (918) INCIDENT

N

NANCHANG UPRISING On August 1, 1927, after nearly a month of preparation, Chinese Communist forces began an attack on the Nationalist arsenal at Nanchang, in Jiangxi Province. Between 16,000 and 20,000 Communist troops participated in the uprising. The goals of the attack on Nanchang were to seize the arsenal, foster a general proletarian uprising, and later drive south toward Guangdong, developing a peasant uprising and a land reform movement. The date marks the birthday of the **People's Liberation Army** (PLA). The Chinese ideographs for August 1 (the characters *ba yi*, for the numbers eight and one) appear on the buttons and insignia of all PLA uniforms and mark military aircraft, vehicles, and naval vessels in China.

The Chinese Communist Party (CCP) central leadership, encouraged by the Soviet Comintern, believed that the workers (proletariat) of the city were sufficiently mobilized and were eager to take part in an urban revolution against the Guomindang government. The CCP center and the Comintern sought to begin a regional revolt intended to spread into a nationwide revolution by initiating action against Nanchang. Some of the Communist commanders in the uprising were secretly CCP members and were members of the Nationalist (KMT) Army. The commander of the Nationalist 20th Army was **He Long**, later one of the marshals of the PLA. The 20th Army took part in the attack, as did five regiments of the 11th Army, which was under the command of **Ye Ting. Zhu De**, who later also became a marshal of the PLA, led two companies of the cadet corps to the uprising. Among the other leaders of the forces at Nanchang were **Zhou Enlai, Liu Bocheng, Nie Rongzhen**, and Zhou Shidi.

The Nanchang rebels formed a "Guomindang Revolutionary Committee" (also initially called the "Guomindang Left Wing") in Nanchang but were surrounded by KMT troops and retreated after only a three-day occupation of the city. Communist forces were then organized into the Ninth Army under Zhu De and the 11th Army under He Long. The forces moved south and east, pursued by KMT forces, attempting to develop the revolt. The next stage of the revolution was to have been the **Autumn Harvest Uprisings** in Henan, Jiangxi, and Guangdong, but that attempt at mass mobilization also failed. By October 1927, the CCP forces retreated into the Jinggang Mountains (**Jinggangshan Revolutionary Base Area**) where, under the leadership of Mao Zedong, they formed the first guerrilla base and the cadre of the **Chinese Workers and Peasants Red Army** (Zhongguo Gong-Nong Hung Jun).

REFERENCES

James Pinkney Harrison, *The Long March to Power: A History of the Chinese Communist Party, 1921–1972* (New York: Praeger, 1972); Immanuel C. Y. Hsu, *The Rise of Modern China*, 5th ed. (New York: Oxford University Press, 1995); Warren Kuo, *Analytical History of the Chinese Communist Party* (Taipei: Institute of International Relations, 1968); Warren Kuo, *A Comprehensive Glossary of Chinese Communist Terminology* (Taipei: Institute for International Relations, 1978); Harrison E. Salisbury, *The Long March: The Untold Story* (New York: Harper and Row, 1985); ZGDBKQS, vol. 2; ZGDSJMCD, vol. 1.

NANJING, RAPE OF (December 1937) As **World War II** began in China with the **Marco Polo Bridge Incident**, on July 7, 1937, **Chiang Kai-shek** began to organize the Nationalist Army for the defense of China. In the south, in the Yangtze River basin, Chiang was going to blockade the Yangtze's lower reaches to protect the cities along the river and to cut off access to the hinterlands of China. The Japanese government, upon learning of these plans, attempted to land troops in Shanghai to protect citizens and property and to evacuate Japanese citizens, precipitating a clash there on August 13, 1937. As Japanese forces began a serious campaign to occupy and control eastern China by first taking Shanghai, which finally fell on December 9, the Nationalist forces defending the city withdrew to Nanjing. The Chinese defenders of Shanghai, 450,000 strong at the start of the battle, lost 60 percent of the force to casualties. They withdrew in disorder, for the most part toward Nanjing, with some soldiers stealing clothing and supplies from civilians to hide their military affiliation. The Japanese, who were supported by naval gunfire and aircraft, attacked with 200,000 troops. Japanese casualties have been estimated at 42,000 in the Shanghai battle.

Nanjing was flooded with refugees from the fighting in Shanghai and Hangzhou in December 1937, as well as Nationalist troops that had deserted their units. Japanese forces moved inland, along the Yangtze, to take the city, which had been the capital of the **Taiping Rebellion** in 1853 and later of the Nationalist government. The defense of the city was undertaken by Tang Shengzhi, a Nationalist leader, who abandoned the city himself on December 12, 1937, as the Japanese forces approached. The Japanese army entered the city of Nanjing on December 13, and for about seven weeks systematically went through the city killing Chinese troops and those civilian males suspected of being Chinese soldiers, brutally assaulting and raping women and young girls and murdering civilians. Most of this carnage was inflicted by hand, in the most brutal ways, as the Japanese had done in the **Port Arthur Massacre** during the **Sino–Japanese War**. The International Relief Committee estimated Chinese deaths in Nanjing at 42,000, many of the people having been hacked to death with swords, bayoneted, or rounded up and burned alive. Of these 42,000, about 30,000 were estimated to be fugitive soldiers, and 12,000 were innocent civilians. The number of female rape victims of the Japanese forces was estimated by these foreign observers to be about 20,000.

REFERENCES

Iris Chang, *The Rape of Nanjing: The Forgotten Holocaust of World War II* (New York: Basic Books, 1997); Immanuel C. Y. Hsu, *The Rise of Modern China*, 5th ed. (New York: Oxford University Press, 1995); F. F. Liu, *A Military History of Modern China: 1924–1949* (Princeton: Princeton University Press, 1956); Barbara W. Tuchman, *Stilwell and the American Experience in China, 1911–45* (New York: Macmillan, 1971); ZGRMJFJZS, vol. 2.

NANJING, TREATY OF (1842) This treaty was the basis for what China today calls the ''unequal treaties'' imposed by foreign military forces on the Qing Court after the **Opium War**, which turned China into a semicolonial state. The treaty was signed in Nanjing on August 29, 1842, after two years of naval and land combat between British forces and troops of the Qing emperor Daoguang. The treaty was signed on the British ship HMS *Cornwallis* on the Yangtze River and approved by Queen Victoria and Daoguang.

The Treaty of Nanjing served as a model for other, similar treaties with Western and foreign powers that eventually created a system of concession areas around China where foreign citizens enjoyed extraterritorial privileges. The main features of the treaty were stipulations that guaranteed to citizens of Britain and China security and protection for their persons and property within each other's borders. The treaty opened five Chinese cities for residence by British subjects for trade and permitted the establishment of consulates in these cities (Guangzhou, Fuzhou, Xiamen, Ningbo, and Shanghai). In addition, the emperor agreed to allow British citizens to trade with whichever merchant they wished, not just individuals designated by the emperor to deal with foreigners. The island of Hong Kong was ceded to Britain ''in perpetuity'' (Hong Kong returned to China by mutual agreement on July 1, 1997). The British merchants who originally delivered opium to the Chinese in Canton in 1840 (leading to the Opium War) were repaid $6 million. The Chinese also agreed to pay for the cost of fighting the Opium War, with interest. Any prisoners who were British subjects were released by China, and the emperor agreed not to punish Chinese who had worked or served with the British. Pursuant to the treaty, derogatory imperial terminology in Chinese court language used when dealing with foreigners was dropped from diplomatic discourse by the Chinese.

In 1843, a supplementary tariff treaty with the British gave Britain "most favored nation" status, so that any additional privileges extended later to other countries by the emperor would also be granted to British subjects.

REFERENCES

Chang Hsin-pao, *Commissioner Lin and the Opium War* (Oxford: Cambridge University Press, 1964); John K. Fairbank, *Trade and Diplomacy on the China Coast* (Cambridge: Harvard University Press, 1953); Gerald Graham, *The China Station: War and Diplomacy, 1830–1860* (New York: Oxford University Press, 1978); Jonathan D. Spence, *The Search for Modern China* (New York: W. W. Norton, 1990).

NANNIWAN EXPERIMENT After the **Second Front Army** moved its forces from western Hunan, making a 3,000-mile **Long March** of its own in 1934, it moved into the Shaanxi-Shanxi-Suiyuan area. **He Long** commanded the army, and **Xiao Ke** and **Wang Zhen**, respectively, were political commissar and commander of the Red Sixth Corps. In 1937, when the **Eighth Route Army** was formed from the **First**, Second, and **Fourth Front Armies**, the Second Front Army converted into the Eighth Route Army's 120th Division. He Long remained division commander. The 359th Brigade of the 120th Division, which was led by Wang Zhen, was assigned an operations area in the Shaanxi-Gansu-Ningxia Province area, which contained some of the worst soil in China, making agricultural production very difficult. Complicating the problem of a lack of arable land, the population base of this operating area was very sparse, making it difficult for the Eighth Route Army's forces to depend on the local populace for supplies and support. In an effort to reach self-sufficiency without putting an unreasonably high burden on the local populace by soliciting food from local farmers, which was a practice common to many Eighth Route Army guerrilla units, between 1941 and 1945, Wang Zhen's 359th Brigade embarked on an experiment to achieve agricultural self-sufficiency in its base area around the base area headquarters in Nanniwan. Using the slogan "a shovel in one hand, a rifle in the other hand," the guerrilla forces of the 359th Brigade, over a period of two years, reached full self-sufficiency in grain production and in raising livestock. They were named a "model unit" for the Eighth Route Army and the **People's Liberation Army** (PLA). Today, the concept of unit self-sufficiency in food production is still practiced in the People's Liberation Army and is the backbone of the PLA logistics system. This practice reduces the tax burden on the state, reduces the military budget, and permits more funds to be used for weapons procurement, training, and modernization of the armed forces. However, it also ties PLA organizations to a specific location and complicates logistics support for deployed forces.

REFERENCES

Xiao Zhaoran, ed., *Zhonggong Dangshi Ciming Cidian* (Beijing: Jiefangjun Chubanshe, 1986); Huang Xiaodong, ed., *Sixiang Zhengzhi Gongzuo Qishi Nian* (Beijing: Guofang

Daxue Chubanshe, 1991); William W. Whitson with Chen-Hsia Huang, *The Chinese High Command* (New York: Praeger, 1973).

NANSHA ISLANDS INCIDENTS (Spratly Islands) The Nanshas, or Spratly archipelagic chain in the South China Sea, may be the site of a major oil source, as confirmed by recent Chinese and international seismic surveys. The question of sovereignty has become a matter of serious concern among the claimants to the islands: China, Vietnam, Malaysia, the Philippines, and Vietnam. Taiwan also occupies one of the islands, Itu Aba.

A January 30, 1980, "white paper" published by the People's Republic of China (PRC) claimed "indisputable sovereignty over the Xisha (Paracel) and Nansha (Spratly) Islands" for China. Since both Beijing and Taipei regard Taiwan as a province of China, the Taiwan garrison on Itu Aba, called Taiping Island by the PRC, reinforces the Chinese claim to sovereignty. China's claims to the islands are based on historic usage by Chinese fishermen as early as 200 B.C.E. and on the 1887 Chinese–Vietnamese Boundary Convention, while Vietnam claims historic links with the islands based primarily on having inherited modern French territory. The Spratly Islands were abandoned by Chinese Nationalists (Taiwan) in 1950, and at the time that South Vietnam fell in 1975, the Saigon regime occupied four islands. Today, Taiwan has a garrison on one island. The successor Hanoi government built up the total number of occupied islands to about 20.

Vietnam remains China's principal protagonist, as exemplified by the March 1988 incident in which the **People's Liberation Army** (PLA) Navy sank three Vietnamese supply vessels. For the PLA Navy, the problem of control of the Spratly Islands is not so much continuing the development of a growing blue-water naval capability but rather one of control of the associated airspace. This problem will not be solved until the PLA Navy either is able to deploy an aircraft carrier or develops an operational air-to-air refueling capability for its naval air arm, thereby ensuring dominance over Vietnam and any other claimant to the Nanshas.

With the perceptions of the Association of Southeast Asian Nations (ASEAN) member countries in mind, China continues to proceed slowly in exercising its claims over the Nanshas. In a statement issued in Manila in 1994, the Chinese Embassy stressed Beijing's "indisputable sovereignty" over the Nansha Islands and adjacent waters. China has contracted a U.S. oil company to drill exploratory wells near the Nanshas and has said it will defend its claimed sites if necessary. In February 1995, China continued to expand its presence in the Nanshas and took over Mischief Reef, a fifteen-square mile set of shoals 150 miles west of the Philippine island of Palawan. China acknowledged building several structures on the barely submerged coral reef, ostensibly as "shelters for Chinese fishermen." China also deployed several ships to the area, believed to be naval vessels. Philippine president Ramos said the Chinese actions were "inconsistent with international law" and with the 1992 declaration on the Nanshas, which was endorsed by China and the other claimants to the islands. As recently as July 1995, the ASEAN Regional Forum conference in Brunei agreed to abide by the law of the sea in negotiating claims to the islands.

REFERENCES

Greg Austin, *China's Ocean Frontier: International Law, Military Force, and National Development* (Sydney: Allen and Unwin, 1998); Rosita Delios, *Modern Chinese Strategy: Present Developments, Future Directions* (New York: St. Martin's Press, 1980); Bruce Swanson, *Eighth Voyage of the Dragon—A History of China's Quest for Seapower* (Annapolis, MD: Naval Institute Press, 1982).

Vance H. Morrison

NANYANG FLEET China's southern fleet operated in the area of Shanghai during the period of the **self-strengthening movement**. A large percentage of the funds allocated for this fleet were diverted to restore the Summer Palace, instead of being used for naval construction, which was the intent of the **Navy Board**. This was one of the four Chinese fleets that were in operation at the time of the **Sino–Japanese War** (1894–1895). The other fleets were the **Beiyang Fleet**, operating out of Port Arthur and Weihaiwei; the Fujian Fleet, operating out of Fuzhou; and the Guangdong Fleet, operating out of Guangzhou. In 1898, after the time of the Sino–Japanese War, the Nanyang Fleet had six German-built cruisers of about 3,400 tons and four old British gunboats of about 400 tons each.

REFERENCE

John L. Rawlinson, *China's Struggle for Naval Development, 1839–1895* (Cambridge: Harvard University Press, 1967).

NATIONAL DEFENSE SCIENCE, TECHNOLOGY AND INDUSTRY COMMISSION. See COMMISSION OF SCIENCE, TECHNOLOGY AND INDUSTRY FOR NATIONAL DEFENSE

NATIONAL DEFENSE WHITE PAPER (1998) China issued its first comprehensive statement of national security strategy and military strategy in its 1998 defense white paper, titled *China's National Defense*, on July 27, 1998, as a confidence- and security-building measure among the countries in the Asia-Pacific region and the Association of Southeast Asian Nations (ASEAN). Earlier, in 1995, China had issued a white paper titled *China: Arms Control and Disarmament*, which described the policies of the nation toward arms control and disarmament regimes. In *China's National Defense*, Beijing provided a description of China's long-term security goals of devoting itself to a modernization drive requiring a peaceful, stable environment. Beijing identified economic security as the main factor in ensuring the success of its future goals, stated its firm intentions to prevent the emergence of an independent Taiwan as a separate political entity from China, reinforced its claims to the South China Sea archipeligoes of the **Spratly** and **Paracel Islands**, and restated the basic principle that the **People's Liberation Army** is the main defender of the Chinese Communist Party and is led by the party. The basic military budget was also set forth in the white paper.

REFERENCES

China: Arms Control and Disarmament (Beijing: Information Office of the State Council of the People's Republic of China, 1995); *China's National Defense* (Beijing: Information Office of the State Council of the People's Republic of China, 1998).

NAVAL AIR FORCE Development of the People's Republic of China (PRC) naval air force began in late 1951, when China followed the recommendation of Soviet air advisers that such a force be established. The Soviet recommendations to the Chinese included a concept for a naval air arm as a defensive force for coastal patrol and control of the littoral. The secondary role of the naval air arm would entail offensive operations, as limited by the range of shore-based aircraft. Thus, in 1952, Soviet instructors began to create a naval air force made up of aircraft provided by the USSR.

The missions of the naval air force were spelled out by its first commander, Vice Admiral Dun Xingyun, on October 8, 1956: "The Chinese Navy Air Force has begun to shape up as a combat force. We have fought cooperatively without brothers in liberating Yijing Island and the Yushan Islands. We have accomplished our missions of protecting the sea routes and fishing fleet, and of defending our territorial air space."

Originally subordinated to the air force, the PRC Naval Air Force was eventually formed into three air divisions—each corresponding to a fleet region (North Sea, East Sea, and South Sea). A naval air school was opened in Qingdao in December 1950, where the pilots initially underwent fifteen months of academic and technical training before being assigned to a division. The first class graduated two years later, and the force became operational in January 1953. The initial organization was two naval air divisions, one of fighters and one of light bombers. Three basic types of aircraft were flown by naval air pilots: Soviet piston, twin-engine TU-2s; MIG-15 jet fighters; and IL-28 twin-jet bombers. By 1958, the naval air force increased its order of battle from 80 aircraft to an estimated 470 aircraft, an order of battle similar to that today and changed only by the introduction of newer aircraft.

By the 1960s, the mission had changed little; the naval air force was a shore-based force of about 500 aircraft with a mission of coastal air defense, offshore reconnaissance, protection of PRC shipping, and offensive operations against offshore islands and hostile shipping. By 1967, the Chinese navy perfected operating aircraft from ships at sea, initially employing a dozen Supper Frelon utility helicopters purchased from France in this role. By 1983, the naval air force consisted of about 200 bombers and antiship strike aircraft, including IL-28 BEAGLE and TU-16 BADGER bombers. The force grew to some 750 aircraft by 1989, including the following aircraft: 500 interceptors of types Jian 5 (MIG-17), Jian 6 (MIG-19), and Jian 7 (MIG-21); about 20 Jian 8 interceptors (modified MIG-21F); about 100 Hong 5 bombers (a licensed version of the Soviet IL-28; 50 Hong 6 bombers (a version of the Soviet TU-16 BADGER); 12 Soviet Be6 MADGE amphibious aircraft; seven HARBIN SH5 amphibious aircraft; 13 French-supplied SUPER FRELON helicopters with more ZHI 8 copies made under licensed production; 26 SHI 8 helicopters (a licensed production copy of the SA 375 DAUPHIN helicopter).

REFERENCES

Jean L. Couhat, ed., *Combat Fleets of the World, 1988/1989* (Annapolis, MD: Naval Institute Press, 1987); *The Military Balance, 1996* (Singapore: Asian Defense Review, 1996); David G. Muller, Jr., *China as a Maritime Power* (Boulder, CO: Westview Press, 1983).

Vance H. Morrison

NAVAL INFANTRY (Marines) The Chinese campaigns to control the offshore islands on China's coast in the 1950s highlighted the need for a naval infantry capable of conducting amphibious operations. In 1958, a People's Republic of China (PRC) naval officer defector reported that, although plans for a Marine Corps had been initiated in the early 1950s, they had been abandoned by 1953. Analysis of Chinese amphibious actions in the 1950s indicates that Soviet-type naval infantry units, made up of regular **People's Liberation Army** (PLA) infantry personnel, were used for the offshore islands campaigns. After the campaigns the units were again subordinated to the shore-based main force armies.

Throughout the 1950s, however, there were persistent reports from the Guomindang government on Taiwan of a large PLA Marine Corps training for a planned invasion of Taiwan. Most of these personnel turned out to be guards at navy installations. Some PLA ground force units did train in amphibious operations, including ''seagoing acclimatization,'' embarkation, debarkation, formation sailing, firing guns from boats, underwater demolition, and beach combat tactics, but little was heard about this force. China developed none of the essential elements required to conduct a large-scale, opposed amphibious invasion. Of special importance, air cover was not available; China had no aircraft carriers, and the range of their combat aircraft from coastal airfields precluded meaningful operations over even destinations as close as Taiwan. In addition, Chinese pilots did little over-water training and lacked that navigational and operational experience.

In the 1970s, the Chinese amphibious force took on a new vigor. Still in service were most of the 30 former U.S. **World War II**–era LSTs and LSMs Landing Ships (Tank) (LSTs) and Landing Ships (Medium) (LSMs). A new, large LST, the YUKAN class, appeared from China's shipyards. Construction of a new class of LSMs began in 1978, and the Chinese experimented with sophisticated, high-speed air-cushion landing vehicles. Most were based in the South Sea Fleet, where formation of a new "landing ship brigade" was announced in 1981. In 1982, the PLA announced that the navy now had a "Marine Corps."

The PLA Navy is estimated to have a force of up to 38,000 naval infantry, reportedly enough to constitute a division, in three elements subordinate to the three major fleet commands. However, PLA leaders acknowledge only a brigade in the South Sea Fleet, headquartered in Zhanjiang. With its limited shipbuilding program and without the necessary attack support ships, there is still no evidence that the PLA Navy is preparing a major amphibious assault capability. Such as it is, its amphibious capability will be of definite value in asserting China's claims in the South China Sea, and there are recent references to the navy as a combined arms force in which land combat units are referred to as a "Marine Corps." In major exercises in the mid-1990s, the amphibious training given to main-force PLA group armies complements that of the Marine Corps. The PLA now defines the Marines' mission as capturing enemy islands, establishing forward operating bases for the navy, and carrying out land operations in support of naval operations.

REFERENCES

Paul H. B. Godwin, *The Chinese Communist Armed Forces* (Maxwell Air Force Base, AL: Air University Press, 1988); Bruce Swanson, *Eighth Voyage of the Dragon—A History of China's Quest for Seapower* (Annapolis, MD: Naval Institute Press, 1982); John R. Young, *The Dragon's Teeth: Inside China's Armed Forces* (New York: Orion Books, 1987).

Vance H. Morrison

NAVY BOARD, QING DYNASTY (1885) After the defeat of the Chinese navy by the French Fleet in the 1884–1885 **Sino–French War**, it became obvious that a national fleet was needed. The French had little trouble defeating piecemeal the locally controlled, decentralized Chinese naval forces. **Zuo Zongtang**, who had been instrumental in the suppression of the **Taiping Rebellion**, sought to convince the Qing Court that China needed a national fleet, with a single navy director. However, Zuo died before his concept could be realized. After his death other Chinese officials also encouraged the Qing Court to establish a centralized board or office to manage naval affairs (Shuishe Yamen).

By September 1885, the Central Board of Revenue of the Qing dynasty put out an order prohibiting the purchase of ships or naval material without prior clearance by itself and the Navy Board. At the suggestion of **Li Hongzhang**, who supported a centralized board for naval affairs, the moneys of the Sea Defense Fund, which had been managed by the northern and southern commissioners of trade, in Tianjin and Nanjing, respectively, were diverted for use by the Navy Board. This provided an annual budget of 4 million taels to the board for naval construction. However, management of the Navy Board and its funds was assigned to the imperial household.

The Navy Board made a series of inspections of China's fleets and coastal fortifications and also conducted naval exercises off China's ports. In addition, sometimes managing to override objections from the Board of Revenue, the Navy Board sought to purchase ships, guns, and material from Britain, Germany, and even the United States. Li Hongzhang also urged the Navy Board to establish a naval academy. However, a great deal of the money allocated for building the navy was actually funneled into the construction and restoration of the Summer Palace and the building of a marble boat there as a tribute to the Qing "dowager empress" Cixi.

Between 1886 and 1889, money allocated for the construction of ships and fortifications was spent by the imperial household on the restoration of the three lakes in central Beijing, Nanhai, Zhonghai, and Beihai, as well as the marble boat and the Summer Palace project. In fact, in 1889, a special fund was established to finish the project at the Summer Palace in time for her 60th birthday in 1894. As much as 100 million taels may have been devoted to build the Summer Palace and marble boat, of which about 10 million taels may have come from funds earmarked for naval construction. By the time of the **Sino–Japanese War** (1894–1895), the Navy Board served almost no useful function, and Li Hongzhang often issued direct orders to the fleets.

REFERENCES

Li Hong Chang, *Memories of Li Hong Chang*, ed. William F. Mannix (Boston: Houghton Mifflin, 1913); John L. Rawlinson, *China's Struggle for Naval Development, 1839–1895* (Cambridge: Harvard University Press, 1967).

NEW FOURTH ARMY (Xinsijun) The New Fourth Army, which evolved from the Fourth Red Army and **Fourth Front Army**, was established in the wake of the **Xi'an Incident** of December 1936 as a force of Communist soldiers to operate in a **United Front** with Nationalist forces along the Yangtze River in the nationalist Third War Zone. It operated in the second United Front with Nationalist forces until that United Front was shattered by the **New Fourth Army Incident** in 1941. Although pressure had been building for Nationalist–Chinese Communist Party (KMT-CCP) cooperation, the **Marco Polo Bridge Incident** and the outbreak of the **Sino–Japanese War** made it urgent that the two parties cooperate better against the Japanese. Despite the fact that Nationalist Army anti-Communist operations continued in southeast and coastal China through September 29, 1937, the period of United Front cooperation dated itself to the Xi'an Incident in 1936. Chinese Communist Party (CCP) histories cite October 12, 1937, as the date for the formal establishment at the New Fourth Army. The army was recruited, with Nationalist concurrence, from remnants of the former Communist guerrillas in the Anhui–Jiangxi–Jiangsu area, once occupied by Fourth Front Army forces operating out of the **Eyuwan Soviet**, between the Yangtze River and Lake Tai (Taihu). Wang Ming, who headed the group in the Communist Party most closely associated with the Communist International in Moscow, was one of the leaders who urged the formation of the United Front. The headquarters of the New Fourth Army moved from Hankou (present-day Wuhan), an area under Wang's influence, on the Yangtze River to Nanchang, Jiangxi Province, in December 1937. The date cited for the establishment of the army headquarters at Nanchang is January 6, 1938. To establish the New Fourth Army, the CCP Central Committee combined units from 13 guerrilla areas in eight provinces (Hunan, Hubei, Jiangxi, Anhui, Fujian, Guangdong, Zhejiang, and Henan). After a meeting between **Ye Ting** and Mao Zedong in Yan'an at the end of 1937, **Xiang Ying** was appointed deputy commander. The First Column of the army was commanded by **Chen Yi**, who later commanded Chinese forces in Korea and became foreign minister of China. The New Fourth Army was organized into four columns centered in Anhui, Fujian, the Zhejiang area, and Hubei. Army operations concentrated primarily in the areas south of the Yangtze River, around the Delta areas, which are fertile. After 1941, the army was forced to move its operations into Anhui, north of the Yangtze. The army eventually expanded its operations north of the Huai River and even into Shandong. Notable leaders in the New Fourth Army included **Peng Dehuai, Xu Huaidong, Li Xiannian, Zhang Aiping, Huang Kecheng**, Wei Guoqing, Ye Fei, and Xu Shiyu.

The New Fourth Army was intended to be an example of a successful anti-Japanese United Front between the CCP and the KMT. Many of the army's leaders were supporters of the United Front Policy. Ironically, those leaders were killed by the Nationalists in the **New Fourth Army Incident** in January 1941.

REFERENCES

Tetsuya Kataoka, *Resistance and Revolution in China: The Communists and the Second United Front* (Berkeley: University of California Press, 1974); Lyman P. Van Slyke, *Enemies and Friends: The United Front in Chinese Communist History* (Stanford, CA: Stanford University Press, 1967); William W. Whitson with Chen-Hsia Huang, *The Chinese High Command* (New York: Praeger, 1973); ZGRMJFJZS, vol. 2; ZGRMJFJZSJB.

NEW FOURTH ARMY INCIDENT (January 4, 1941) After the **Hundred Regiments Campaign**, the Japanese counterattacked against the **Eighth Route Army**'s operating areas, inflicting very heavy casualties on the army, part of which moved south of the Yangtze River into the New Fourth Army operating area. The New Fourth Army, meanwhile, was expanding its operations around the Yangtze River delta and the food-producing and industrial areas. As the New Fourth Army began to enlarge its operations, Guomindang (KMT) generals wanted to force the New Fourth Army units, as well as the units of the Eight

Route Army regrouping south of the Yangtze, back to operating areas north of the river.

The Guomindang leadership, in response to the pressure of the Nationalist generals, issued a directive on December 9, 1940, setting a deadline of December 31, 1940, for the movement of Eighth Route Army forces to Anhui, north of the Yangtze River. New Fourth Army units were to have crossed the river to Anhui by January 31, 1941. However, New Fourth Army leaders delayed the move and even sought to develop popular mass support to remain south of the river, while attempting to renegotiate with the Guomindang, arguing that the Guomindang commander in Anhui was hostile to Communist forces.

On January 4, 1941, seven divisions of Nationalist troops entrapped about 9,000 of the New Fourth Army troops in the vicinity of Maolin, Zhejiang Province. Some 3,000 New Fourth Army troops were killed in heavy fighting with the Nationalist forces between January 7 and 13, 1941. The remainder of the 9,000 troops were captured. **Xiang Ying**, who supported the **United Front**, was killed. This incident punctuated the slow disintegration of the United Front between the Chinese Communist Party (CCP) and the Guomindang. The CCP Central Committee sent Liu Shaoqi to the North Yangtze River area to reorganize in January 1941. A new headquarters was established north of the Yangtze River at Yancheng on January 29, 1941, with a reorganized New Fourth Army under the command of **Liu Shaoqi** and **Chen Yi**.

REFERENCES

Guofang Daxue, *Zhongguo Jiefangjun Zhanshi Jianbian* (Beijing: Jiefangjun Chubanshe, 1992); Tetsuya Kataoka, *Resistance and Revolution in China: The Communists and the Second United Front* (Berkeley: University of California Press, 1974); Jonathan D. Spence, *The Search for Modern China* (New York: W. W. Norton, 1990); Lyman P. Van Slyke, *Enemies and Friends: The United Front in Chinese Communist History* (Stanford, CA: Stanford University Press, 1967); William W. Whitson with Chen-Hsia Huang, *The Chinese High Command* (New York: Praeger, 1973).

NIAN REBELLION (1852–1868) The Nian Rebellion was a regional problem in the area north of the Huai River in the Shandong–Jiangsu–Henan–Anhui border area. The term "Nian" referred to the nature of the rebels, who were mobile bands or groups of bandits. "Nian" rebels or bandit groups had existed in the area north of the Huai River for over 50 years. The bandit groups included members of secret societies, like the White Lotus sect, as well as smugglers. The **Taiping Rebellion** also had a causative effect on the Nian Rebellion, insofar as the Nian gangs had organized themselves and built defensive areas in response to the Taiping expeditions west and north of Nanjing.

In 1852, the leaders of 18 separate Nian bandit groups met and declared that an Anhui landlord and smuggler, Zhang Luoxing, was their leader. Zhang was later elected to head the Nian alliance, and Nian forces organized into five "banners," designated by different colors. The Qing dynasty had officially declared them as rebels much earlier, in 1851. Nian forces were probably never stronger than 30,000 to 50,000 troops, including cavalry. However, they had a significant impact on the Qing dynasty because they operated in an area that served important lines of communication, including the Grand Canal, between Beijing and the Taiping forces in Nanjing and Shanghai, impeding the movement of Qing forces.

In 1865, the Nian rebels were able to ambush and kill the Qing general who was chasing them, Senggelinqin, an ethnic Mongol. The Qing dynasty then appointed **Zeng Guofan**, who orchestrated the Taiping defeat, as commander of military affairs to suppress the Nian. Although Zeng had already disbanded his own **Xiang Army**, he depended on troops from **Li Hongzhang**, another central figure in the Taiping defeat. Using foreign-purchased arms and armored gunboats, the troops suppressed the Nian by 1868, executing any Nian who survived the battles.

REFERENCES

Chiang Siang-tseh, *The Nian Rebellion* (Seattle: University of Washington Press, 1954); Elizabeth Perry, *Rebels and Revolutionaries in North China, 1845–1945* (Stanford, CA: Stanford University Press, 1980).

NIE RONGZHEN (1899–1996) Nie Rongzhen was born in Sichuan Province in 1899. He was appointed a marshal in the **People's Liberation Army** (PLA) in 1955. He came from the family of a wealthy landowner and attended middle school in Chongqing, where he was a classmate of **Deng Xiaoping**. Nie was a participant in the May 4 Movement in 1919, after which he was sent to France as part of a work-study program in Paris. He worked in the Schneider-Creuzot arsenal and weapons factory, at the Renault automobile factory, and in the Thompson electrical equipment factory. Nie joined the Communist Youth League in 1922. He also majored in Chemical Engineering at the Universite de Travail, Charleroi, Belgium. The Communist Party sent him to Germany in 1923 and to the University of the Red Army in Moscow in 1924. Nie's early experience in major defense-related industries provided him a deep appreciation for technology and the defense industry, which he later brought to bear when he ran the National Defense Industry Office and the National Defense Science and Technology Commission (NDSTIC) in the 1950s and 1960s (see **Commission of Science, Technology and Industry for National Defense**).

Nie Rongzhen was perhaps the best educated of the early Red Army leaders. He took part in the **Nanchang Uprising** and the **Guangzhou Uprising**. For a short time Nie did underground work, trying to organize coal miners in the Tangshan area south of Beijing and Tianjin. He moved to the **Jiangxi Soviet** in the **Jinggangshan Revolutionary Base Area** after 1931 and was a political commissar under **Lin Biao** during the **Long March**. When the **Eighth Route Army** formed in 1937, Nie was both deputy commander and political commissar under Lin Biao in the 115th Division. He established a guerrilla base area in the Wutai mountain range east of **Yan'an**, where his forces operated against the Japanese. During the **Civil War**, Nie's forces were involved in the capture of Shijiazhuang, capital of Hebei Province, Beijing and Tianjin. He was mayor of Beijing from 1949 to 1951. He later commanded both the Beijing and Tianjin Garrison Commands. Nie Rongzhen was a chief of the PLA General Staff. Notably, Nie, on behalf of **Zhou Enlai**, actually delivered the warning to the United States through the Indian ambassador in Beijing that, if American forces advanced toward the Yalu, the PLA would enter the **Korean War**.

Nie's influence in the State Science and Technology Commission was tremendous. His son-in-law, Ding Henggao, for a number of years was the head of the Commission on Science, Technology, and Industry for National Defense (COSTIND), the successor organization to the NDSTIC. Nie Rongzhen's daughter, Nie Li, retired from the PLA as a lieutenant general. Her last post was the COSTIND deputy director.

REFERENCES

Wolfgang Bartke, *Who's Who in the People's Republic of China* (Armonk, NY: M. E. Sharpe, 1981); John Frankenstein and Bates Gill, "Current and Future Challenges Facing Chinese Defense Industries," *The China Quarterly*, no. 146 (June 1996): 394–427; JFJJLZ; Tong Bai, ed., *Junren Shouce* (Beijing: Changzheng Chubanshe, 1987).

NINE ONE EIGHT (918) INCIDENT (September 18, 1931) This is considered by Chinese historians one of the most serious, large-scale incidents of Japanese aggression in northeast China. It is also known as the "Mukden Incident."

On September 18, 1931, at 10:30 P.M., a small group of Japanese military officers stationed in Shenyang (Mukden) sought to preempt an order from Tokyo calling for more restraint in dealing with the occupation of Manchuria. To create an incident to which they could respond, they blew up a section of railroad track in the northern suburbs of the city. The track section was part of the north-south line in the near Liutiaohu (Willow Twig Lake) and was not far from a Nationalist military barracks, China's Northeast Barracks, near Shenyang. Following the explosion, intermittent fighting broke out between Chinese and Japanese troops. One hour later, the Japanese Kwantung Army fired on the Nationalist barracks with artillery. Japanese forces ignored a request for restraint from their own consul in Shenyang and mounted an attack on the Chinese military barracks. The Japanese forces followed this by capturing the walled city of Shenyang. Meanwhile, the Japanese commander in Korea independently ordered his forces to advance into southern Manchuria in support of the Kwantung Army.

By September 19, Japanese forces began attacks on a number of cities along the rail line in Manchuria, including Changchun and Siping. Fighting spread throughout Jilin and Liaoning Provinces. Seeking to avoid a larger-scale conflict, **Chiang Kai-shek**, already preoccupied with fighting the Communist Red Army in southern China, ordered Nationalist forces under the command of **Zhang**

Xueliang to withdraw from the area and position themselves south of the Great Wall. The withdrawal of Nationalist forces left all of Manchuria under Japanese control.

REFERENCES

George M. Beckmann, *The Modernization of China and Japan* (New York: Harper and Row, 1962); Dorothy Borg, *The United States and the Manchurian Crisis of 1933–1938* (Cambridge: Harvard University Press, 1964); Immanuel C. Y. Hsu, *The Rise of Modern China*, 5th ed. (New York: Oxford University Press, 1995); Jonathan D. Spence, *The Search for Modern China* (New York: W. W. Norton, 1990); ZGDSJMCD, vol. 1; ZGJSSL, vol. 3.

NINGDU CONFERENCE (Ningtu Conference) (October 1932) The criticism of **Mao Zedong** and his emphasis on small-unit dispersion and peasant-based guerrilla operations reached its nadir at the Ningdu Conference, held in mid-October 1932 in the town of Ningdu, Jiangxi Province. After the Third **Encirclement Campaign**, where Communist forces achieved modest successes through conventional operations, this conference marked a period in the **People's Liberation Army**'s (PLA) history where conventional, mobile warfare was the principal tactic emphasized by PLA leaders (the PLA continued to use, to a lesser extent, guerrilla operations and relied on the peasantry as a base of popular support for logistic activities, but the focus of Communist military leaders shifted somewhat in favor of modern, maneuver warfare). Mao was relieved of leadership positions at the conference, and despite the restoration of Mao to the central leadership of the PLA in 1935, at the **Zunyi Conference**, it was not until after the **Hundred Regiments Campaign**, in 1941, that the PLA returned to primarily a Maoist, guerrilla base-oriented strategy. Both **Liu Bocheng** and **Peng Dehuai** attacked Mao for self-glorification at the conference and accused him of pedantry because his notions of tactics and campaign strategy were based on ancient historical writings. After the Ningdu Conference, **Zhou Enlai** replaced Mao as the commissar of the Red Army, Mao was expelled from all of his posts, and many of Mao's supporters were also expelled or persecuted. **Xiao Jingguang**, for instance, was removed from his party positions and sentenced to five years in prison by Zhou Enlai. **Lin Biao**, a traditional Mao ally, may have temporarily abandoned Mao at this juncture, since Lin was given command of a corps after the conference.

The emphasis on standardized tactics and conventional battlefield operations was rewarded during the battles against the Nationalists during their Fourth Encirclement Campaign, where Communist forces captured large amounts of weapons and ammunition, radios, equipment, and prisoners. Entire Nationalist units defected to the PLA. It was not until the Fifth Encirclement Campaign, where the Nationalists brought more forces to bear more effectively, that the decisions of the Ningtu Conference were modified, at the Zunyi Conference, and Mao was restored to leadership positions. This debate within the PLA over the extent to which emphasis should be placed on guerrilla tactics versus conventional tactics and over the role of the Communist Party's political commissars within units started at the **Gutian Conference** in 1929, came to a head at the Ningdu Conference, and was, to a certain extent, reversed at the Zunyi Conference. Nonetheless, this tension between guerrilla warfare and conventional warfare is the major doctrinal conflict in the PLA even today.

REFERENCES

Laszlo Ladany, *The Communist Party of China and Marxism, 1921–1985: A Self-Portrait* (Stanford, CA: Hoover Institution Press, 1988); William W. Whitson with Chen-Hsia Huang, *The Chinese High Command* (New York: Praeger, 1973); Xiao Zhaoran, *Zhonggong Dangshi Ming Cidian* [A Concise Dictionary of the Chinese Communist Party's History] (Beijing: Jiefangjun Chubanshe, 1986); ZGDBKQS, vol. 1.

NORTH CHINA FIELD ARMY The North China Field Army, really the "Fifth Field Army" in the **Field Army System**, traces its military lineage to the 343d Brigade, 115th Division of the **Eighth Route Army**. It operated on the North China Plain. The leaders most closely associated with the North China Field Army are **Nie Rongzhen, Bo Yibo, Peng Zhen, Luo Ruiqing, Geng Biao**, Yang Chengwu, and Yang Dezhi. Many of these people were also closely associated with **Lin Biao**, even throughout the time that the North China Field Army was activated in 1949, and even back into the Eighth Route Army period (as part of the 115th Division), it was led by Nie Rongzhen.

During the **Anti-Japanese War**, after 1937, the **People's Liberation Army** (PLA) element that formed the North China Field Army operated out of a base area in the Jin-Cha-Zhi Border region (the areas of North China around Hebei Province and Beijing, comprising the Tianjin area, reaching east toward the Shandong Peninsula, the Chahar area, and the Zhifeng area, which reaches into what is now Inner Mongolia.

The units of the 115th, including those that later formed the North China Field Army, took part in the **Hundred Regiments Campaign**, which was fought in their area of operation. By the end of **World War II**, as the Anti-Japanese War that started in China in 1937 expanded, the forces that formed the North China Field Army took over Shanxi Province, fighting a hard campaign in Taiyuan. During the **Civil War**, they participated in the **Ping-Jin Campaign** to secure Beijing and Tianjin.

Few of the North China Field Army's leaders or units fought in the **Korean War**. The 19th Army, led by Yang Dezhi, was sent into Korea in February 1951, and Yang served as a deputy commander of the **Chinese People's Volunteers**. He stayed in Korea until 1954, after which he became the commander of the Beijing military region.

By the time the military regions were established, the Field Armies were disestablished, but the Field Army System retained its influence in Chinese politics.

REFERENCES

Fang Tian et al., *Si Ye Zuihou Yizhan* [The Last Battle of the Fourth Field Army] (Beijing: Guofang Daxue Chubanshe, 1995); William W. Whitson with Chen-Hsia Huang, *The Chinese High Command* (New York: Praeger, 1973).

NORTHEAST CHINA FIELD ARMY. See FOURTH FIELD ARMY; NORTHEAST DEMOCRATIC UNITED ARMY

NORTHEAST DEMOCRATIC UNITED ARMY (1945–1948) The Northeast Democratic United Army (Dongbei Minzhu Lianjun) is a forerunner of the **Fourth Field Army**. When the Soviet Union declared war on Japan and initiated combat operations into China on August 8, 1945, the Chinese Communist Party Central Committee decided to send units into northeast China (Manchuria) in greater strength. **People's Liberation Army** (PLA) units were sent to the northeast from the **Eighth Route Army** and the **New Fourth Army** to reinforce the units of the Northeast United Anti-Japanese Army already operating in Manchuria.

The need for the Central Committee to further reinforce Manchuria became more urgent in the fall of 1945, as the United States began to transport Nationalist troops into Manchuria, both to control surrendering Japanese forces and to offset the Soviet and Chinese Communist military presence in the area. In September 1945, the Communist Party Central Committee organized the Northeast Bureau with **Peng Zhen** as its general secretary. Then on October 31, 1945, the Central Committee declared the PLA forces in Manchuria to be the ''Northeast People's Autonomous Army'' (Dongbei Renmin Zizhikun) with **Lin Biao** as commander and concurrently political commissar and **Luo Ronghuan** as first political commissar. To increase troop strength in the region, in November 1945, the Central Committee sent into the northeast over five divisions from the Shandong area (the First, Second, Third, Sixth, and Seventh Divisions with elements of the Fifth Division), three divisions from the New Fourth Army, four brigades from the Shaanxi–Gansu–Ningxia area, including the 359th Brigade, troops from **Yan'an**, and troops from the **Anti-Japanese Military and Political College**.

By January 1, 1946, the Communist Party Central Committee had brought its military strength in northeast China to 270,000 soldiers. At that time, the name of the forces was changed to the ''Northeast Democratic United Army.'' (Whitson translates the term as Northeast Democratic Allied Army, so some references use this term, but People's Liberation Army histories translate the term as ''Northeast Democratic United Army.'') The Northeast Democratic United Army fought the initial battles of the PLA campaign to control Manchuria, including the **Sungari River Campaign** and the **Siping Defensive Battles**. In January 1948, while the Northeast Democratic United Army was engaged in the **Siping Offensive Campaign**, the Central Committee redesignated Manchuria the Northeast Military Region and renamed the army the ''Northeast Field

Army'' (Dongbei Yezhanjun). Lin Biao remained as commander and political commissar, and Luo Ronghuan remained as first deputy political commissar. Chen Yun, later China's economic planner, was also a deputy political commissar. In early 1949, the **Northeast China Field Army** was converted into the Fourth Field Army, with Lin Biao still in command.

REFERENCES

Fang Tian et al., *Si Ye Zuihou Yizhan* [The Last Battle of the Fourth Field Army] (Beijing: Guofang Daxue Chubanshe, 1995); William W. Whitson with Chen-Hsia Huang, *The Chinese High Command* (New York: Praeger, 1973); ZGDBKQS, vol. 1.

NORTHERN EXPEDITION (BEIFA, 1926–1928) After the Canton Coup (March 20, 1926), the Soviets were more careful about what money or arms were sent into China to assist the Nationalists, fearing the funds would be used to attack the Chinese Communist Party (CCP). At the same time, the Nationalist Army had grown considerably in size, to over 85,000 men, causing a drain on the coffers of the Nationalist government of 3 million yuan a month. Compounding the drain on revenues caused by the size of the army and the reduction of military aid by the Soviets was an embargo on arms for China by the West.

In response to these pressures, the Nationalist government sought to accomplish two objectives: to unify the country under its rule and to capture arsenals in North China at Shanghai, Nanjing, and Hanyang. To do this, the central government decided to follow a plan advocated by Sun Yat-sen, a ''northern expedition to unify China.'' This meant breaking the stronghold that regional **warlords** held on separate parts of the country and forcing them to pay taxes and support the central government. The first action was an attack on Hunan warlord Tang Shenzhi by forces under Marshal Wu Peifu, who supported the Nationalist cause. Generals Li Congren and Li Chisen, from Jiangxi, assisted Wu. Meanwhile, other parts of China remained under strong warlord control. Sun Zhuanfang controlled the southern part of the country, including Shanghai. The Shandong area was controlled by Zhang Congchang. Shanxi was under the control of **Yan Xishan. Feng Yuxiang** controlled most of northern China. The northeast, including Beijing and Manchuria, was controlled by **Zhang Zuolin**. (See warlords for a more complete listing.)

Chiang Kai-shek was appointed commander in chief of the Nationalist Army forces on June 9, 1926, with the broad mission of overthrowing the warlords and stopping support for the warlords by foreign powers. Some Chinese Communist leaders were part of the expedition and contributed to its success, notably, **Ye Ting, Ye Jianying**, and **Zhou Enlai**, all of the 24th Division, Fourth Corps. The initial campaign in the Northern Expedition succeeded in gaining control of Wuhan, Hankow, Hanyang, and Wuchang, giving the Nationalists control of Hunan and Hubei Provinces. Chiang Kai-shek then turned his forces east on Nanchang against Sun Quanfang, gaining control of the Yangtze River area by November. Concurrently, a campaign by General He Yingqin gained control of

Fujian and began to move into Zhejiang. By March 1927, Chiang was moving forces farther north to take Shanghai and Nanjing.

The Nationalist troops attacked the American, British, and Japanese consulates in Nanjing, killing seven people. In reaction, U.S. and British gunboats opened fire, killing 15 Chinese troops. Chiang had to travel to Shanghai to calm the foreign legations, which had 16,000 troops of their own in the city. Chiang arrived on March 26, 1927, to settle the ''Nanjing Incident.'' In a further drive north, Chiang's Nationalist forces captured Zhengzhou and Xuzhou, attacking the forces of Feng Yuxiang. Feng joined forces with the Nationalists and turned the attack north again. In the spring of 1928, forces of Yan Xishan (Shanxi Province), Chiang, and Feng attacked Beijing from three axes, causing a withdrawal by force of Zhang Zuolin. Zhang Zuolin withdrew from Beijing in a manner that allowed the forces of Yan Xishan to take Beijing instead of the forces of Feng Yuxiang, an old enemy of Zhang's. A drive on Tianjin and Jinan was halted by Chiang to avoid conflict with Japanese forces sent there to protect Japanese nationals in that concession area. By June 1928, the completion of the Northern Expedition was announced by Chiang and other warlords in a temple in Beijing's Western Hills.

REFERENCES

Hsi-sheng Ch'i, *Warlord Politics in China, 1916–1928* (Stanford, CA: Hoover Institution Press, 1977); F. F. Liu, *A Military History of Modern China: 1924–1949* (Princeton: Princeton University Press, 1956); Lucian W. Pye, *Warlord Politics: Conflict and Coalition in the Modernization of Republican China* (New York: Praeger, 1971).

NUCLEAR PROGRAM The first Chinese nuclear explosion was detonated on October 16, 1964, from an enriched uranium fission device similar to that exploded by the United States at Hiroshima. The weapon was developed with Soviet assistance by a team headed by Deng Jixian, an American-trained physicist who had earned a doctoral degree from Purdue University. Marshal **Nie Rongzhen** was responsible for both the nuclear weapons program and a parallel missile program. China's first nuclear ballistic missile test was conducted on December 28, 1966, using a medium-range (1,100 kilometers) Dongfeng 2 (CSS-1) ballistic missile. The yield was about 20 kilotons. On June 17, 1967, China successfully tested a thermonuclear device. The development and successful test of an intercontinental ballistic missile (ICBM) by China waited until 1972, when the Dongfeng 3 (CSS-2) silo-based system was detonated with a three-megaton warhead. The range of this system, which China has provided to Saudi Arabia, is about 2,700 kilometers. To date China has conducted 26 nuclear tests and may have stockpiled about 1,400 nuclear weapons. China has also developed a submarine-launched ballistic missile capability but, as of 1998, has not deployed submarines operationally. The only **Broad Ocean Missile Launch** by China as a test of its intercontinental ballistic missile program took place on

May 16, 1980. This test also involved the deployment of an 18-ship naval task force to support the test.

China's nuclear weapons policies in 1998 still supported the concept that the nation must be prepared, if necessary, to fight a major nuclear war. Strategists in China believe that the "nuclear retaliatory capabilities of medium-sized powers can prevent great powers' interference in a local, limited war." The responsibility for China's nuclear forces lies with the Second Artillery Corps of the PLA, which in the **Taiwan Straits Crisis of 1995/1996** demonstrated that it controls not only strategic ballistic missiles but also the shorter-range Dongfeng series, with the Dongfeng-15 and Dongfeng-11 (called the M-9 and M-11 in the West). The tactical battlefield missiles like the Dongfeng-11 and 15 have been incorporated into PLA war-fighting doctrine as "deep-strike" weapons intended to break up enemy troop concentrations, logistical hubs, and reserve forces.

China's strategic nuclear program continues to modernize the force in the late 1990s. The PLA is likely to make a three-stage, solid propellant missile, the Dongfeng (DF)-41, with a 12,000-kilometer range carrying an 800-kilogram payload, operational around the turn of the century. The current land-mobile strategic missile system in China, the DF-5, will be replaced by the land-mobile, solid-propellant, intermediate-range (IRBM) DF-31, which will see initial operational capability just after the turn of the century. This system may owe some of its technology to assistance from Russian missile scientists hired by China after the collapse of the Soviet Union. The Second Artillery, as the command element for China's nuclear forces, has about 90,000 troops. China's nuclear-capable forces range from an aging fleet of H-6 (former Soviet Tu-16) bombers with a range of 1,650 miles, to the DF-5 intercontinental ballistic missiles (ICBM) with a range of 13,000 miles. As of 1998, China had about 15–20 ICBM systems deployed and 70–80 IRBMs. The number of its shorter-range systems, such as the DF-11 and DF-15, is more difficult to estimate. A future goal of China's nuclear and strategic missile program is to develop multiple, independently targeted reentry vehicles (MIRV) for its ICBM forces.

REFERENCES

Paul H. B. Godwin, "From Continent to Periphery: PLA Doctrine, Strategy and Capabilities towards 2000," *The China Quarterly*, no. 146 (June 1996): 464–487; John W. Lewis and Xue Litai, *China's Strategic Seapower: The Politics of Force Modernization in the Nuclear Age* (Stanford, CA: Stanford University Press, 1994); John W. Lewis and Xue Litai, *China Builds the Bomb* (Stanford, CA: Stanford University Press, 1988); Larry M. Wortzel, ed., *China's Military Modernization: International Implications* (Westport, CT: Greenwood Press, 1988).

O

ONE TWO EIGHT INCIDENT (January 28, 1932) On January 28, 1932, in response to weeks of serious anti-Japanese demonstrations and boycotts among the Chinese populace in Shanghai that developed after the Mukden Incident (see **Nine One Eight (918) Incident**), the Shanghai Municipal Council declared a state of emergency in that city. Troops of the Nationalist 19th Route Army were deployed around the city to protect the foreign concession areas from popular unrest and antiforeign incidents.

Japan, which had sent troops to Shanghai aboard ships as early as January 18, 1932, landed several thousand marines at Shanghai in three columns on the night of January 28. The Japanese marines advanced to establish protective perimeters around the international settlement area that housed Japan's foreign concessions, along the Huangpu Creek, and to occupy the rail stations in the city. When the Japanese marines encountered elements of the 19th Route Army, which was responsible for security and garrison duties in the Nanjing–Shanghai area, they exchanged fire with the Chinese soldiers in the Chapei District of Shanghai. In retaliation for this incident, the senior Japanese naval officer ordered the Chapei District, which was primarily inhabited by Chinese workers, to be bombed by Japanese aircraft. By the beginning of February, the Japanese had landed 7,000 more marines in Shanghai, and the Guomindang reinforced its 19th Route Army with the Fifth Corps. By February 7, Japan had moved a force of three infantry divisions, 80 ships, and 30 aircraft in Shanghai. Japanese forces there reportedly totaled about 90,000 men, while the Nationalist Army defended with about 50,000 personnel. In fierce battles on March 1–3, 1932, Chinese forces suffered over 10,000 casualties but managed to contain the Japanese forces within the Shanghai area. Under strong pressure from foreign countries, including the United States and England, Japan agreed to a cease-fire, and an armistice went into effect in Shanghai on May 5, 1932. Under the terms of the agreement, the Nationalist government was forced to accept a neutral zone around Shanghai and to withdraw its military forces.

REFERENCES

George M. Beckmann, *The Modernization of China and Japan* (New York: Harper and Row, 1962); James Crowley, *Japan's Quest for Autonomy: National Security and Foreign Policy, 1930–1938* (Princeton: Princeton University Press, 1966); F. F. Liu, *A Military History of Modern China, 1924–1949* (Princeton: Princeton University Press, 1956); ZGDBKQS.

OPERATION ICHIGO. See ICHIGO, OPERATION

OPIUM WAR (1839–1842) In 1838, Chinese emperor Daoguang appointed Lin Zexu, a Confucian scholar, as imperial commissioner to Canton (Guangzhou), with a mission to end the trade in opium. Lin began a campaign to suppress opium use among Chinese in the Canton area in 1839. He used a combination of social pressure, arrests, and confiscation of the drugs in his campaign. His measures against British opium traders were also harsh.

In an attempt to pressure the foreign opium traders to stop sales, Lin ordered all the staff and servants in Canton to leave foreign employ. He then blockaded all foreign merchants and traders in Canton in their factories, insisting that they turn over all the opium. The senior British official in China, superintendent of foreign trade Charles Elliott, was among those subject to the Chinese blockade. At the end of a six-week period, the foreign traders turned over about 3 million pounds of raw opium to Lin Zexu, which Lin destroyed. The majority of foreigners were then allowed to leave Canton. British merchants withdrew from Canton to Hong Kong Island, which they used as a base of trade. The British Parliament, at the same time, authorized Admiral George Elliott, a cousin of Charles Elliott, to seek reparation from China or to hold Chinese ships and cargo in custody if the Chinese refused to compensate British merchants for the opium destroyed by Lin Zexu.

Admiral George Elliott commanded a force of 16 warships, four armed steamships, about 4,000 troops, and 28 transports. The British fleet arrived off Canton in June 1840 and began to blockade China's main ports along the coast. The bulk of the fleet moved north and arrived at Beihe, on the approaches to the city of Tianjin, where Elliott began negotiations with Qi Shan, a secretary to the emperor, in August and September 1840. The negotiations were later moved to Canton. In the end, the Chinese agreed to cede Hong Kong to the British, paid 6 million Mexican silver dollars to the British merchants in indemnities, permitted the British direct contact with the Qing Court, and reopened trade in Hong Kong. The emperor was furious with his secretary for agreeing to these terms. British foreign secretary Lord Palmerston was also angry that Charles Elliott had given up Zhoushan Island in Canton in return for Hong Kong. The British insisted on a new agreement that was to be signed by the emperor, not a negotiator.

A new negotiator was sent out, Sir Henry Pottinger. By August 1841, when Pottinger reached China, fighting had already broken out between British troops and local Chinese militia around Canton. Pottinger took the British fleet north and captured the ports of Xiamen, Ningbo, and Zhoushan. In 1842, landing British forces, he also severed all of China's main river and canal links from north to south. By June 1842, British forces, reinforced from India, captured Shanghai, Zhenjiang was taken by July, and by August 5, 1842, British forces were in position to attack Nanjing. Representatives of the Qing Court signed the **Treaty of Nanjing** on August 29, 1842, and Emperor Daoguang accepted the treaty in September. In December 1842, the Treaty of Nanjing was ratified by Queen Victoria.

REFERENCES

Wolfram Eberhard, *A History of China* (Berkeley: University of California Press, 1977); John K. Fairbank, *Trade and Diplomacy on the China Coast* (Cambridge: Harvard University Press, 1953); John W. Garver, *Foreign Relations of the People's Republic of China* (Englewood Cliffs, NJ: Prentice-Hall, 1993); Jonathan D. Spence, *The Search for Modern China* (New York: W. W. Norton, 1990).

OYUWAN SOVIET. See EYUWAN SOVIET

P

***PANAY* INCIDENT** (December 12, 1937) The USS *Panay*, a U.S. Navy gunboat assigned to the "China Station," was attacked and sunk by Japanese warplanes off Shanghai on December 12, 1937.

The *Panay* had been built especially for service on the Yangtze River and on the coast of China. On November 24, 1937, *Panay* assumed duty as China Station's ship for the U.S. Asiatic Fleet in Nanjing. *Panay*'s mission was to protect American lives and property in China. The American Embassy was, at that time, in Hankou, having shifted there from Beijing after war broke out between China and Japan following the **Marco Polo Bridge Incident** in July 1937. *Panay* was instructed by the fleet to maintain communications with the U.S. ambassador in Hankou and with the American consulate in Nanjing.

When Japanese forces stormed Nanjing on December 10, 1937, serious artillery duels broke out around the city between Japanese and Nationalist forces. Since the United States was neutral in the war at that time, *Panay*, with other U.S. Navy gunboats, the *Meiping, Mei-Hsia*, and *Meian*, moved upriver to avoid the fighting. Japanese troops boarded the *Panay* on the river at 9:40 A.M. and were told by the *Panay*'s captain that the ship was moving away to avoid the artillery.

After noon on the same day, December 12, the *Panay* was attacked by Japanese warplanes, bombed, and strafed. *Panay* was hit by a bomb, *Meiping* was set afire, and *Mei-Hsia* and *Meian* were damaged. Six Japanese biplanes attacked again, seriously damaging *Panay*, which began to sink. The *Meian* was beached, and the crew evacuated. The British gunboats HMS *Bee* and *Ladybird*, also patrolling the Yangtze River, came to the assistance of the American ships and, with another American gunboat, the USS *Oahu*, evacuated the wounded to Shanghai, arriving on December 17.

The United States formally protested the attack to Japan by diplomatic note, demanding an expression of regret and an indemnity from Japan, calling the incident a "disregard of American rights by the Japanese." The U.S. State Department asked for assurance that American property in China would not, in the future, be subject to attack or interference by Japan. The Japanese responded with a note of regret, claiming the attack was accidental and explaining that the commander of the aircraft had been removed from his duties.

REFERENCES

W. A. Angin, "Compiled Records of the Commander in Chief, United States Asiatic Fleet," Shanghai, December 1938, papers of Milton E. Miles, Hoover Institution Archives, Hoover Institution on War, Revolution, and Peace, Stanford, CA; Thomas A. Bailey, *A Diplomatic History of the American People* (Englewood Cliffs, NJ: Prentice-Hall, 1980); Dorothy Borg, *The United States and the Manchurian Crisis of 1933–1938* (Cambridge: Harvard University Press, 1964).

PARACEL ISLANDS. See XISHA ISLANDS DEFENSIVE CAMPAIGN

PEIPING–TIANJIN CAMPAIGN. See BEIPING–TIANJIN CAMPAIGN

PEIYANG ARMY. See BEIYANG ARMY

PEIYANG FLEET. See BEIYANG FLEET

PEKING CONVENTION (1860) (Beijing Convention) The Peking Convention was signed in 1860 between China and Great Britain after British troops attacked Beijing and leveled the Summer Palace (Yuanmingyuan). The Convention of Peking ended a post-**Opium War** engagement that started with the seizure in Canton of the *Arrow*, a Chinese-owned, Hong Kong-registered ship (see **Arrow War**). The British acted because the Chinese emperor had ignored the provisions of the **Treaty of Tianjin** (1858), refusing to allow a British ambassador to be resident in Beijing, nor would the emperor agree to exchange ratification of the Treaty of Tianjin in Beijing. Working together, British and French forces occupied the **Dagu forts**, protecting the approaches to Tianjin, and Anglo–French forces marched to Peking and occupied parts of the city. The emperor finally agreed to ratification and also ceded to Britain the peninsula of Kowloon. Among other provisions, Tianjin was also opened as a treaty port.

REFERENCES

Wolfram Eberhard, *A History of China* (Berkeley: University of California Press, 1977); John K. Fairbank, *Trade and Diplomacy on the China Coast* (Cambridge: Harvard University Press, 1953); Jonathan D. Spence, *The Search for Modern China* (New York: W. W. Norton, 1990).

PEKING RAID. See RAID, SOVIET COMPOUND IN BEIJING

PENG DEHUAI (1901–1974) Peng Dehuai was one of the most respected and experienced combat leaders of the **People's Liberation Army** (PLA). His purge and subsequent mistreatment at the hands of **Mao Zedong** and the party are one of the tragic examples of the Chinese Communist Party's (CCP's) inability to respond to inner-party criticism in the ideologically charged period of the 1960s (a problem, perhaps, for all Leninist political parties). Peng Dehuai was born in 1901 in Xiangtan, Hunan Province. As a child he ran away from home, supporting himself through manual labor. During a catastrophic famine in Hunan in 1919, he joined the army. Within a year he was made a lieutenant and was involved in an assassination attempt on the governor-general of Hunan, Fu Liangzuo. Although Peng was arrested, little is known of the terms of his imprisonment. When the **Northern Expedition** took place in 1926, he joined the Nationalist Army with the rank of major. In 1927 or 1928, Peng joined the CCP and took an active part in founding the Hunan Soviet. He later joined Mao Zedong and **Zhu De** at the Red Army in the **Jinggangshan Revolutionary Base Area**. In 1930, Peng joined forces with the Red Army under Zhu De. During the **Long March**, Peng Dehuai commanded the lead units of the advanced guard. At the beginning of the **Anti-Japanese War**, Peng was appointed deputy commander of the **Eighth Route Army**, where he carried out guerrilla warfare in east and south Shanxi Province. By the end of the Anti-Japanese War, Peng held the post of deputy commander in chief of the PLA and commander of the **First Field Army**.

Peng commanded the **Chinese People's Volunteers** in Korea in 1951, remaining in Korea throughout the war. In 1954 he was recalled from Korea and appointed a marshal of the PLA, minister of national defense, and vice chairman of the National Defense Council. At the **Lushan Conference** in July and August 1959, after making an inspection trip of conditions in China, Peng criticized Mao Zedong for the disastrous economic policies of the "Great Leap Forward." Peng was purged during the Lushan Conference for criticizing Mao, the party line, and the system of people's communes. He suffered further during the **Cultural Revolution** and was officially rehabilitated only posthumously in 1978. Peng Dehuai died on November 29, 1974.

REFERENCE

Wolfgang Bartke, *Who's Who in the People's Republic of China* (Armonk, NY: M. E. Sharpe, 1981).

PENG ZHEN (1899–1997) Peng Zhen is an interesting figure who was a central actor during the Revolutionary War, the **Civil War**, the **Korean War**, and the **Cultural Revolution**. Although his career was spent almost exclusively in the Chinese Communist Party (CCP) organization department, it was tied intrinsically with the **People's Liberation Army** (PLA).

Peng Zhen was born in Shanxi Province, a poor area that had a lot of coal mines but little agricultural development. His career in the Chinese Communist

Party began early, since he was deeply influenced by leftist schoolteachers. He began to work in the labor and youth movements organized by the CCP in Shanxi and joined the Communist Party in 1926, after the **Northern Expedition**.

Peng began party organization work in the North China Bureau, which included the provinces of Shanxi, Hebei, and Chahar and parts of Inner Mongolia. He worked closely with **Bo Yibo** in 1929, organizing Communist Party cells and support in "White areas" (areas controlled by the Nationalists). He also was closely associated with party officials like **Zhou Enlai** and **Liu Shaoqi**, with whom he established a particularly close relationship. His first direct military experience probably came in mid-1937. After the **Anti-Japanese War** began, he brought a group of students and workers to join 2,000 Communist guerrillas operating in the Wutai Mountains, between Hebei and Shanxi, under the leadership of **Nie Rongzhen**. They established a **Revolutionary Base Area** in Fouping, and Peng Zhen became secretary of the Communist Party's North China Bureau. From 1938 to 1941, Peng also was the political commissar for Nie Rongzhen's forces in the "Jin-Cha-Zhi" Base Area.

For some time, between 1938 and 1942, Peng Zhen was concurrently head of the Central Communist Party school in **Yan'an** and the political commissar for the forces at Fouping under Nie. By 1945, Peng was a Politburo member, along with Liu Shaoqi, Zhou Enlai, and, of course, **Mao Zedong**. When **World War II** ended, and **Lin Biao** moved his forces into the northeast for the **Manchurian Campaign**, Peng Zhen served as his political commissar. In 1947, when Lin's forces incurred serious casualties in the fifth offensive across the **Sungari River** in an attempt to take Siping, both Lin Biao and Peng Zhen came under heavy criticism by the Communist Party. Peng was recalled from Manchuria and replaced by **Luo Ronghuan**. In 1966, during the Cultural Revolution, while he was mayor of Beijing, Peng Zhen came under serious attack by Red Guards and the Communist Party Cultural Revolution Group for having encouraged Lin Biao to use conventional tactics at Siping instead of "People's War" tactics. Peng Zhen was attacked also for his alleged efforts to act on behalf of "revisionist intellectuals" in Beijing during the Communist Party's 1957 Anti-Rightist Campaign. He was purged as an "anti-Party element" in the Cultural Revolution but in 1979 was rehabilitated by Deng Xiaoping and restored for a short time to his position as mayor of Beijing. He retired in 1988 as the leader of the National People's Congress. Peng died on April 26, 1997.

REFERENCES

Lowell Dittmer, *Liu Shao-ch'i and the Chinese Cultural Revolution: The Politics of Mass Criticism* (Berkeley: University of California Press, 1974); Richard Curt Kraus, *Class Conflict in Chinese Socialism* (New York: Columbia University Press, 1981); Laszlo Ladany, *The Communist Party of China and Marxism, 1921–1985* (Stanford, CA: Hoover Institution Press, 1988); James R. Townsend, *Politics in China* (Boston: Little, Brown, 1974); William W. Whitson with Chen-Hsia Huang, *The Chinese High Command* (New York: Praeger, 1973).

PEOPLE'S LIBERATION ARMY (PLA) The People's Liberation Army is the collective term for the armed forces of the People's Republic of China. It includes the ground forces, or army, with a strength of about 2.1 million men and women; the PLA Navy (PLAN), with a strength of about a half million sailors; and the PLA Air Force (PLAAF), with a strength of about a half million personnel. The People's Liberation Army is controlled by the Chinese Communist Party's Central Military Commission (CMC). In addition to the ground forces, navy, and air force (which includes China's air defense forces), the PLA includes the strategic rocket forces, or Second Artillery (Er Pao), and the People's Armed Police.

Various paramilitary forces must also be considered as augmenting or included in the other People's Liberation Army forces that are controlled by the Central Military Commission. These other forces include the People's Militia, a loosely constructed, lightly armed organization in time of war or disaster; the People's Armed Police, about 1 million men strong, which operates under the joint control of the Central Military Commission of the Communist Party and the Ministry of Public Security; and vestiges of the Production Construction Corps (PCC), a militia-like organization that still operates state farms in remote areas of western China (Xinjiang) and on the Sino–Russian border in Manchuria.

The People's Liberation Army dates its foundation to the **Nanchang Uprising** on August 1, 1927, when about 30,000 Communists and disaffected, Communist-led elements of **Chiang Kai-shek**'s Nationalist Army began an insurrection in that city to take over the Nanchang Arsenal. Although the insurrection and subsequent **Autumn Harvest Uprisings** failed, Communist guerrilla units established themselves in **Revolutionary Base Areas** and, after the **Long March** from Jiangxi Province to Shaanxi Province, headquartered themselves in **Yan'an**. There, as the **Eighth Route Army** and **New Fourth Army**, the PLA fought the Japanese until 1945, when the **Civil War** broke out in China. After a three-year series of campaigns, the People's Liberation Army succeeded in driving the Nationalist Army off the mainland, and the Nationalist government withdrew to Taiwan, where it remains today.

REFERENCES

Defense Intelligence Agency, *Handbook on the Chinese Armed Forces* (Washington, DC: Defense Intelligence Agency, 1976, 1984); ''Special Issue: China's Military in Transition,'' *The China Quarterly*, no. 146 (June 1996).

PEOPLE'S WAR (Renmin Zhan Zheng) People's War is a general strategy and system of organization for China's armed forces that is essentially defensive in nature. It depends on the mass mobilization of the vast Chinese populace, the use of main force combat units of the military to engage attackers, and the use of local militia forces and reserves in a supporting role and to conduct guerrilla warfare. There are three basic characteristics to People's War: the mobilization of the populace both as a deterrent measure against attack and a defense in case of attack; the conduct of a protracted war in order to take advantage of terrain, the time it takes to mobilize the populace, and the size of China's population; and the fact that because of inferior weapons China tended to fight a defensive battle. People's War is best viewed as a national strategy designed to cope with threat and attack. In the event of an attack on China, however, the basic tactics used in the strategy were essentially designed to fight a guerrilla war against a superior attacking force, absorbing attacks from that force while using China's superiority in manpower to overcome the enemy. The entire populace was expected to contribute to the war effort, either as guerrillas or by assisting with supply, transport, and other logistical support. At decisive times, when force could be concentrated to China's advantage, conventional tactics were used. People's War tactics call for ''luring the enemy deep'' into the defended area, trading space for time, and then encircling and eliminating an attacker on terms favorable to Chinese forces. One feature of People's War is that, as practiced against the Japanese in **World War II**, it did not concentrate on controlling specific points of terrain or cities as its main objective. Instead, it focused on using mass-mobilized militia and reserves to harass and weaken an enemy in a war of attrition until the main military forces of the enemy could be completely destroyed.

With the resolution of the disputes over the demarcation of Sino–Soviet border between 1989 and 1997 and the improvement in relations with Russia, this term has been modified to provide for the incorporation of modern weapons and technology. In 1985, China significantly reduced the size of the military and began to focus on a more active defense, using joint operations (ground, naval, air, and strategic missile forces). This has been interpreted as a strategy of **People's War under Modern Conditions**. But it remained an essentially defensive strategy adapted to use more modern technology. This is also called a ''Local Wars Strategy,'' meaning that China expects to fight only on its periphery, in defense of its sovereignty.

REFERENCES

Paul H. B. Godwin, ed., *The Chinese Defense Establishment* (Boulder, CO: Westview Press, 1983); Ellis Joffe, *The Chinese Army after Mao* (Cambridge: Harvard University Press, 1987); Mao Zedong, *Selected Works*, vol. 1: ''A Single Spark Can Start a Prairie Fire,'' pp. 117–128 (Beijing: Foreign Languages Press, 1975), vol. 2: ''Problems of Guerrilla War against Japan,'' pp. 79–112, and ''On Protracted War,'' pp. 113–194 (Beijing: Foreign Languages Press, 1965).

PEOPLE'S WAR UNDER MODERN CONDITIONS The concept of modernizing the **People's War** strategy to permit the **People's Liberation Army** (PLA) to fight a ''people's war under modern conditions'' took place in the late 1970s and the 1980s. It reflects a recognition in the leadership of the PLA that improvements in weaponry, command and control, and training and the changed economic and industrial infrastructure in China called for a changed defense strategy. Instead of allowing enemy forces to penetrate deep into China's interior, a characteristic of the concept of ''luring the enemy deep'' in the ''People's War'' strategy, the PLA was prepared to engage and defeat an enemy close to the border. The reliance on a war of attrition in a protracted war was also modified in the strategy of ''People's War under modern conditions,'' which placed an emphasis on winning early battles and combat engagements in order to influence the course of a war. Whereas classic ''People's War'' doctrine called for mobile warfare with a fluid line of defense, the new doctrine placed more emphasis on positional defense to protect key terrain and lines of communications, giving China's defensive strategy more balance. This meant that cities would be defended rather than abandoned to allow PLA forces to retreat into the countryside. Finally, the improved nuclear deterrent force in China led to a change in how China thought about dealing with a nuclear attack against its territory. In ''People's War'' theory, China believed that it could survive a nuclear attack and then defeat a conventional force by ''luring the enemy deep'' into China. Improved nuclear forces permitted China's strategists to use the threat of retaliation to deter nuclear attack.

Along with the change to a strategy of ''People's War under modern conditions,'' China's strategic thinkers also began to investigate a **local war doctrine**, which amounted to a realization that the PLA would probably fight along the periphery of China. This shift reflected the realization that a direct attack designed to occupy and control the heartland of China by traditional adversaries, Russia, Japan, and the Western powers (particularly, the United States) was not likely.

The most recent doctrinal literature in Chinese military operational art evolves from the PLA's exploration of ''high-technology warfare under modern conditions.'' Based on its understanding of the 1990–1991 Gulf War between a U.S.-led coalition and Iraq, the PLA is experimenting with equipment and strategies to take advantage of a ''revolution in military affairs'' and, within its own limitation, employ information warfare and information operations in its military doctrine. The changes in the PLA's strategic views are consistent with a general trend toward taking advantage of computers and automated information distribution systems to link ''shooters'' (weapons and weapon systems) with ''sensors'' (radars, intelligence systems, and other means of acquiring intelligence about the battlefield).

REFERENCES

Ellis Joffe, *The Chinese Army after Mao* (Cambridge: Harvard University Press, 1987); Li Jijun, *Junshi Zhanlue Siwei* [Military Strategic Thought] (Beijing: Junshi Kexue Chubanshe, 1996); Li Qingshan, *Xin Junshi Geming yu Gao Jishu Zhanzheng* [The Revolution in Military Affairs and High Technology Warfare] (Beijing: Junshi Kexue Chubanshe, 1995); Charles D. Lovejoy, Jr., and Bruce W. Watson, eds., *China's Military Reforms* (Boulder, CO: Westview Press, 1986); Nan Li, ''The PLA's Evolving Warfighting Doctrine, Strategy and Tactics, 1985–1995: A Chinese Perspective,'' *The China*

Quarterly, no. 146 (June 1996): 443–463; Michael Pillsbury, ed., *Chinese Views of Future Warfare* (Washington, DC: National Defense University Press, 1997); Yue Shuiyu, *Sunzi Bingfa yu Gao Jishu Zhanzheng* [Sun Tzu's Military Thought and High-Technology Warfare] (Beijing: Guofang Daxue Chubanshe, 1998); Zhu Youwen, *Gao Jishu Tiaojian Xia de Xinxi Zhan* [Information Warfare under High-Technology Conditions] (Beijing: Junshi Kexue Chubanshe, 1994).

PING-JIN CAMPAIGN. See BEIPING-TIANJIN CAMPAIGN

PINGXINGGUAN, BATTLE OF (September 25–26, 1937) The **People's Liberation Army** (PLA) **Eighth Route Army**'s 115th Division was committed against the Japanese army in late September 1937 to seal a major pass between Hebei and Shanxi Provinces at Pingxingguan. The defense of the pass was the responsibility of approximately 65,000 soldiers of the **warlord Yan Xishan**, who was operating under the control of the Nationalist government. Yan's son-in-law, Yan Aiyuan, commanded the Nationalist troops. The 115th Division, in a coordinated operation designed to ambush the Japanese, was then commanded by **Lin Biao** (whose name is often dropped from many of the People's Republic of China (PRC) military histories of the event because of his alleged 1971 coup attempt against **Mao Zedong**).

On September 25, 1937, the 115th Division positioned itself before dawn to ambush a Japanese column from the 21st Brigade of the Fifth Division in the pass. By late in the afternoon, the 685th, 686th, and 687th Regiments of the 115th Division had cut off approximately 3,000 Japanese troops in the Pingxingguan Pass. An Independent Regiment sealed the area to prevent Japanese reinforcement. The 344th Brigade of the 115th Division had been assigned a reserve mission. The plan called for the Nationalist troops to assist the 115th Division as the Japanese counterattack the ambush force. However, when the Japanese counterattack came, Yan Xishan's Shanxi Nationalist troops failed to provide the agreed support. Lin Biao therefore ordered a withdrawal of the 115th Division on September 26, under heavy Japanese pressure, including air attacks.

The net result was that the Japanese 21st Brigade lost about 1,000 killed. The Communist forces also captured an artillery piece, 20 heavy machine guns, 1,000 rifles, 200 vehicles, and 53 horses. PLA military histories credit this battle as the psychological turning point in **World War II** that convinced PLA leaders that the Japanese were not invincible. It also demonstrated to the Communist leaders that Nationalist forces were willing to let the Communists do battle and die on the battlefield, without the support promised by Nationalist commanders in the plan of operations.

REFERENCES

William W. Whitson with Chen-Hsia Huang, *The Chinese High Command* (New York: Praeger, 1973); ZGDBKQS, vol. 2; ZGRMJFJZH, vol. 2.

PORT ARTHUR (Lushan) In July 1945, in conversation with Nationalist foreign minister **T. V. Soong**, Stalin suggested that he wanted an alliance with China, stating he needed use of the Chinese ports of Dalian and Lushun for 30 years in case Japan restored its forces. "We could strike from there," Stalin told Soong. The Soviets were granted the access. After the end of **World War II**, an occasional Nationalist Chinese presence was permitted in Lushun, but the base was kept solely in Soviet hands. By 1948, 14 Soviet submarines, 5,000 naval personnel, and 20,000 Soviet ground troops occupied the Liaodong Peninsula.

The Soviet presence in Lushun later played an important role in the development of the **People's Liberation Army** (PLA) Navy. The Chinese established there the Democratic Naval Academy for future Chinese naval officers. By 1948, 300 cadets were reported to have graduated and received additional training aboard Soviet naval vessels in Lushun, in Dalian, and at Vladivostok.

In January 1950, **Mao Zedong** visited Moscow and signed the Sino–Soviet Treaty of Friendship, Alliance, and Mutual Assistance, in which it was stipulated that Soviet control over Lushun and the Changqun Railroad into the Liaodong Peninsula would end in 1952, not 1975, as had previously been agreed by the Nationalist government. In 1952, despite Soviet agreement to leave the Liaodong Peninsula, the Chinese requested that the Soviets remain in the area until a peace treaty with Japan had been signed. The ongoing **Korean War**, with U.S. involvement in Korea and U.S. actions in the Taiwan Strait as factors, probably influenced the Chinese request. After the death of Stalin in 1953, the Soviets gradually gave up their special position in Manchuria, ending joint control of the main railway line.

The memoirs of Nikita Krushchev provide some detail into how Port Arthur was returned to China. While the Sino–Soviet partnership was quite tense, Krushchev recounts the 1954 meeting with Mao, where the USSR agreed to return the treaty ports to China in May 1955. The Soviet premier states that Mao, while strongly desirous of the ports, did not want the Soviets to leave China, claiming that the United States might take advantage of such a move and attack China. Krushchev countered that the Soviet navy could provide adequate protection against such an eventuality from Vladivostok. The Soviets withdrew from the Port Arthur naval base in 1955, issuing this statement: "During the joint use of Port Arthur . . . Chinese and Soviet forces have stood side by side on the naval defense front in vigilant defense of peace in the Far East."

REFERENCES

John K. Fairbank, *The United States and China* (Cambridge: Harvard University Press, 1983); Sergei N. Goncharov, John W. Lewis, and Xue Litai, *Uncertain Partners: Stalin, Mao, and the Korean War* (Stanford, CA: Stanford University Press, 1993); David G. Muller, Jr., *China as a Maritime Power* (Boulder, CO: Westview Press, 1983); Bruce Swanson, *Eighth Voyage of the Dragon—A History of China's Quest for Seapower* (Annapolis, MD: Naval Institute Press, 1982).

Vance H. Morrison

PORT ARTHUR MASSACRE (November 21–24, 1894) Port Arthur, on the Liaodong Peninsula, today the location of the city of Dalian, fell in one day after attack by 18,000 Japanese troops of the Second Army, during **World War II**. According to Japanese records, three-quarters of the 12,000 Chinese in uniform who fought there were new recruits. Between November 21 and 24, 1894, Japan had taken control of the city and surrounding forts. According to the diary of a Japanese soldier, when Japanese forces entered the city, they found the head of a Japanese soldier on a stake. Japanese forces reacted brutally, murdering civilians, male and female, who were fleeing. They engaged in rape and acts of sexual mutilation of men, women, and children. The death toll from this Japanese attack was put in the hundreds or thousands.

REFERENCES

James V. Davidson-Houston, *Russia and China: From the Huns to Mao Tse-tung* (London: R. Hale, 1960); Stewart Lone, *Japan's First Modern War: Army and Society in the Conflict with China, 1894–95* (New York: St. Martin's Press, 1994).

Q

QIAN XUESEN (1912–) Born in Shanghai in 1912, Qian Xuesen graduated from Jiaotong University in 1934 and went to the United States to study one year later. He studied at Massachusetts Institute of Technology and California Institute of Technology, where he began work on aerodynamics, jet propulsion, and aircraft design. During **World War II**, Qian worked for the U.S. Army Air Forces as the director of the rocket section of the U.S. National Defense Scientific Advisory Board. He was made a colonel of the board and worked under the supervision of General Henry (''Hap'') Arnold. Qian was one of the individuals sent by the United States to dismantle the German rocket production center in Peenemunde and ship its equipment back to the United States. He returned to Shanghai and married the daughter of a Chinese general in 1947 but returned to the United States to become the director of the Guggenheim Jet Propulsion Laboratory at the California Institute of Technology from 1949 to 1955. Although Qian attempted to return to China in 1950, he was prevented from doing so under U.S. law after the outbreak of the **Korean War**. In 1955, Qian and his family were allowed to return to China in exchange for the return of Americans detained by Beijing. Qian proved to be instrumental in the development of China's missile and space program and was a senior official of the **People's Liberation Army**'s National Defense Science, Technology, and Industry Commission (NDSTIC), the forerunner of the **Commission of Science, Technology, and Industry for National Defense** (COSTIND).

REFERENCES

Wolfgang Bartke, *Who's Who in the People's Republic of China* (Armonk, NY: M. E. Sharpe, 1981); John W. Lewis and Xue Litai, *China Builds the Bomb* (Stanford, CA: Stanford University Press, 1988).

QINGDAO, GERMAN OCCUPATION OF (1898) As the **Boxer** movement began to develop, and antiforeign, anti-imperialist feeling began to strengthen in China, German missionaries in the port city of Qingdao, on the Shandong Peninsula, were the target of an attack by the secret society. In response, Germany used the attack on its missionaries to occupy the city of Qingdao. The German government also claimed the mining and railway rights on the peninsula around Qingdao at that time. The Germans retained concessionary over Qingdao and the peninsula until after World War I. Germany's concessionary rights over Qingdao were granted to Japan after the war in return for Japan's support of the Allied cause against Germany based on a secret 1917 agreement among Japan and England, France, and Italy.

REFERENCES

John W. Garver, *Foreign Relations of the People's Republic of China* (Englewood Cliffs, NJ: Prentice-Hall, 1993); Hou Chi-ming, *Foreign Investment and Economic Development in China, 1840–1937* (Cambridge: Harvard University Press, 1965).

QIU SHOU QI YI (September 1927) **Mao Zedong** led peasants and miners in the Hunan-Jiangxi border area in an uprising, forming the First Division of the First Workers and Peasants Revolutionary Army. The uprisings started on September 8, 1927, and eventually the organization became the **Chinese Workers and Peasants Red Army**. See AUTUMN HARVEST UPRISINGS.

REFERENCE

Mao Zedong, *Selected Works*, vol. 1 (Beijing: Foreign Languages Press, 1975).

QUEMOY. See JINMEN

R

RAID, SOVIET COMPOUND IN BEIJING (April 6, 1927) Chinese police and paramilitary gendarmes under the direction of Nationalist General Zhang Zuolin, the former **warlord** leader of Manchuria, conducted a raid on the offices of the Soviet military attaché in Beijing on April 6, 1927. The raid was part of a wider action taken by **Chiang Kai-shek** and the Nationalist government to purge the Guomindang Party (KMT) of Communist and Soviet Comintern influence. Earlier, on March 20, 1926, in the **Canton Coup**, Chiang had arrested Soviet advisers to the **Whampoa Military Academy**. The raid on the attaché's office resulted in the seizure of hundreds of documents, which established that the Soviet military attaché office was the directing authority for Soviet agents working as military and political advisers and instructors among Chinese revolutionary groups and the KMT. Many of the documents were published in English and Chinese, creating an international storm of inquiry and protest into the actions of the Comintern and Soviet agents. The Beijing government's position on why it conducted the raid was that the Soviet Embassy compound was harboring Li Dazhao and other Chinese Communist revolutionaries that had conspired against the Nationalist state, violating the Sino-Soviet Agreement of 1924 and international law. The Soviets denounced the raid as a breach of international law.

REFERENCES

Conrad Brandt, Benjamin Schwartz, and John K. Fairbank, *Documentary History of Chinese Communism* (Cambridge: Harvard University Press, 1952); Arif Dirlik, *The Origins of the Chinese Communist Party* (London: Oxford University Press, 1989); Michael Y. L. Luk, *The Origins of Chinese Bolshevism, 1920–1928* (London: Oxford University Press, 1990); C. Martin Wilbur and Julie Lien-ying How, *Documents on Communism, Nationalism and Soviet Advisers in China, 1918–1927: Papers Seized in the Peking Raid* (New York: Columbia University Press, 1956).

RAILROADS AND ROAD NETWORK The industrialization of war was a nineteenth-century phenomenon. Weapons manufacture was transformed into an industry that manifested itself in China with places like the **Fuzhou** and **Jiangnan Arsenals**. The industrialization of sea and river transport was manifest in steamships, like those in the **Beiyang Fleet**. But it was significant changes in the use of fossil fuels that permitted a more rapid, massive movement of armies, equipment, and food stocks that began to transform warfare in the world. Perhaps the largest scale manifestation of this was in the U.S. Civil War, which showed how modern militaries that depend on rail transport can support broad maneuver warfare with the necessary logistic stocks. Like the steamship and the manufacture of weapons in arsenals, the railroad also spread to China in the nineteenth century. The first railway in China was built in 1876 between Shanghai and Wusong, in the Jiangsu Province area. However, because of its perceived disruption of the landscape, it proved so controversial that it was torn up and moved to Taiwan in 1877. The Qing Dynasty withheld permission for building more railways after this until, in 1886, a rail line was constructed along the rim of the Taihang Mountains, in the Hunan-Hebei area, to support coal mines run by the Kailan Mining Corporation. This was the only effort at rail construction for the remainder of the century.

The shock of defeat after the **Sino–Japanese War** (1894–1895) both weakened the Qing Dynasty and led to stronger competition among the Western powers for concession areas. This trend was exacerbated by the **Boxer Rebellion**. In 1903, a Beijing-Shenyang (Mukden) line was built running a distance of about 840 kilometers. This was followed by a Beijing-Hankow (Wuhan) line (1208 kilometers) in 1905, a Tianjin line in 1912, and a Beijing-Kalgan-Datong-Baotou line built between 1909 and 1923. A great deal of the construction was done by European countries, with China retaining ownership and administrative control of the railroads. France, between 1901 and 1910, built a rail line of narrow gauge between Haiphong (Vietnam) and Yunnan Province. The northeast was developed mainly by Russia, but after Japan's defeat of Russia in 1905, much of the rail system came under Japanese control. When Japan took over parts of Manchuria in the **Nine One Eight Incident**, Tokyo further developed the rail lines in China. By 1949, only about 22,000 kilometers of track were laid, but only half were operable after the **Civil War**.

Rail lines proved to be more of a burden than a blessing for the Japanese during **World War II** in China. They were absolutely necessary to supply the forces Japan had deployed in China, but so much time, effort, and manpower had to be devoted to the protection of the lines, which were open to interdiction by guerrilla forces, that both Nationalist and Communist forces were able to pin the Japanese down along the lines and defeat them in detail through maneuver. During the Civil War, the Nationalist forces suffered the same fate at the hands of the Communists. The **Manchurian Campaign** shows how Nationalist forces made themselves targets for the Communists by protecting the rail lines and avoiding maneuver warfare. After the Communists took over, with the establishment of the People's Republic of China, great emphasis was placed on the railroads. A significant portion of the **People's Liberation Army** (PLA) was assigned to railway construction, and the entire nationwide rail network today has a parallel system of management still controlled by the PLA. China has 33 locomotive and rolling stock factories, all under the control of the Ministry of Railways, and 54,000 kilometers of track, all of which can be commandeered by the PLA. As an example, in 1979, when China began to build up the troop concentrations for the **Self-Defensive Counterattack against Vietnam**, the PLA depended very heavily on night movement of forces by rail to conduct a rapid, secret troop deployment.

The road network in China, which only now is developing well, is modeled on the German Autobahn and U.S. Interstate Highway system. Although China has over 1 million kilometers of total road network, only 170,000 kilometers are paved. This is also changing, and today an excellent, multi-lane, paved highway runs the length of China from Beijing to Guangzhou, a road system runs from Beijing to Shanxi, and a developing system can be found in the southwest. As in the rail system, there are provisions for the PLA to control road use and road traffic, a means to ensure that the transportation network serves national defense.

REFERENCES

William H. McNeill, *The Pursuit of Power* (Chicago: University of Chicago Press, 1980); T. R. Treagar, *China: A Geographical Survey* (New York: John Wiley and Sons, 1980); T. R. Treager, *A Geography of China* (Chicago: Aldine Publishing Co., 1965).

RED ARMY COLLEGE (Hongjun Daxue) The Red Army College, a forerunner to the Chinese **People's Liberation Army** (PLA) National Defense University and cornerstone of the PLA's professional education system, was established in Ruijin, Jiangxi Province, in August 1933, to train military cadres. The school ran separate courses for company command-level officers, for division commanders, and for training staff officers. During the **Jiangxi Soviet** period a Russian adviser, Li De (Otto Braun), taught at the college. Military subjects offered at the school included tactics, operations, military history, weaponry, staff procedures, and the construction of military fortifications. The college was reestablished at Baoan, in northern Shaanxi Province, after the **Long March** (1934–1935). In October 1936, it combined with **Zhang Guotao**'s **Fourth Front Army**-subordinated Red Army College. Once the PLA moved to its base in **Yan'an**, the college was renamed the **Anti-Japanese Military and Political College** (Kangda).

REFERENCE

Warren Kuo, *A Comprehensive Glossary of Chinese Communist Terminology* (Taipei: Institute of International Relations, 1978).

RED BASES. See REVOLUTIONARY BASE AREAS

RED GUARDS As the **Cultural Revolution** broke out in 1966, Red Guard organizations of young people, mostly students, formed throughout China. They were idealistic, anti-Establishment, young men and women, often in their teens and 20s, who responded to calls by **Mao Zedong** to criticize Chinese Communist Party (CCP) leaders. The first such organization to form was the Qinghua University Affiliated Middle School Red Guard Unit, which was organized in May 1966. This group of young students between the ages of 12 and 17 sent a poster to Mao in July 1966, drawing his attention. As Red Guard organizations formed, encouraged by radicals in the Communist Party and the Communist Youth League, Mao Zedong took advantage of the movement to strengthen his own position. On August 18, 1966, Mao reviewed a mass rally of Red Guards in Tiananmen Square, which formally recognized the establishment of Red Guard organizations and gave them a paramilitary flavor. Mao was accompanied by **Lin Biao**, and both appeared in military uniform. By the end of 1966, over 13 million Red Guards had converged on Beijing from all over China. They then fanned out around China and conducted sometimes violent campaigns to root out establishment power holders accused of ''following the capitalist road.'' Once the Red Guards seized an institute, they often killed or maimed its leaders, whom they condemned as ''bourgeois reactionaries.'' Red Guards units also fought each other in arguments over ideological purity, sometimes in armed battles. After 1967, the situation in China was so chaotic that the PLA had to step in to restore order in some cities. The Red Guards were then ''sent down'' to the countryside to ''labor with the workers.'' They form part of a ''lost generation'' who were either denied a formal education or received a late education. The Cultural Revolution officially ended with the arrest of the **Gang of Four** after Mao Zedong's death and the rehabilitation of **Deng Xiaoping** in 1976.

REFERENCES

Parris Chang, *Radicals and Radical Ideology in China's Cultural Revolution* (New York: Columbia University Press, 1973); Lee Hong-yung, *The Politics of the Chinese Cultural Revolution: A Case Study* (Berkeley: University of California Press, 1978); Laszlo Ladany, *The Communist Party of China and Marxism, 1921–1985* (Stanford, CA: Hoover Institution Press, 1988).

RED VERSUS EXPERT These two terms, though not diametrically opposed to each other, constitute the historical tension in emphasis on the requirements for leaders in the **People's Liberation Army** (PLA). ''Expert'' denotes a competent, tactically proficient, well-trained army of disciplined, professional, small-unit commanders. ''Red'' connotes the peasant-based, revolutionary guerrilla tradition of local forces steeped in Communist political ideology, responsive to the theories of **Mao Zedong**, which emphasized small-unit, hit-and-run tactics and the use of terrain to the advantage of indigenous forces that traded geographical space for time. The role of the political commissar (PC) is especially critical in the "Red" Military as the transmitter of political rectitude. The PC holds a parallel and coequal position in major military units. Although at one time PCs could overrule commanders, the role of the commander, especially in combat, is better defined today.

REFERENCE

Ellis Joffe, *Party and Army: Professional and Political Control in the Chinese Officer Corps, 1949–1964* (Cambridge: Harvard University Press, 1965).

REPUBLICAN REVOLUTION This is the period marked by the fall of the Qing dynasty. It actually began with the unrest of the Railway Protection Movement in late 1910 and early 1911. The Republican Revolution of 1911, however, is officially marked by the date of the Wuchang Uprising: October 10, 1911 ("Double Ten Day"), which is celebrated by the Nationalist government on Taiwan as national day. See WUCHANG UPRISING.

REFERENCES

Jonathan D. Spence, *The Search for Modern China* (New York: W. W. Norton, 1990); Mary C. Wright, ed., *China in Revolution: The First Phase* (New Haven, CT: Yale University Press, 1968).

REVOLUTIONARY BASE AREAS After the failed **Autumn Harvest Uprisings** and the successes of Guomindang agents against the Communists in Shanghai and Guangzhou, Communist forces withdrew into mountainous regions in South China. In the area of the **Jiangxi Soviet**, generally headquartered in the Jinggang Mountains (**Jinggangshan**) around Ruijin, Jiangxi Province, Mao Zedong had the strongest influence. The borders of Jiangxi, Fujian, and Zhejiang Provinces meet at this point. **Gao Gang**, in the Baoan area of Shaanxi, not far from **Yan'an**, organized another Soviet Revolutionary Base Area. The **Eyuwan Soviet**, in the Anhui, Henan, Hubei border area, northeast of Wuhan, was also a major revolutionary base area. The Eyuwan Soviet survived repeated **Encirclement Campaigns** by the Nationalists until, in 1932, Zhang independently decided to move west to Sichuan. In addition, there were three other Revolutionary Base Areas northwest of the Jinggang Mountains where Communist forces also organized regular and **militia** military units and developed part-subordinated government systems. The Soviet revolutionary base areas operated from 1927 through 1934, when they were abandoned under Nationalist pressure as the **Long March** began.

REFERENCES

JGRMJFJZS, vol 1.I; Jonathan D. Spence, *The Search for Modern China* (New York: W. W. Norton, 1990).

REVOLUTIONARY MILITARY COMMISSION, PEOPLE'S The People Revolutionary Military Commission was established by the Chinese Communist Party Central Committee on September 27, 1949, as the supreme military command authority of the state of China. Its first chairman was **Mao Zedong**. The vice chairmen were **Zhu De**, **Liu Shaoqi**, **Zhou Enlai**, **Peng Dehuai**, and Cheng Qian. The commission was abolished by the Communist Party in 1954, with Mao Zedong still as the chairman.

REFERENCE

Warren Kuo, ed., *A Comprehensive Glossary of Chinese Communist Terminology* (Taipei: Institute of International Relations, 1978).

ROAD NETWORK. See RAILROADS AND ROAD NETWORK

RUIJIN CONFERENCE Following a preliminary meeting in September 1931, the First Party Conference of the Central Soviet Area met at Ruijin, Jiangxi Province, in November 1931. The Ruijin Conference removed **Mao Zedong** from any position of direct influence over the army, organizing a Central Soviet Revolutionary Military Committee under the chairmanship of **Xiang Ying**. **Zhou Enlai** was appointed head of the Chinese Communist Party (CCP) Central Committee Military Department, and **Zhu De** was made chairman of the military counsel. Mao was named premier of the Executive Council of People's Commissars. From this point, for four years until the **Zunyi Conference** in 1935, the **People's Liberation Army** (PLA) was dominated by Soviet-influenced military professionals. The Nationalists defeats of Communist forces during the **Encirclement Campaigns** brought a reversal of the Ruijin decisions and forced the PLA to embark on the **Long March**. Mao finally evened the score during the **Cultural Revolution** against many of those who had opposed him.

REFERENCES

Laszlo Ladany, *The Communist Party of China and Marxism, 1921–1985* (Stanford, CA: Hoover Institution Press, 1988); William W. Whitson with Chen-Hsia Huang, *The Chinese High Command* (New York: Praeger, 1973).

S

SACO. See SINO-AMERICAN COOPERATIVE ORGANIZATION

SECOND ARTILLERY CORPS (The Chinese Strategic Rocket Forces) The Second Artillery Corps comprises the strategic rocket forces of the **People's Liberation Army** (PLA) and numbers about 90,000 personnel. It functions under the training direction of the **General Staff Department**, but takes its mission guidance and direction from the **Central Military Commission of the Chinese Communist Party**. It is called the Second Artillery Corps (*di er pao*) by most Chinese officers and was given that name by **Zhou Enlai**. In most formal Chinese language references, however, it is listed as the *Di Di Zhanlue Daodan Budui* (ground-to-ground strategic missile troops). China's nuclear program had its initial success in 1964, and Beijing, taking the lead from the establishment of the strategic missile forces in the United States and the Soviet Union in 1960, formally established its own strategic missile forces in 1966. Later, with the establishment of a submarine launched ballistic missile (SLBM) program (the first Chinese SLBM test was in 1982), release authority for all strategic missiles and nuclear weapons came under the CMC. The Second Artillery Corps retains its own identity as primarily a surface-to-surface missile force, but it has operational control for all land, air, and sea-based nuclear missiles. In wartime, release authority for the use of the weapons, whether tactical nuclear weapons or strategic weapons, is under the control of the CMC.

In function, according to Li Shuiqing of the Academy of Military Science, the Second Artillery Corps functions most like the rocket forces of France. Some assets are dedicated to intercontinental deterrence, and other missile systems are dedicated to shorter and intermediate-range targets on China's periphery. The total Chinese nuclear warhead inventory is estimated at about 330–350, with about 200 of the warheads at the disposal of the Second Artillery. The firing units of the Second Artillery are organized into missile divisions, each of which has three launch regiments under its control according to *Jane's Strategic Weapons Sys-*

tems. Early ballistic missile systems in China were liquid-fueled, requiring a great deal of preparation for firing and often fixed firing sites. In the late 1980s, however, China developed a road-mobile, solid-fuel system, the *Dongfeng* (East Wind) 21, or CSS-5 as it is referred to in its NATO designation. The *Dongfeng* (DF)-21 has a range of about 2,000 kilometers. In about the year 2000, however, a road-mobile system with a range of as much as 8,000 kilometers, the DF-31, is expected to become operational in China. The DF-31 may also be capable of carrying multiple, independent reentry vehicle (MIRV) warheads. A more formidable strategic system under development is the DF-41, which is expected to have a range of about 12,000 kilometers and be mobile. The shorter range systems of the Second Artillery Corps, like the DF-15 (M-9) missile that was fired in the **Taiwan Strait Crisis of 1995/1996**, has a 600-kilometer range and can carry a 90-kiloton nuclear warhead. The DF-11 (M-11) missile, which China supplied to Pakistan, was based on the Soviet "Scud-B." It has a range of about 280 kilometers and can carry a 90-kiloton nuclear warhead. China has also supplied a missile system to Saudi Arabia, the DF-3 (called the CSS-2 by NATO). The CSS-2 is an older, 1971-vintage, liquid-fueled missile with a range of about 2,800 kilometers and is capable of carrying a 1- to 3-megaton nuclear warhead.

REFERENCES

Defense Intelligence Agency, *Handbook of the Chinese People's Liberation Army* (Washington, DC: DIA, 1984); *Jane's Strategic Weapons Systems*, Issue 28, September 1998; John Wilson Lewis and Hua Di, "China's Ballistic Missile Programs: Technologies, Strategies and Goals," *International Security*, Vol. 17, No. 2 (Fall 1992), pp. 5–40; ZGDBKQS.

SECOND FIELD ARMY The Second Field Army was formed from the forces of the **Fourth Front Army** and one corps of the Third Army of the **First Front Army**. Many of its leaders, such as **Deng Xiaoping, Liu Bocheng, Xu Xiangqian**, and **Peng Dehuai**, emerged as the dominant figures in China. The Red 11th Corps, comprising the Red 31st, 32d, and 33d Divisions, is the antecedent of the Fourth Front Army and, in **People's Liberation Army** (PLA) military lineage, the Second Field Army. The same Communist rebels who led the **Huangan-Macheng Uprising** during the **Autumn Harvest Uprisings** in Hunan and Hubei Provinces and later moved into the **Dabieshan** (Dabie Mountains) to form the **Eyuwan Soviet** were the nucleus of the Second Field Army.

A portion of the Red Third Army, which later developed into the First Front Army, also moved into the area developed by the Communist leaders who organized the Eyuwan Soviet. This was led by Peng Dehuai. Deng Xiaoping, who operated a Communist guerrilla force in Guangxi Province designated the Red 7th Corps, emerged as a seminal figure in Chinese history and is one of the leading cadre associated with the Second Field Army. Deng was instrumental in organizing peasant support for the forces that conducted the **Hundred Regiments Campaign**, which is also part of the Second Field Army lineage.

During the **Civil War**, in the campaign to control Central China, the PLA elements that formed the Second Field Army operated under Liu Bocheng and Deng Xiaoping. In fact, these forces were instrumental in ensuring the Communist victory over the Nationalist Army in the **Huai-Hai Campaign**.

When the PLA formed the **Field Army System** in 1949, Liu Bocheng was the commander of the Second Field Army, and Deng Xiaoping was its political commissar. The Second Field Army had subordinate armies in it (the Third, Fourth, and 18th Armies), each of which had three subordinate corps, and each corps had three divisions. This basic triangular organizational structure for combat forces is maintained by the PLA today, but the PLA corps are now called "group armies." The PLA has no "armies" or "field armies" in its present organization in the sense that these terms are used in U.S. doctrine.

The leaders of the Second Field Army, which had moved into parts of Sichuan, Yunnan, and Guizhou Provinces after the Civil War, dominated the military and important party and government posts in that area. Four corps of the Second Field Army, the 12th, 15th, 16th, and 60th, participated in the **Korean War**.

REFERENCES

William W. Whitson with Chen-Hsia Huang, *The Chinese High Command* (New York: Praeger, 1973); ZQDBKQS.

SECOND FRONT ARMY The Second Front Army was organized in western Hunan Province and emerged from a rebel guerrilla and bandit group headed by **He Long**, who later became a Marshal of the **People's Liberation Army**. Its history includes taking part in the **Nanchang Uprising** on August 1, 1927, the anniversary of the establishment of the People's Liberation Army, as the 20th Corps of the National Revolutionary Army, part of the Nationalist Army. At that time, He Long and **Ye Ting** led the 20th Corps together. From 1928 to 1930, during the **Autumn Harvest Uprisings**, the 20th Corps, which had left the Nationalist Army, fought as the Red Second Corps, led by He Long. In 1934, the Red Second Corps joined forces with the Red Sixth Corps, led by **Xiao Ke**, and **Xiao Jingguang**'s Red 29th Corps to form the Second Front Army.

Soon after its formation, the Second Front Army came under heavy military pressure by the Nationalist Army, which was then conducting its "bandit Suppression" or **Encirclement Campaigns**. The start of the **Long March** by the **First Front Army** in October 1934 drew the Nationalist units away from He Long's base areas, allowing him to expand his forces. In fall 1935, however, the Nationalists again began to move forces into position to encircle He Long's Second Front Army, which had grown in strength to over 25,000 soldiers under arms. In November 1935, the Second Front Army began its own 3,000-mile "Long March" to the west, finally linking up with the **Fourth Front Army**, led by **Zhang Guotao**, in June 1936 in the Gansu–Sichuan provincial border area. The strength of the Second Front Army was reduced by about half as a consequence of the rigors of the march. In July 1936, He Long set out to march north again, seeking to cross the grasslands of Gansu Province and to link up with the First Front Army. By October 1936, when the Second and First Front Armies finally joined forces and linked up in Huining, Gansu Province, He Long's force was reduced to about 10,000 soldiers from its original strength of 25,000. The Second Front Army finally joined the **Mao Zedong**-led **Yan'an** base area and the First Front Army in Shaanxi Province in November 1936, at about the time of the **Xi'an Incident**.

When the **Eighth Route Army** formed in 1937, He Long's Second Front Army was converted almost intact into the 120th Division of the Eighth Route Army. One of its roles was to provide security and screening forces to protect the Yan'an base area. He Long's forces fought in the **Hundred Regiments Campaign** and secured the approaches to Yan'an in northern Shanxi and Suiyuan Provinces. Other elements of He Long's 120th Division, in the Shaanxi–Gansu–Ningxia provincial border area, were isolated in poor, agricultural areas and were pressed heavily by Japanese forces in the **Three-All Campaign**. The 120th Division's 359th Brigade, led by **Wang Zhen**, achieved essential agricultural self-sufficiency in the **Nanniwan Experiment**, setting the model for future People's Liberation Army logistics systems.

After the defeat of Japan in 1945, the forces of the former Second Front Army were converted into the Northwest Field Army. In 1949, when the **Field Army System** developed, they were converted into the **First Field Army**.

REFERENCES

William W. Whitson with Chen-Hsia Huang, *The Chinese High Command* (New York: Praeger, 1973); Xiao Zhaoran, *Zhonggong Dangshi Ming Cidian* [A Concise Dictionary of the Chinese Communist Party's History] (Beijing: Jiefangjun Chubanshe, 1986); ZGDBKQS, Vol. 1; ZGRMJFJZS, Vol. 2.

SECOND REVOLUTIONARY WAR. See WORLD WAR II

SELF-STRENGTHENING MOVEMENT This movement characterizes an effort that lasted over 30 years during the late nineteenth century, designed to strengthen China by mastering a number of foreign skills and subjects, primarily military in nature. The ''self-strengthening movement'' owes its name to a series of essays written by the scholar Feng Guifen to to one of the most influential grand councillors of the Qing Dynasty, **Zeng Guofan**.

The essence of the movement was to make China into a modern state with a military force capable of responding to and resisting foreign domination. Between 1861 and 1865, a series of diplomatic and military projects was launched around China. In Beijing a foreign office was established, the Zongli Yamen (Tsungli Yamen), to deal with foreign businesses, diplomats, and matters of state. The Zongli Yamen also established a language school. In military affairs, an effort began to import machine tools and to establish arsenals and shipyards. The **Jiangnan Arsenal** was established in 1865 to make cannon and small arms, and the Nanjing arsenal was established in 1867. Industrial projects were developed, including the Tianjin Machine Factory in 1870 and the Kaiping Coal mines in 1877. The self-strengthening movement also included an effort to make China a credible maritime power. A dockyard was established in Fuzhou in 1866; the Chinese Merchants Steam Navigation Company was founded in 1872; and the Beijing Fleet was organized in 1888. The effort was unsuccessful, for it failed to prevent the defeat of China by Japan in 1895. It is also significant because the establishment of regional arsenals was a factor in the rise of **warlords**.

REFERENCES

Albert Feuerwerker, *China's Early Industrialization* (Cambridge: Harvard University Press, 1958); John L. Rawlinson, *China's Struggle for Naval Development, 1839–1895* (Cambridge: Harvard University Press, 1967).

SHIMONOSEKI, TREATY OF (1895) **Li Hongzhang** was sent to Japan in 1895 to negotiate a treaty ending the **Sino-Japanese War**. In the Treaty of Shimonoseki, China ceded the island of Taiwan, the Pescadores, and the Liaodong Peninsula (where Lushun, or Port Arthur, is located) to Japan. China also agreed to recognize the independence of Korea, ending its status as a Chinese tributary state, and opened four treaty ports to Japan. China also agreed to pay indemnities to Japan, but Russia, Germany, and France later pressured Japan to return Port Arthur for an indemnity of 30 million taels of silver. China had already paid Japan an indemnity of 200 million taels of silver to end the war.

REFERENCES

John K. Fairbank, ed., *The Chinese World Order: Traditional Chinese Foreign Relations* (Cambridge: Harvard University Press, 1968); Rhoads Murphy, *The Treaty Ports and China's Modernization* (Ann Arbor: University of Michigan Press, 1970).

SINO-AMERICAN COOPERATIVE ORGANIZATION (SACO) The Sino-American Cooperative Organization, or SACO (pronounced "socko," as in a hard punch), was a combined organization run by the U.S. Office of Strategic Services (OSS) and the Nationalist Chinese government's Bureau of Investigation and Statistics. SACO trained thousands of Chinese to act as guerrillas in the fight against Japan in **World War II**. The organization also proved to be a very effective intelligence collection arm for directing U.S. naval and air forces against the Japanese. The director of SACO was the head of Nationalist intelligence, espionage, and counterespionage, Lieutenant General **Dai Li (Tai Li)**. The deputy director of SACO was U.S. Navy commander (later captain and rear admiral) **Milton Edward "Mary" Miles**. Since SACO was a secret organization, officially Miles' cover for being in China was as the U.S. naval observer, Chongqing. Miles was to act concurrently as chief of U.S. Strategic Services, Far East. Dai Li was the chief of the Bureau of Investigation and Statistics of the National Military Council of the Chinese government. This national Nationalist organization was a rough equivalent of the U.S. Federal Bureau of Investigation (FBI).

The genesis of SACO seems to have been a series of informal discussions between Miles, his supervisor on the staff of the Navy Department in Washington, Captain (later Admiral) Willis J. ''Ching'' Lee, and a representative of Dai Li who operated as one of the Chinese military attachés in Washington.

In addition to Miles, the original group, trained by the navy and the Office of Strategic Services (the OSS, wartime forerunner of the Central Intelligence Agency), consisted of seven officers: Daniel J. Heagy, Theodore J. Wildman, Robert J. Dormer, Clarence P. Taylor, Lashley H. Mann, K. A. Mann, and Terrence J. O'Neil, all of whom deployed to China in May 1942, after Miles' arrival there. Captain Jeff Metzger was also a SACO member and often functioned as Miles' representative in the United States. In October 1942, Navy Lieutenants Ray Kotrla and Edward S. Gilfiller were added to the organization. One month later two marines, Gunnery Sergeant Way Holand and Major John H. Masters, and navy aviation machinist Willie D. Flournoy arrived in Chongqing and were added to the U.S. Naval Observer Group. On January 1, 1943, six photointerpreters arrived to assist in bomb damage assessment and targeting: Lieutenant Colonel Gregory A. Williams, Lieutenant Commander David D. Dwight, Lieutenant Colonel Bankston T. Holcomb, Jr., marine gunner Boyd Jackson, and Lieutenants John R. Horton and Merrill R. Stewart.

None of these early people officially belonged to SACO at the time of their arrival. The SACO organization was officially established by a formal bilateral memorandum signed in Washington, D.C. on April 1, 1943, and countersigned in China on July 4, 1943, by Dai Li. Even without formal establishment, however, SACO began to function and to establish active guerrilla training cells and units.

SACO Unit One was established in Hengyang, Henan Province, on February 1, 1943, by seven of the original members of the U.S. Naval Observer Group. It started out by training members of Dai Li's ''Loyal Patriotic Army,'' which was a local militia-like group, in the use of small arms, guerrilla tactics, demolitions, and intelligence collection. Unit One later moved to Huizhou. It eventually trained about 7,000 Chinese guerrillas.

Unit Two was established in Nanyo, Hunan Province, in June 1943, with a guerrilla-training mission. Not long after that, SACO Unit Three was established in the area of Loyang, Henan Province, north of the city of Lin Ru, also with a primary mission of training guerrillas. SACO Unit Four departed Chongqing on November 18, 1943, with 12 charcoal-burning trucks, 80 Chinese, and 12 Americans to establish a weather-monitoring and -reporting station in the Gobi Desert. It arrived at Shan Ba, Suiyian (now Huhehaode, capital of Inner Mongolia) on January 18, 1944. Unit Four personnel also worked with forces under Nationalist general Fu Zuoyi, training guerrillas in addition to carrying out the weather mission.

SACO Unit Five was established in Nanning, Guizhou Province, in August 1944. Miles used some of the members of Unit Two to set up the Nanning organization. Unit Five trained over 1,000 Chinese to be stay-behind assassins in the Nanning area after the Nationalist troops and the American air forces had to abandon that base in the face of **Operation Ichigo**.

SACO Unit Six was established as a coast-watcher unit in Zhangzhou and Hua'an, Fujian Province. Miles, accompanied at times by Dai Li, had made his own dangerous, personal reconnaissance in this area in 1942, when the SACO organization was first conceived. Unit Six conducted beach reconnaissance in support of future potential amphibious operations and gathered intelligence in the area. Farther up the coast, in Dengfeng, Zhejiang Province, SACO established Unit Seven on September 19, 1944, to train guerrillas and collect intelligence. Unit Eight was also established with a beach reconnaissance and coastal surveillance mission on August 21, 1944, at Qingdian, Zhejiang Province. It later moved to Yuhu when threatened by Japanese forces.

SACO Unit Nine was set up in Chongqing as the school of intelligence and counterespionage. It was established with assistance from the U.S. Treasury Department, the Office of Naval Intelligence, the FBI, and several U.S. police departments, including the New York City Police. It also taught countersabotage and clandestine communications. Unit 10, a guerrila training unit, was established in Guiyang, Guizhou Province.

In addition to the numbered guerrilla training and intelligence collection units, SACO had its own medical unit, accounting unit, and aerology section, which took balloon meteorological readings at 100 locations around China in support of General **Claire Chennault**'s 14th Air Force. SACO also trained a large number of coast-watchers and provided them with communications equipment to provide naval and shipping information to submarines and naval air groups.

SACO's integral intelligence section controlled the activities of Units One, Six, Seven, and Eight and also controlled the aerology (weather) unit. It totaled 435 U.S. Navy and Marine Corps officers and enlisted personnel and, among its four subordinate numbered units, trained 16,000 guerrillas.

Along the Yangtze River, on the Gan River at Changsha, Hunan Province, and on the Minjiang and Wuning Rivers, SACO established and operated its Riverine Unit. This organization was tasked to block Japanese supply lines into Hankou (the city is now Wuhan). At the end of the war Unit 13 moved to Hankou and was redesignated as Naval Group, Central China. Miles also established Unit 14, which functioned as the combat intelligence provider for Chennault's 14th Air Force.

SACO is notable for several reasons. First, its guerrillas inflicted a number of serious causalities on the Japanese and effectively interdicted their supply lines. Second, it provided valuable intelligence to U.S. air, naval, and ground forces operating against Japan. Finally, when the Office of Strategic Services (OSS) evolved into the Central Intelligence Agency (CIA), it provided the basis for the long-term, Cold War cooperation between the CIA and the intelligence services of the Republic of China and later Taiwan, which was important in the 1950s–1970s. On the negative side of the operation, because of SACO's close identification with Dai Li and the Nationalist Intelligence, OSS officers were never trusted by the Chinese Communist Party or by **People's Liberation Army** forces in **Yan'an**. Thus, the **Dixie Mission** was probably doomed from the start, since **Mao Zedong**'s people viewed all OSS personnel as agents of Dai Li. This probably led to the Communists' resistance to suggestions by the Dixie Mission and General **Wedemeyer** that the United States begin a program to train Communist troops as anti-Japanese guerrillas.

REFERENCES

Milton E. Miles, *A Different Kind of War: The Little-Known Story of the Combined Guerrilla Forces Created in China by the U.S. Navy and the Chinese during World War II* (New York: Doubleday, 1967); Milton E. Miles, "U.S. Naval Group China," *United States Naval Institute Proceedings* 72, July 7, 1946; Ray Olin Stratton, *SACO—The Rice Paddy Navy* (Pleasantville, NY: C. S. Palmer, 1950); U.S. Navy, *Official History of the United States Naval Observer Group, Chongqing, China*, from Milton E. Miles Papers, Hoover Institution Archives, Hoover Institution on War, Revolution, and Peace, Stanford, CA.

SINO-FRENCH WAR. See FRANCO-CHINESE WAR

SINO-INDIAN BORDER WAR (1962) After the **Longju Incident** along the **McMahon Line** in August 1959, China continued to observe the McMahon Line as the de facto Sino–Indian border. Chinese forces pulled back from disputed territory in that month to avoid a clash. In the western sector of the border, where China and India had clashed in October 1959, Indian forces, under New Delhi's orders, began a "forward policy" designed to push the disputed boundaries out to geographical features favorable to, and claimed by, India.

In the east, on the McMahon Line, Indian forces had avoided patrolling within two miles of the border after 1959. But in December 1961, Indian forces were ordered to move forward and to begin patrolling again along the disputed line. During the first six months of 1962, the Indian army was to establish 24 new border outposts. In June 1962, a platoon of Indian Assam Rifles moved forward of the McMahon Line, as marked about four miles north of the line to the Thagla Ridge, which they treated as the border. Indian troops established a position there on June 4, even though India's own maps showed the ridge to be in Chinese territory. On September 8, a Chinese force advanced on the Thagla post, attempting to pressure the Indians to withdraw. Beijing also issued a diplomatic protest on September 16, complaining about the presence of the Indian troops.

The Indian government took the position that, when Henry McMahon drew the line in 1944, he intended that it run along the line of the highest ridges. They argued that the Thagla Ridge was the dominant terrain feature and, therefore, should be the border. The Chinese reaction, according to India, was the implementation of a central policy by Beijing to advance the border into Indian territory. India began a buildup of forces that was logistically insupportable and military dangerous, while Chinese forces increased their strength and weaponry along the border using a road system that would support fire and even seven-ton vehicles. Through September, there were skirmishes around the Thagla area, in the course of which forces of both sides took casualties. By October, faced with bad weather and unfavorable terrain, the Indians attempted a buildup of forces.

The Indian army was ordered to carry out Operation Leghorn on October 10, 1962, designed to push back the Chinese. Chinese intelligence was aware of the operation, and Beijing, on October 6, issued a warning to India in the form of a diplomatic note. Meanwhile, the **People's Liberation Army** (PLA) concentrated superior forces and artillery in the area. On the night of October 19, Chinese troops assembled in assault positions. On the morning of the twenty-ninth, they attacked the Thagla Ridge, wiping out the Indian Seventh Brigade and taking prisoner Indian brigadier Dalvi.

In the western sector of the border, along the Galwan River, the PLA launched a simultaneous attack against Indian forces in the Chip Chap River valley. The Chinese government declared these to be self-defensive counterattacks to clear Indian troops out of Chinese territory. American supplies, meanwhile, began to flow into India, with about 20 tons of military equipment a day arriving in India. England also sent in military equipment.

An Indian counterattack along the McMahon Line was beaten back by Chinese forces in November, while Indian forces also built up in the west but were trounced by Chinese forces. By November 21, 1962, China announced a unilateral cease-fire. The Chinese simultaneously announced a December 1, 1962, withdrawal to positions 20 kilometers behind the "line of actual control" that existed between China and India on November 7, 1959, reviving a formula that was used to defuse the crisis in that year. Indian figures indicate that 1,383 troops were killed, 1,696 were missing in the operation, and 3,968 were captured by Chinese forces. The Chinese, having incurred far fewer losses, were left in control of the Aksai China Plateau.

REFERENCES

Deng Liqun et al., eds., *Zhongguo Renmin Jiefangjun* (Beijing: Dangdai Chubanshe, 1994); John W. Garver, *Foreign Relations of the People's Republic of China* (Englewood Cliffs, NJ: Prentice-Hall, 1993); Neville Maxwell, *India's China War* (Bombay: JAICO Press, 1971); Gerald Segal, *Defending China* (London: Oxford University Press, 1985).

SINO–JAPANESE WAR (1894–1895) The Sino–Japanese War was fought over Korea and China's efforts to keep Korea as a tributary and buffer state. In the 1880s, two Chinese leaders, **Yuan Shih-kai** and **Li Hongzhang**, managed Korean affairs, with Li in Beijing and Yuan in Korea. These officials urged Korea to develop stronger links with Western powers to counter Japan's attempts to annex the peninsula. In 1894, a pro-Japanese Korean politician was assassinated in Shanghai. At the same time, a domestic rebellion against the Korean king broke out on the peninsula.

Japan professed to be interested in a limited campaign to suppress the rebellion on the Korean Peninsula in order to strengthen Korea's independence. However, as early as July 17, 1894, Japanese military planners met secretly to complete a campaign plan for the conquest of the Korean Peninsula and parts of China. Japan and China were obligated by the 1885 **Treaty of Tianjin** to notify each other of military moves or exercises on the Korean Peninsula, and each party was permitted to send an equivalent force to maintain a balance of influence in Korea.

By 1894, Japan's army had seven infantry divisions, the Imperial Guards division, stationed in Tokyo, the First Division, also in Tokyo, and the Second through Sixth Divisions, which were all stationed on the island of Honshu. The Japanese military staff planners continued to meet between July 17 and August 5, sketching out a plan. The attack plan produced by August 5 envisioned a two-phase war against Korea and China. In the maritime phase, the Japanese navy was to seize control of the Yellow Sea and the Gulf of Zhili (the Bohai Gulf), guaranteeing safe passage for Japanese forces to mainland China. Then the Japanese Fifth Division was to attack in Korea, to restrain Chinese forces there and to gain a foothold on the peninsula. The Japanese navy and ground forces were then to move to seize the port of Weihaiwei, on the coast of the Shandong Peninsula. At Weihaiwei, several more Japanese divisions were to land, with a mission to move inland to defeat Chinese troops and, if possible, to drive on Beijing. The Japanese also developed an alternative plan, in the event that the Japanese fleet was unable to defeat the Chinese fleet, which was considered a worthy opponent because China had been building it during the years of the **self-strengthening movement**. According to the alternative plan, if the Japanese navy was unable to control the maritime theater and the Yellow Sea, Japan's ground forces were to expel China from the Korean Peninsula and seize control there.

Although the entire campaign plan was not complete until August 5, Japan actually declared war on China on August 1, 1894. A five-hour battle had already taken place between Chinese and Japanese forces at Songhuan, five miles south of Seoul, on July 29, involving 3,000–4,000 troops. Japanese forces claimed to have inflicted 500 casualties on the Chinese, losing only 82 killed and wounded on their own side. The Japanese navy, under Togo Heihachiro, also sank a British ship, the *Kowshing*, which was carrying Chinese troops to Korea. The remainder of the Japanese Fifth Division was moved to Korea, where it landed at Pusan between August 2 and 6, 1894. Attacking north, the division reached Seoul by August 19. Japan's military successes were not based on superior troop strength but on better training than that of the Chinese and on superior artillery strength.

On the maritime front, Japan believed that the Chinese **Beiyang Fleet** was avoiding a decisive engagement. On September 17, 1894, in the Battle of the Yalu River, Japan sank part of a Chinese naval fleet off the Korean coast near the Yellow Sea. The rest of the Chinese ships retreated to Port Arthur (Dalian).

On the Korean Peninsula, the Chinese forces pulled back to the Yalu River. On October 23, 1894, Japan sent a regiment of troops across the Yalu and bridged the river. The Japanese Third Division then crossed the Yalu into China, continuing the attack. Chinese forces retreated to Jiuliancheng. The Japanese attacked that city on October 26, driving the Chinese forces farther back and seizing the city. By November 1, the Japanese Third and Fifth Divisions, constituting the First Army, controlled all of southern Manchuria on the Korea–China border. The Japanese forces established a civilian government occupation office in China.

The Japanese Second Army, comprising the First, Second, and Sixth Divisions, was dispatched to strike Port Arthur and, if possible, to continue the attack to Beijing. Port Arthur was occupied by November 6. Many accounts of the capture of Port Arthur say that Japanese troops were guilty of committing serious atrocities, recounting the massacre of as many as 60,000 Chinese soldiers and civilians. Women and children were allegedly raped and sexually mutilated by Japanese troops. The **Port Arthur Massacre** was the worst Japanese military atrocity in China before the December 1937 **Rape of Nanjing** by Japanese forces. The Second Army then regrouped at Port Arthur, preparing to attack Weihaiwei after the New Year. The Chinese fleet still had not left port at Weihaiwei, because Li Hongzhang feared losing it.

In the international arena, Russia and Britain warned Japan about the offensive. France, however, was sympathetic to the Japanese and did nothing to restrain Tokyo. Despite the ongoing hostilities, Japan and the United States signed a treaty in Washington, D.C., on November 22, 1894, although Congress expressed some reservations about the treaty because of the war.

Up to January 1895, Japanese ground and naval forces had encountered only Chinese regional forces, which they quickly overwhelmed. In late January 1895, the Japanese Second Army landed on the Shandong Peninsula to march to Weihaiwei, with the objective of neutralizing the Chinese fleet and securing the peninsula. The attack on Weihaiwei started in January 30. By February 7, Japanese forces were in control of the city, but access to the port was being denied by shelling from the Chinese Beiyang Fleet. On February 3, the Japanese fleet entered the bay, engaging the Chinese ships. The Chinese naval commander, Admiral Ding, surrendered his fleet to the Japanese on February 12 and committed suicide by poisoning himself. Japan returned some of its ground forces to Port Arthur by March 5, and Japanese forces then occupied the area between Niuzhang and Yingkou. Japan also moved against Taiwan (Formosa) and occupied the Pescadore Islands (Diaoyu Islands) on February 20, 1895.

Li Hongzhang was sent to Japan to negotiate the **Treaty of Shimonoseki**. In the negotiations, Japan asked for cession of the Liaoning Peninsula, cession of Taiwan and the Pescadores, indemnity for Japan's war expenses, Chinese recognition of Korea as an independent country, and the negotiation of a new Sino–Japanese commercial treaty opening the Chinese domestic market to Japan. Li agreed to the terms, and the Treaty of Shimonoseki was signed on April 17,

1895. Japanese troops, at that time, continued to occupy Weihaiwei. Triple intervention by Germany, France, and Russia on April 23, 1895, forced Japan to abandon its claims in the treaty on the Liaoning Peninsula. These countries had significant commercial interests there, which they did not want to lose to Japan. On May 30, 1895, Japan's Cabinet agreed to abandon the claims on the Liaoning Peninsula, accepting instead an additional 30 million taels of silver from China as an indemnity payment.

REFERENCES

Stewart Lone, *Japan's First Modern War: Army and Society in the Conflict with China, 1894–95* (New York: St. Martin's Press, 1994); John L. Rawlinson, *China's Struggle for Naval Development, 1839–1895* (Cambridge: Harvard University Press, 1967); Tai Sung-an, *The Sino-Soviet Territorial Dispute* (Philadelphia: Westminister Press, 1973).

SINO–SOVIET CONFLICT. See ZHENBAO ISLAND CLASH

SINO–VIETNAMESE WAR. See VIETNAM, SELF-DEFENSIVE COUNTERATTACK AGAINST

SIPING, DEFENSIVE BATTLE OF (February–May 1946) By mid-March 1946, after completing the approach march for the **Manchurian Campaign**, moving north up the rail line from Shanhaiguan, where the Great Wall meets the sea, Nationalist forces had occupied Shenyang. They were moving toward Changchun, farther north in Liaoning Province. The Nationalist advance was slowed by a strong, although eventually unsuccessful, defense of the city of Siping by Communist forces.

In January 1946, the Communists and Nationalists had reached a cease-fire agreement, one component of which gave the Communists nominal control of the northeast (Manchuria). By early March 1946, however, 11 Nationalist infantry divisions had gathered in the Shenyang area, from which they began operations to clear areas around the region of Communists. The goal was to establish Nationalist control over cities to the north of Shenyang, especially Siping, which was a major rail junction, Harbin, and Changchun. Communist forces in Manchuria at the time operated under the command of **Lin Biao**, with **Peng Zhen** as political commissar. Having already ''liberated'' the cities of Harbin, Changchun, and Qiqihar, the Communist Party Central Committee decided to concentrate its defenses at Siping, using the northwest–southeast rail line passing through the city as a defensive boundary feature to block Nationalist advances. The initial meeting engagement there was on March 17, 1946, during which the Communist forces stopped the Nationalist troops.

In early April 1946, the Nationalist New First Army, led by General **Sun Liren**, moved forward toward Siping. Between April 8 and 14, 1946, the Nationalist 71st Army, minus the 88th Division, and the New First Army (these armies were really about the strength of a corps in U.S. terminology, about

40,000 troops) began their advance on Siping. The New First Army advanced up the north–south rail line from the south toward the city, while the 71st Army attacked northward on the west flank. However, the 71st Army was stopped well south of the city by Communist counterattacks carried out by the First, Second, and Third Divisions, and units of the Seventh Column of Lin Biao's forces. The New First Army was halted outside Siping by the Communists.

A second Nationalist attack on the city of Siping was initiated from April 16 by the New Sixth Army, reinforced by the 88th Division, 71st Army, and the 195th Division of the 52d Army. During early May, these reinforcing Nationalist forces advanced to the east of Sun Liren's New First Army in an attempt to envelop Siping from the east flank. Lin Biao, leading the Chinese forces, counterattacked with the 359th Brigade, which moved south from the Harbin area, and the Third Column, which moved into the battle from the east but failed to blunt the second Nationalist offensive. Finally, on May 18, 1946, Nationalist forces began a second assault on Siping, supported by air attacks and artillery. As Lin Biao's forces retreated to the west, Nationalist forces occupied Siping on May 19, 1946.

Communist losses in the battle, based on **People's Liberation Army** histories, were about 8,000 troops of some 110,000 defenders. The Nationalists, who used some of the troops from the American-trained **ALPHA Force**, had attacked with about 70,000 troops and lost approximately 10,000 men.

REFERENCES

Trevor N. Dupuy, *The Military History of the Chinese Civil War* (New York: Franklin Watts, 1969); Liu Qi et al., *Liao-Shen Zhanyi Qin Li Ji: Yuan Guomindang Jiangling de Huiyi* [Diaries of the Liaoning-Shenyang Campaign: Original Recollections of the Guomindang Generals] (Beijing: Wenshi Chubanshe, 1996); ZGDBKQS; ZGRMJFJZS, vol. 3.

SIPING, OFFENSIVE CAMPAIGN AGAINST (Spring 1947–March 1948) By spring 1947, the Nationalists had massed eight corps-strength units in northeast China, including the 13th, 52d, 60th, 71st, and 93d Armies, the New First Army, the New Sixth Army, and several independent divisions. (A Chinese army, whether Nationalist or Communist, is equivalent to a U.S. corps: two to five divisions, with a strength of 30,000 to 45,000 troops.) Nationalist General Du Yuming was the major commander, with a main objective of holding on to the important population centers and defending strategic rail links in the heavily industrialized northeast. This defense focused on the cities of Changchun, Siping, Shenyang, and Jilin. Du had about 250,000 troops under his command.

Communist forces in northeast China at the time were much stronger but more dispersed. Their numbers, about 460,000, included part of the civil populace, which was organized into support units. **Lin Biao** was the Communist commander; the political commissar was **Peng Zhen**. Lin's main combat force was about 100,000-strong, divided into four columns. In April and May 1947, Lin Biao initiated a general **Northeast Offensive Campaign** that led to a series of battles centered around Siping, the main rail junction for lines from the north, south, east, and west. Lin's First Column was led by Li Tianyu (who went on to be commander of Guangzhou Military Region, a commander of forces in Korea, and deputy defense minister). The Second Column was led by Liu Zhen, the Sixth Column was headed by Yang Guofu, and the Artillery and Support Column was led by Zhu Rui.

As the campaign kicked off, Lin moved the four main columns, which were his most mobile, south in a broad feint. This induced Du Yuming, the Nationalist commander, to reinforce the city of Shenyang. By May 13, 1947, the First and Second Columns and the First and Second Independent Division of the Communist forces moved south across the Sungari River (see **Sungari River Campaign**) toward the cities of Huaide and Changchun. Huaide was defended by the Nationalist New First Army's 30th Division and the 17th Defense Regiment. After capturing Huaide, the Communist force set up a blocking position along road and rail lines south of Changchun, isolating the city of Siping, which lay farther south, from any reinforcement from the north. Meanwhile, southeast of Siping, by May 14, the Third Column of Lin Biao's forces attacked against the Nationalist 184th Division, 60th Army, which was defending the Shenyang-Jilin rail line. This attack was supported by a division of the Communist Fourth Column and another independent division. By May 24, this column secured the Shenyang-Jilin rail line, and it began a drive west, attacking the Nationalist 71st Army, which defended Siping. To the northeast of Siping, on May 13–14, the Communist Sixth Column and two independent divisions attacked the Nationalist New 38th Division. By May 30, 1947, the New 38th Division had been defeated, and troops of the Communist Sixth Column had marched southwest toward Siping. While the Communist First Column blocked the rail line south of Siping, preventing reinforcement of the Nationalist troops by units in Shenyang, another Communist column, composed of an independent cavalry division and an infantry division, drove on the city of Siping from the northwest.

Lin Biao's forces were assembled and in position to attack Siping by May 22–23, but, instead of attacking, Lin tried to conduct a siege. This tactical error allowed the Nationalist defenders time to organize the city's defenses and to coordinate air attacks on massed Communist forces using aircraft provided to the Nationalist government by the United States during the war against Japan. The 29,000 Nationalist troops defending Siping were also assisted by the city's civilian populace, which numbered about 100,000. The Communist assault on the city began on June 2, 1947, but the advance was slow, primarily because Lin failed to mass his forces and did not use all of the artillery at his disposal. Lin Biao's forces did not penetrate Siping's defenses until June 9. When they finally entered Siping, the Communists faced heavy street fighting, which delayed their advance to the city center. It was actually the Communist Sixth Column's 17th Division, attacking from the south with the First Column, that first penetrated to the heart of Siping, securing the center of the city on June

21. However, in the northeast section of Siping, where the Nationalist 87th and 88th Divisions mounted a strong defense, the Communist forces were held off by continuous Nationalist air strikes.

Nationalist forces in Shenyang, to the south of Siping, began to regroup as soon as elements of the Communist forces left their blocking positions on the rail lines between Siping and Shenyang in late June. By June 24, the Nationalist New Sixth Army and the 52d Army had fought their way north to within 70 kilometers of Siping, supported by the 53d Army. These Nationalist units had the advantage of close air support by P-51 fighter/ground attack aircraft and B-25 bomber aircraft, which had been supplied by the United States. Also on June 24–25, Nationalist forces in the Changchun area succeeded in attacking south, further threatening Lin Biao's troops around Siping. On June 28, 1947, facing heavy Nationalist pressure, Lin Biao withdrew his forces from Siping northward, back across the Sungari River. Communist forces lost about 40,000 troops during this offensive against Siping, for which Lin Biao was severely criticized. Lin's political commissar, Peng Zhen, was recalled and replaced. Both men were later criticized during the **Cultural Revolution** for their actions in Siping.

During summer 1947, Nationalist forces reorganized into four army groups (equivalent to U.S. field armies), while more divisions were brought north to reinforce the defense of the Shenyang–Siping–Changchun rail line. The Nationalist commander, Du Yuming, was also replaced by General Chen Cheng. In what proved to be a disastrous policy mistake, Chen Cheng refused to employ any of the Japanese-trained, former **Manchukuo** Army troops that had joined the Nationalists in the field. This caused many of the former Manchukuo soldiers to defect to the Communist cause. The Communists also took the time they had to rest north of the Sungari River to reorganize and to begin a political campaign of land reform to gain the support of the local populace, a conscription campaign to build new forces, and a campaign of "Red terror" to eliminate Nationalist supporters in the local populace.

In late September 1947, the Communists again initiated another offensive. This time they had nine columns (of roughly corps or army size), comprising 27 divisions, 10 more independent divisions, and two cavalry divisions at their disposal. In a marked difference from earlier campaigns in Manchuria, the Communists also employed a range of tactics, including both conventional maneuver warfare and guerrilla operations supported by a freshly mobilized local populace. The Communist forces attacked all along the Shenyang–Siping–Changchun rail corridor from the north, east, and west, driving Nationalist forces out of the cities of Changchun and Siping for a time. This campaign again turned into a major battle centered around control of Siping and lasted until early November, when Lin Biao again withdrew his forces to safer grounds north of the Sungari River. By cutting land lines of communication, he had forced the Nationalists to resupply Changchun by air and to airlift reinforcements into the city. **Chiang Kai-shek** also flew to Shenyang on October 8 to personally inspect the defenses of that city and to shore up flagging morale in the Nationalist ranks. By the end of this second offensive, Lin Biao succeeded in controlling 15 cities in the area and in breaking Nationalist control of the rail line from Siping to Changchun.

In a third campaign against the Changchun–Siping–Shenyang rail corridor, Lin Biao began a winter offensive in mid-December 1947. Instead of driving directly on the cities of Changchun and Siping, as he had before, Lin Biao moved his forces south of Shenyang and concentrated on capturing the rail line south to Yingkou, on the Liaodong Bay, in the northern Bohai Gulf. This cut the line of communication to Beijing. In doing so, Lin Biao gave Communist forces the freedom to operate in, and control large parts of, Liaoning Province and prevented Nationalist reinforcements from coming north. Lin then moved on Siping between January 31 and March 1948, eventually taking the city with forces of the First, Third, Seventh, Eighth, and Tenth Columns on March 15, 1948.

REFERENCES

Trevor N. Dupuy, *The Military History of the Chinese Civil War* (New York: Franklin Watts, 1969); Liu Qi et al., *Liao-Shen Zhanyi Qin Li Ji: Yuan Guomindang Jiangling de Huiyi* [Diaries of the Liaoning-Shenyang Campaign: Original Recollections of the Guomindang Generals] (Beijing: Wenshi Chubanshe, 1996); William W. Whitson with Chen-Hsia Huang, *The Chinese High Command* (New York: Praeger, 1973); ZGDBKQS, vol. 1; ZGRMJFJZS, vol. 3.

SOONG, T. V. (Soong Tse-wen; AKA Soong Tse-Ven and Song Ziwen) (December 4, 1894–1971) Although not a military figure, T. V. Soong had significant influence over Chinese military history, especially during **World War II**, in his position as **Chiang Kai-shek**'s representative in Washington, D.C. He was the brother of Chiang's second wife, Soong Mei-Ling. Another sister, Soong Ch'ing-Ling (Qingling), was the wife of Sun Yat-sen, the Nationalist president of China, and was a strong opponent of the Guomindang who held important positions in the People's Republic of China. A third sister, Soong Ai-Ling, was married to the head of the National Ministry of Industry and Commerce, H. H. Kung.

T. V. Soong was brought into the Sun Yat-sen government by his older sister, Soong Qingling, and managed the salt administration in Guangdong and Guangxi Provinces, which then comprised the parts of Nationalist China under Sun's control. He later managed the Central Bank and used this position to finance the 1926 **Northern Expedition**. In September 1925, T. V. Soong was named minister of finance of the Nationalist government. Chiang Kai-shek married T. V. Soong's sister Soong Mei-Ling on December 1, 1927, after Chiang retired from the government. T. V. Soong financed the establishment of the Revenue Guards, which later became the Nationalist New First Army, which fought in Burma in World War II. They also fought in Shanghai against the Japanese as part of the Nationalist 19th Route Army. During the **Xi'an Incident** in 1936, T. V. Soong took part in the negotiations with **Zhang Xueliang** to secure the release of Chiang Kai-shek and the cooperation agreement between the Communists and Nationalists in the Second United Front.

After representing Chiang in Washington in 1940 and 1941, where he succeeded in obtaining loan credits for China's fight against the Japanese, Soong was made minister of foreign affairs. He secured a credit loan of $500 million from the United States and signed the Lend-Lease Agreement. When Major General **Patrick J. Hurley** visited China in September 1944, as President Roosevelt's personal representative, Soong voiced strong opposition to Hurley's attempts to negotiate a Nationalist-Communist coalition government. Soong negotiated the Sino–Soviet Treaty of Friendship and Alliance in 1945, which granted Mongolia independence but gave the Soviet Union rights to Port Arthur (Dalian) and Chinese railways under the framework of the **Yalta Agreement**. T. V. Soong was declared a war criminal by the Communists in 1949 and established a residence in New York. He remained in the United States until his death in 1971.

REFERENCES

Sterling Seagrave, *The Soong Dynasty* (New York: Harper and Row, 1985); Barbara W. Tuchman, *Stilwell and the American Experience in China, 1911–45* (New York: Macmillan, 1971).

SOUTH CHINA SEA NAVAL EXPEDITION. See BROAD OCEAN MISSILE LAUNCH

SOVIET BASE AREAS This term is almost synonymous with a **Revolutionary Base Area**, except that it refers to the type of administration and government. The soviets were the secure areas under the control of Communist forces that were administered and governed by Communist Party cadre. The term "soviet" actually means a council or body of delegates and dates back to the Russian revolutions of 1905 and 1917. In 1920, the Communist International (Comintern) advocated the establishment of soviets in colonial areas, especially China and India. In July 1922 the Second Chinese Communist Party (CCP) National Congress adopted a platform calling on workers to join with peasants forming in soviets for self-defense and liberation. The first CCP soviet area was established in Guangdong in 1927. The **Jiangxi** Soviet was established at a CCP National Soviet Congress on November 7, 1931, held in Ruijin in the Jinggangshan Mountains. It was called the Soviet Republic of China but may also be called the **Jinggangshan Revolutionary Base Area**. Between 1927 and 1949, the Communists established as many as 40 separate soviets and Revolutionary Base Areas.

REFERENCE

Warren Kuo, *Analytical History of the Chinese Communist Party* (Taipei: Institute of International Relations, 1968).

SPRATLY ISLANDS. See NANSHA ISLANDS

STILWELL, JOSEPH W. (1883–1946) General Joseph Warren Stilwell was a U.S. Army officer whose career was almost intrinsically tied to China throughout his time in the military. General Stilwell was born on March 19, 1883, in Palatka, Florida. He was the son of Benjamin W. Stilwell (a lawyer and physician) and the former Mary A. Peene. Stilwell's father and mother both came from families with roots in the United States going back to the 1630s. He graduated from the U.S. Military Academy at West Point, New York, on June 15, 1904, as an infantry second lieutenant. Stilwell first went to China in November 1911, traveling en route home from a tour of duty in the Philippines. His wife preceded him back to the United States, while Stilwell continued on to Shanghai, which he reached on November 23 of that year. Stilwell was a lifetime diarist, but made little record of the events of the 1911 revolution. For the 17 days he spent in China he concentrated on his impressions of the foreign presence in Shanghai and on what he saw as the tradition-bound, superstitious, xenophobic nature of the Chinese people. After learning of the fighting in Guangzhou over the 1911 revolution, Stilwell immediately made his way there. He traveled up the West River for 200 miles on a British gunboat, the *Sandpiper*, and witnessed the beheading of prisoners by Chinese soldiers, after which, in some cases, the soldiers ate the hearts and livers of those they executed. He returned to Manila on December 9, 1911.

Between 1913 and 1916, Stilwell taught English and history at West Point. He was promoted to captain in September 1916 and major on August 25, 1917, at Camp (now Fort) Lee, Virginia. In December 1917, Stilwell went to France to work gathering intelligence for the American Expeditionary Force, where the army hoped to take advantage of his capabilities as a French linguist. He remained in Europe through 1919, where he took part in campaigns as a staff officer and intelligence planner. Stilwell was sent back to China as a language officer in Beijing from 1920 to 1923. He lived with his family in a compound in the city, mingling with the diplomatic community and perfecting his language abilities while he traveled around the country to gain first-hand impressions of how the Chinese people lived and thought. Later, Stilwell served again in China as an officer in the U.S. 15th Infantry, where he was stationed in Tianjin from 1926–1929. He returned to China once more, where he served as U.S. military attaché in the American legation in Beijing from 1935 to 1939.

In January 1942, Stilwell was appointed a lieutenant general and ordered to China by President Roosevelt. Getting there was no easy task. His logistical supply link, the transport *Normandie*, burned at the dock in New York on February 9, the same day that Japanese forces crossed the Salween River in Burma, putting pressure on Rangoon, the capital. Stilwell flew from Miami, Florida, with a staff of officers, arriving in India on February 25, 1942, just as the British forces in the Far East were about to collapse under Japanese pressure. Stilwell spent a week in India coordinating with the British commanders of the theater, and he left Calcutta for Chongqing, where his headquarters was established on March 3. He was the head of the American forces there in **World War II** and was commander of the China–Burma–India Theater. En route, in a stop at Lashio, Burma, he met **Chiang Kai-shek** and Madame Chiang, who were conferring with Chinese military commanders. Stilwell's first military campaign was to attempt to relieve the pressure on Rangoon, with the Chinese Fifth and Sixth Armies. However, he had to fight constantly with Chiang Kai-shek for supplies and reinforcements. Stilwell received air support for his campaign from the Royal Air Force and from **Claire Chennault**'s **Flying Tigers** (the American Volunteer Group, as it was officially known). Chennault had a far better relationship with Chiang than Stilwell, and the two Americans were at odds over resources and strategy throughout the war. Stilwell, the infantryman, believed that the key to beating the Japanese was through land forces and campaigns to defeat and destroy troops and control terrain. Chennault sought to draw supplies for an airpower campaign designed to cut Japanese supply lines with the eventual goal of forcing their withdrawal.

Stilwell and Chiang Kai-shek never trusted each other. Stilwell believed that Chiang was corrupt, making all of his decisions on the basis of private interests, the interests of his family, and those of his client generals. Throughout the war, Stilwell had to keep continuous pressure on Chiang to ensure that U.S. Lend-Lease supplies and loans were actually reaching the Chinese forces for which they were intended. Because he believed that eventually he would again fight the Communists, which he had tried to crush in successive **Encirclement Campaigns** before World War II, Chiang never fully supported Stilwell's strategy for a decisive land campaign against the Japanese.

On August 7, 1944, Stilwell was appointed a general (four stars). The intent of President Roosevelt was to ensure that Stilwell could command all of the U.S. and Chinese armed forces in the Allied war effort against the Japanese. The poor relationship with Chiang, however, frustrated all of Stilwell's attempts to conduct a land war. Stilwell believed that Chiang wanted to avoid any fight with the Japanese, waiting for American forces to do that, while he preserved his Nationalist Army to fight the Communists. Chiang's refusal to commit the Chinese **Y-Force** against the Japanese in 1944 convinced Stilwell that the prosecution of the war with Chiang Kai-shek was impossible. Convinced of the corrupt nature and weakness of the Nationalist government, Stilwell eventually recommended that, after the war, the United States should get out of China because a Communist victory over the corrupt and inefficient Guomindang (KMT) was virtually certain. Stilwell was recalled from China in October 1944 by President Roosevelt under pressure from Chiang Kai-shek and the British. General **Wedemeyer** replaced Stilwell as chief of staff to Chiang Kai-shek and commander of American forces in China. The Burma–India Theater was split away from the China Theater and put under the command of General Sultan. Wedemeyer, however, was no more successful than Stilwell in forcing Chiang to commit Chinese forces in decisive battle. In the end, only the defeat of

Germany permitted the United States to focus on the Pacific and force the retreat of Japanese forces from southeast China, something that Stilwell had never accomplished. Stilwell was appointed commanding general of the 10th Army, under MacArthur, in preparation for the invasion of Japan. Stilwell was present on the battleship *Missouri* on September 2, 1945, in Tokyo Bay, when Japan surrendered. On October 18, 1945, Stilwell returned to the United States and was reassigned as commander of the Sixth Army, headquartered in the Presidio of San Francisco, California. On October 3, 1946, he had surgery for stomach problems and was found to have advanced, metastatic stomach cancer. General Stilwell died on October 12, 1946.

REFERENCES

Charles F. Romanus and Riley Sunderland, *The United States Army in World War Two. The China–Burma–India Theater: Stilwell's Command Problems* (Washington, DC: Government Printing Office, 1956); Charles F. Romanus and Riley Sunderland, *The United States Army in World War Two. The China–Burma–India Theater: Stilwell's Mission to China* (Washington, DC: Government Printing Office, 1953); Barbara W. Tuchman, *Stilwell and the American Experience in China, 1911–45* (New York: Macmillan, 1971).

STRATEGIC ROCKET FORCES. See NUCLEAR PROGRAM; SECOND ARTILLERY CORPS

SU YU (1907–1985) Su Yu was born in Hutong County, Hunan Province, in 1907. He joined the Communist Youth League in 1926, which led to his dismissal from school for agitation. He joined a force under **Ye Ting** and took part in the **Nanchang Uprising** on August 1, 1927. He later moved to the **Jiangxi Soviet**, where he was a company commander under **Chen Yi**. This lifelong association with Chen Yi continued through the **Civil War**, where he was part of Chen Yi's **Third Field Army** during the **Huai-Hai Campaign**. Su Yu was vice minister of defense under **Lin Biao** and later was instrumental in the organization of the National Defense Industries Office, where he managed the defense industrial establishment in cooperation with **Nie Rongzhen**. He was at one time also the mayor of Nanjing. Su Yu's daughter, Su Huining, is a **People's Liberation Army** (PLA) colonel working in the Commission of Science, Technology, and Industry for National Defense (COSTIND), a follow-up organization to the National Defense Industries Office. In 1997, she was married to one of Chen Yi's son's, Chen Xiaolu, who is a retired PLA colonel.

REFERENCES

Author interview with the family of Su Yu, Beijing, China, June 1996; Wolfgang Bartke, *Who's Who in the People's Republic of China* (Armonk, NY: M. E. Sharpe, 1981); Tong Bai, ed., *Junren Shouce* (Beijing: Changzheng Chubanshe, 1987).

SU ZHENHUA (1909–1979) Su Zhenhua is of the generation of early leaders associated with **Peng Dehuai** and **Deng Xiaoping** who were seen as potential challengers by **Mao Zedong**. Su was an architect of a naval tactic that mirrored his experience as a guerrilla leader after he moved into the **People's Liberation Army** (PLA) Navy. He was purged in 1967, but reemerged from obscurity in 1972 as deputy commander of the PLA navy.

Su Zhenhua was born in 1909 or 1911 (sources differ on the date) in Hunan Province. In 1928 he took part in the **Autumn Harvest Uprisings** and subsequently served in a unit commanded by Peng Dehuai. He moved with Peng to the guerrilla base in the **Jinggangshan Mountains**. Su took part in the **Long March** and, like many Long March veterans, attended the **Anti-Japanese Military and Political Academy**. In 1937, he was made political commissar of the 343d Brigade, 115th Division, **Eighth Route Army**, serving under **Lin Biao**. He took part in the defeat of the Japanese in **Pingxingguan**, Shanxi Province. After serving as political commissar of the Training Regiment of the 115th Division, he served as political commissar of Shanxi–Chahar–Hebei Military District. In 1949 he was a political commissar of a corps in the **Second Field Army**, associated with Deng Xiaoping. He stayed in southwest China, primarily in Guizhou Province, through 1953.

In February 1953 Su Zhenhua was appointed deputy political commissar of the PLA. He became political commissar of the navy in August 1957. In 1958, Su commanded the PLA naval units in the Taiwan Strait during the crisis with the Nationalist government on Taiwan. Su championed a tactic of close combat, using small, fast patrol craft in the navy, that attacked at a distance of 200 meters. In 1967, at the height of the **Cultural Revolution** factionalism, Su was accused of planning a coup d'état against Mao together with He Long, and he disappeared. Unlike some Cultural Revolution victims, Su made an early appearance in March 1972 as deputy commander of the PLA Navy.

Su served as an alternate member of the Politburo from 1973 to 1977; as chairman of the Revolutionary Committee of Shanghai Municipality in 1976; and as first political commissar of the PLA Navy in 1977. Su Zhenhua died on February 7, 1979.

REFERENCES

Wolfgang Bartke, *Who's Who in the People's Republic of China* (Armonk, NY: M. E. Sharpe, 1981); William W. Whitson with Chen-Hsia Huang, *The Chinese High Command* (New York: Praeger, 1973).

SUN LIREN (Sun Li-jen) (1900–1990) General Sun Liren was one of the most active military leaders of the Nationalist Army in **World War II**, although there is some debate as to his effectiveness. Sun was described by U.S. Army General **Joseph W. Stilwell**, during the campaign on the Irriwadi Front in Burma in 1942 as ''sulky and often hesitant and weak.'' Among many Nationalist Chinese, however, he was known as ''the ever-victorious general.'' Sun

Liren rose to the position of commander-in-chief of the Nationalist Army on Taiwan. He was born in Anhui Province in 1900 and attended Qinghua University. In 1922, he was a member of China's national basketball team. Sun later earned a civil engineering degree at Purdue University in the United States in 1924 and enrolled in the Virginia Military Institute (VMI), from which he graduated in 1927. Sun returned to China after a tour of Europe in 1928 and became the deputy brigade commander of the student brigade at the Guomindang (KMT) Nanjing Central Party School. By 1930, Sun became a colonel and a regimental commander. He became closely associated with **Chiang Kai-shek**'s brother-in-law, **T. V. Soong**, and commanded a brigade of troops organized by Soong in Changsha in the early 1930s.

Sun's brigade formed the nucleus of the Nationalist New 38th Division in 1942, with Sun as the divisional commander. Historians Charles F. Romanus and Riley Sunderland credit Sun Liren as a ''brilliant commander'' who kept his division in Burma intact as a disciplined combat unit even under the harshest conditions. Stilwell, however, complained that Sun would not act decisively. This criticism seemed to hold true in 1946, during the **Civil War**, when Sun commanded the Nationalist New First Army in northeast China opposing the Communists in the **Sungari River Campaigns** and the **Siping Campaigns**. He was defeated in both campaigns primarily because he was hesitant to commit forces and insisted on keeping his troops close to cities and rail or road systems instead of maneuvering against his enemy. As Nationalist forces were being defeated in Manchuria, Sun was removed from command and shifted to Nanjing as deputy commander in chief of the Chinese Army and commander of army training. After the Nationalists moved to Taiwan, Sun was made commander of the (Nationalist) Chinese Army and was promoted to general in 1951. Sun competed for authority, however, with Chiang's son, Chiang Ching-kuo, who ran the political commissar system in the Army. Allegedly, Sun attempted to abolish the political commissar system in the Nationalist Army in 1954, and he was relieved of his post and placed under house arrest in Taichung, Taiwan, in 1955. No public trial was held, but Sun was alleged to have taken part in acts that involved ''communist agents.'' Sun regained his freedom in 1988, after Chiang Ching-kuo's death, and died at the age of 90.

REFERENCES

Charles F. Romanus and Riley Sunderland, *The United States Army in World War Two. The China–Burma–India Theater: Stilwell's Command Problems* (Washington, DC: Government Printing Office, 1956); Shen Ke-tong, *Sun Liren Zhuan* [The Story of Sun Lijen], 2 vols. (Taipei: Taiwan Student Press, 1998); Barbara W. Tuchman, *Stilwell and the American Experience in China, 1911–45* (New York: Macmillan, 1971).

SUNGARI RIVER CAMPAIGNS (1946–1947) The Sungari River Campaigns were part of the larger offensive to control the Shenyang-Siping-Changchun rail corridor in Liaoning Province (see **Siping, Offensive Campaign**

Against). In a broader sense, these were part of the major Communist offensive that took place between 1946 and 1948 for the control of Manchuria, which culminated in the **Liaoning-Shenyang** (Liao-Shen) **Campaign** of 1948.

Summer 1946 found Communist forces under the command of **Lin Biao** concentrated in eastern Heilongjiang and in Jilin province, Manchuria. Lin Biao's forces were primarily distributed north of the Sungari River, under continuous pressure from Nationalist forces. In an initial offensive, which Lin Biao described as an "offensive-defensive campaign" or "active defense," the First and Sixth Columns of the Communist Army moved south across the Sungari River in November 1946. This was actually more of a major feint, designed to relieve pressure on the Communist defensive force around Linjiang, a city in southwest Jilin Province, which was under attack by the Nationalists. The result of this move was to catch Nationalist forces in a pincerlike envelopment, inflicting heavy casualties and breaking the Nationalist offensive against Linjiang.

On January 6, 1947, the Communists again crossed the Sungari River to conduct offensive operations against Nationalist forces. The Communist First Column, led by Li Tianyu, sent its Third Division to attack the Nationalist garrison at Qidainu. Meanwhile, the First and Second Divisions established positions where they could ambush any Nationalist reinforcements moving east toward Qidainu to relieve the garrison. On January 7, 1947, the First and Second Divisions attacked the relief column, wiping out two regiments of the Nationalist 50th Division. Having eliminated the relief force, the Communists then attacked and destroyed the garrison force in Qidainu. On January 17, 1947, the Communist forces withdrew across the Sungari River. In Communist military histories, this is called the First Sungari River Offensive although the Nationalist histories refer to it as the second.

On February 21, 1947, Lin Biao's forces launched another major offensive across the Sungari River designed to protect the flank of his base area from Nationalist forces. Using the same "defensive-offensive" tactic he had previously employed, Lin Biao attacked the city of Chengzejie, west of the Sungari, with forces of the First, Second, and Sixth Columns. The bulk of the First Column established an ambush between the city of Quitai and Chengzijie, forcing the withdrawal of Nationalist troops from the area. After destroying large sections of the rail line, Lin Biao again withdrew his forces north across the Sungari when the Nationalists began to regroup for a counterattack.

A fourth foray across the Sungari River was initiated on March 7, 1947. In this thrust, Lin Biao's forces eliminated the fighting strength of the Nationalist 88th Division, whose commander, General Chen Mingren, later successfully defended Siping. Interestingly, Chen Mingren later defected to the Communist side and fought to drive the Nationalists out of Guangxi Province as the commander of the Communist 21st Army. Chen had come over and turned his entire army into the Communist 21st Army in May 1949, ironically while fighting against Lin Biao's forces in South-Central China. This fourth offensive across the Sungari, in addition to destroying a Nationalist division, captured weapons and ammunition that helped to increase the fighting capacity of the Communist forces. A fifth offensive strike across the Sungari River in May 1947 turned into the defensive battle of Siping, as Lin's forces drove on that city (see Siping, Defensive Battle of).

REFERENCES

William W. Whitson and Chen-Hsia Huang, *Civil War in China, 1946–1950* 2 vols. (Washington, DC: Office of Military History, U.S. Army, 1967); ZGDBKQS; ZGRMJFJZS, vol. 3.

T

TAI LI. See DAI LI

TAIPING REBELLION (1850–1864) The Taiping Rebellion was an indigenous Chinese, anti-Manchu (Qing dynasty), millenarian movement that promised to bring to China a "kingdom of heavenly peace" (*Taiping Tianguo*), eliminating historical clan and communal conflict in Chinese society. The movement grew from its area of origin, in Guangxi Province, to the point that its adherents controlled large areas of East-Central China, including the Yangtze River delta around Hangzhou, Suzhou, and Shanghai. The Taiping capital was in Nanjing, and its military forces once threatened Beijing and Tianjin. Qing dynasty forces, alone, failed to suppress the movement. As a result, a coalition of forces from the Manchu regime; foreign-dominated, but Chinese-manned and financed, private militia; and military forces from France and England joined together to fight the Taiping rebels. The ensuing conflict, which, combined with the experience of the **Arrow War**, crystallized in the Shanghai area in the period 1860–1864 had the effect of demonstrating to Qing officials the effectiveness of Western military organization and modern weapons. The **self-strengthening movement**, from the period 1860–1894, is viewed by many scholars as having its genesis in the Taiping Rebellion and the foreign military actions that helped the Qing dynasty and local Shanghai officials and merchants suppress the Taiping rebels.

The leader of the Taiping Rebellion was a teacher from Guangdong Province of Hakka minority descent, Hong Xiuquan. Hong was born in 1814 and studied for the imperial scholar's examination that would have permitted him to be appointed an official of the dynasty and exempted him, as a scholar, from physical punishment. Hong Xiuquan attempted to pass the imperial examinations twice in the early 1830s but failed both attempts, the latest of which took place in 1836. After failing the prefectural examination in that year, Hong apparently suffered a serious mental breakdown, and he left to study in the city of Guangzhou. While he was in Guangzhou, Hong apparently met and came under the religious influence of a Christian missionary, believed to have been an American Protestant missionary, Edward Stevens, who was preaching at the Seamen's Friends Society in Whampoa. In 1837, after failing his third attempt at the imperial examination, Hong Xiuquan again suffered some form of delirium and had a dream or vision in which he believed himself to have been directed by God to purge the world of evil and demons and to establish a society based on peace and brotherhood. Later, in 1843, Hong Xiuquan failed the imperial examination for a fourth time and at this point began to read Christian texts. He decided that in his earlier dream he had been the son of God and the younger brother of Jesus Christ, appointed to spread a message and establish a heavenly kingdom on earth.

Hong Xiuquan established his movement and began to proselytize among Hakka minority people in the rural area of Guangxi Province, west of Guilin. Between 1848 and 1849, in order to protect itself during a clan conflict with another minority in a neighboring village, the Hakka followers of Hong also developed the capability to defend themselves. The movement attracted as many as 10,000 followers and adherents, including former soldiers and secret society members, who sought to join Hong's Christian-based community and to destroy the Qing dynasty, which Hong called wicked, corrupt, and deceitful.

Qing forces were sent by the emperor to put down the rebel forces west of Guilin in December 1850. They were beaten by the Taipings, who by then had begun to drill troops, manufacture weapons, and establish a military organization that included a segregated women's force. After this victory over the government forces, Hong Xiuquan declared himself to be the "Heavenly King of the Taiping Tianguo." When larger government forces returned to suppress the Taipings, Hong's forces moved to the mountains of the Guangxi-Hunan border. They eventually seized the city of Yongan, south of Guilin, and, having grown in strength to 60,000 people, moved to attack Guilin itself. Hong failed to capture the capital of Guangxi and continued to campaign north, into Hunan, where he tried to capture Changsha. Despite a two-month siege of that city, the Taiping forces failed again. In December 1852, however, the Taiping rebels succeeded in seizing the city of Yuezhou, north of Changsha on Dongting Lake. They captured large amounts of arms, gunpowder, and cash and increased their strength in Yuezhou, moving farther north to capture Hankou in late December 1852 and Wuchang in January 1853 (these two cities are now the metropolis of Wuhan). Hong's Taiping forces continued to range north, eventually capturing Nanjing in March 1853, a city with a Manchu population of over 40,000. Hong Xiuquan executed all Manchus in the city who had not been killed in the battle. The Manchu survivors were burned, stabbed, or drowned by the Taiping forces. Hong established Nanjing as the capital of the Taiping Heavenly Kingdom, while his forces ranged Tianjin and Beijing and as far west as Chengdu. However, the Taiping forces never succeeded in controlling these regions. Much of Jiangsu and Zhejiang, however, came under the control of the Taipings.

Despite the Christian basis for the religion, Western countries were not sympathetic toward the Taiping Kingdom. Hong Xiuquan's forces destroyed trade and commerce, undermining the profitability of the newly won concession areas in Shanghai and Ningbo. The reaction to the Taiping Kingdom was the development of a coalition of Chinese merchants and bankers, Qing officials, foreign traders and mercenaries, and foreign military forces—all of whom sought to preserve the existing order for their own reasons. The French-led **Ever-Triumphant Army** emerged from the Zhejiang Artillery Corps, supported by the French fleet. In Shanghai, **Frederick Townsend Ward** and **Henry Burgevine**, two Americans, with the help of Qing officials such as **Li Hongzhang**, formed the **Foreign Arms (Rifle) Corps**, which eventually turned into the **Ever-Victorious Army**. With support from the English, after Townsend's death in 1862, the Ever-Victorious Army was led by a British army royal engineer, Lieutenant Colonel **Charles Gordon**, who because of his exploits in putting down the Taiping Rebellion, came to be known as "Chinese Gordon" in the British army. Meanwhile, on the Chinese side, **Zeng Guofan**, a Qing official who controlled the area along the Xiang River in Hunan, along which the Taiping rebels had advanced to take control of Nanjing, organized the **Xiang Army**. Eventually, the combined pressure of the foreign-led mercenary and militia armies, the Xiang Army, British and French forces, and other Qing dynasty troops defeated the Taiping Rebellion and recaptured Nanjing in July 1864. According to Zeng Guofan's account of the seizure of Nanjing, none of the rebels surrendered, and those who were not killed in the battle for the city committed suicide, many by self-immolation.

The Taiping Rebellion is significant because it illustrates the effect of the combined pressures of bureaucratic inefficiency and corruption in the Qing dynasty and foreign, imperialist intervention in China for trading rights on the ruling dynasty. The Qing were collapsing, but some of its officials, especially Li Hongzhang and Zeng Guofan, were seeking to modernize the country and its military forces. Concurrent with the Taiping Rebellion, the Qing was also faced with the Arrow War, the **Nian Rebellion** in Central China, and the Anglo-French led expedition to Beijing that captured the **Dagu forts** at Shanghai and burned the **Yuanmingyuan**, or Summer Palace. These pressures kept the Qing rulers from reacting to the rebellions. At the same time, the Western powers, both because of the profits they earned from trade with China and because they had exacted concession and port rights, had no desire to see fundamental dynastic or political change in the country—thus, the coalition of forces that developed against the Qing in Shanghai.

REFERENCES

Charles G. Gordon, *General Gordon's Private Diary of His Exploits in China* (London: Simpson Low, 1885; New York: Kraus Reprint, 1971); J. S. Gregory, *Great Britain and the Taipings* (New York: Praeger, 1969); Teng Ssu-Yu, *The Taiping Rebellion and the*

Western Powers: A Comprehensive Survey (Oxford: Clarendon Press, 1971); Douglas Hurd, *The Arrow War: An Anglo-Chinese Confusion, 1856–1860* (London: Collins Press, 1967); Franz Michael, *The Taiping Rebellion: History and Documents*, 3 vols. (Seattle: University of Washington Press, 1966).

TAIWAN STRAIT CRISIS OF 1958 Sino–U.S. talks conducted in Geneva over bilateral recognition and the return of U.S. prisoners of war from the **Korean War** broke down by 1957. When the U.S. ambassador in Geneva, U. Alexis Johnson, left for another diplomatic assignment in December 1957, no U.S. replacement was assigned. By June 1958, after requesting the continuation of talks and the assignment of a new U.S. ambassador, Beijing decided to use military force to demonstrate to the United States why it must deal with China. The Chinese Communist Party **Central Military Commission** met over a period of two months, May through July 1958, to decide on a means to bring military force to bear on the situation.

Beijing first initiated a strong propaganda campaign in its internal and external media, calling for the "liberation of Taiwan." This campaign lasted through June and July. Following a July 1958 visit to Beijing by Soviet premier Khrushchev, during which plans for Chinese military action reportedly were withheld from Khrushchev by **Mao Zedong**, the **People's Liberation Army** (PLA) Air Force began to deploy to forward airfields. On August 23, the PLA began shelling Quemoy Island (**Jinmen**), firing more than 30,000 rounds at the island on the first day of the campaign. The United States, meanwhile, rushed six aircraft carrier battle groups to the area and sent in combat aircraft and transports. U.S. ships even escorted Nationalist vessels resupplying the offshore islands of China up to the three-mile territorial limit. Other U.S. support flowed in, including atomic-capable, eight-inch howitzers (which were deployed on the offshore islands) and air-to-air missiles. Taiwan and the mainland fought seven air battles between August 23 and the end of October. China, meanwhile, began to receive surface-to-air missiles (the SA-2) from Moscow. In the end, no invasion of Quemoy or Matzu was mounted. China and the Soviet Union split over assessments of whether the United States was willing to risk war with the socialist camp, and U.S. support continued to flow more strongly into Taiwan.

REFERENCES

Kenneth W. Allen et al., *China's Air Force Enters the 21st Century* (Santa Monica, CA: Rand Corporation, 1995); Gordon H. Chang, "To the Nuclear Brink: Eisenhower, Dulles, and the Quemoy-Matsu Crisis," *International Strategy* 12, no. 4 (Spring 1988): 96–123; John W. Garver, *Foreign Relations of the People's Republic of China* (Englewood Cliffs, NJ: Prentice-Hall, 1993).

TAIWAN STRAIT CRISIS OF 1995/1996 In response to an invitation to Taiwan president Lee Teng-Hui to visit and speak at Cornell University in New York, his alma mater, Chinese authorities began a series of military, propaganda, and diplomatic moves intended to show their displeasure to Taiwan, Washington, and the world. Beijing's objective was to discourage other countries from developing stronger, informal ties with Taiwan. After a near-unanimous vote in the U.S. Congress and Senate in favor of granting a visa to President Lee, permitting him to visit and speak at Cornell University, President Clinton approved the visa. The Chinese ambassador to the United States, Li Daoyu, who failed to predict for his government that a visa would be approved, was recalled to Beijing for consultations, and, at the same time, China instituted a series of moves designed to show its pique to Washington. In November and December 1995, Beijing conducted a series of military exercises opposite Taiwan in Fujian Province that included the firing of ballistic missiles, aerial bombardment exercises, and amphibious landings supported by naval gunfire. Taiwan's first democratic, contested presidential election was scheduled for March 1996, and Beijing kept up military and political pressure against the island, fearing that a democratically elected president on Taiwan would draw greater international recognition.

In March 1996, on the eve of the Taiwan elections, Beijing again began a series of military exercises off Taiwan as a means to intimidate the electorate on the island and to warn voters not to support an independence candidate. Although Beijing expected Lee Teng-hui to win the election, another goal of the exercises was to pressure Lee to avoid an immediate call for the independence of Taiwan from the mainland of China. In addition to holding land, sea, and air exercises, including airborne drops by paratroopers, the strategic rocket forces of the **People's Liberation Army**, the Second Artillery, established closure zones at sea north and south of Taiwan into which they fired M-9 (CSS-6, or Dongfeng 15) ballistic missiles as a warning of a possible strike against Taiwan in the event of a declaration of independence. As a show of resolve, the United States sent a naval force of two aircraft carrier battle groups, the *Independence* and the *Nimitz*, as well as command and control ships, to monitor the situation, positioning them southeast of Taiwan. The exercises concluded without serious incident prior to the Taiwan elections, and the U.S. carriers were withdrawn.

REFERENCES

Greg Austin, ed., *Missile Diplomacy and Taiwan's Future: Innovations in Politics and Military Power* (Canberra: Strategic and Defense Studies Center, Australian National University, 1997); James R. Lilley and Chuck Downs, eds., *Crisis in the Taiwan Strait* (Washington, DC: National Defense University Press, 1997); Office of Naval Intelligence, *Chinese Exercise Strait 961: 8–25 March 1966* (Washington, DC: Office of Naval Intelligence, 1966).

TAKU FORTS. See DAGU FORTS

TAN ZHENG (1902–1988) Tan Zheng was one of the 10 officers appointed generals when military ranks were established in the **People's Liberation Army** (PLA) in 1955. He was also among the senior PLA officers and Communist

Party leaders who were attacked by leftist elements in the party, branded as a "rightist," and purged from all party and military posts during the **Cultural Revolution**.

Tan Zheng was born in 1902 in Xiangxiang County, Hunan Province, and had only a middle school education. He took part in the **Nanchang Uprising** in 1927 and later became the director of the Political Department of the 12th Red Army. By 1932, Tan Zheng was made the director of the Political Department of the First Red Army, which by 1934 evolved into the **First Front Army** in the **Jinggang Mountains** (Jingganshan). Tan participated in the **Long March** during 1934 through 1935. He was appointed deputy director of the **Eighth Route Army**'s Political Department in 1938, a position he held until 1945. He was concurrently the director of the Political Department of the Shanxi–Suiyuan and Shaanxi–Gansu–Ningxia Military Regions, in the heart of the Eighth Route Army's operating area.

When the **Civil War** broke out in 1946, Tan Zheng was sent to Manchuria with the **Northeast Democratic United Army**. Through 1949, Tan Zheng continued to serve in the north and northeast parts of China, and in 1949 he was part of the **Fourth Field Army** when it was formed. He was then moved to Central China, where he was director of the Central China Military District. During 1949–1950, Tan Zheng continued as a deputy political commissar and Political Department director in the Fourth Field Army. By 1952, Tan was appointed political commissar of the Fourth Field Army and the Central-South Military Region.

In 1952, Tan Zheng was brought to Beijing, where he was appointed deputy director of the **General Political Department**. He seems to have spent most of his career closely associated with **Lin Biao** and **Luo Ronghuan**. After Tan was appointed to the rank of general in 1955, he was made director of the General Political Department, a post he held until 1965, when he was moved out of the military to become vice governor of Fujian Province. He was purged in 1967 during the Cultural Revolution and restored as a party member in 1975. Tan Zheng died on November 6, 1988.

REFERENCE

Wolfgang Bartke, *Biographical Dictionary and Analysis of China's Party Leadership, 1922–1988* (New York: K. G. Saur, 1990).

TAO ZHU (1906–1969) Tao Zhu represents an example of the career pattern and life of many of the founding leaders of the People's Republic of China (PRC). Tao started out as a military leader and political commissar. After the establishment of the PRC in 1949, he moved into civil government and party positions. As **Mao Zedong** became ascendant in the early 1960s, Tao returned to military leadership positions. Caught up in factional struggles during the **Cultural Revolution**, Tao was purged and, like many of China's early leaders, died under persecution by the Red Guards at the instigation of the Communist Party during the height of the Cultural Revolution.

Tao Zhu was born in 1906 in Hunan Province. He participated in the **Northern Expedition** in 1926 and in the 1930s in a raid in Xiamen, Fujian Province, which freed 18 arrested Chinese Communist Party (CCP) cadres from Guomindang (KMT) control. He was imprisoned in Nanjing from 1935 to 1938. During the years of the **Anti-Japanese War**, he organized and led guerrilla units in Hubei Province. In 1948, Tao merged his guerrilla units with the forces under **Li Xiannian** to form the Fifth Division of the Fourth Army. He moved twice in the next year. In 1948 Tao was identified as deputy director of Shenyang Military Control Commission, but by the following year he was named deputy director of the Political Department of the **Fourth Field Army**. In December 1949 he was appointed chairman of Wuhan Military Control Commission. In 1950, as political commissar of Guangxi Military District, Tao was appointed a member of Central-South China Military and Political Council and second political commissar of South China Military Region. Moving into civil government, Tao was appointed a member of the Central-South China Administrative Council in 1952. He served as vice chairman of the people's government of Guangdong Province from 1953 to 1955 and governor and secretary of the Guangdong Province Communist Party Committee from 1955 to 1957. In 1962, Tao reassumed his military role as political commissar of Guangzhou Military Region. As the Cultural Revolution started, Tao served as adviser to the Cultural Revolution Group under the CCP Central Committee, moving to Beijing in 1966. He was also appointed director of the Propaganda Department of the CCP Central Committee in that year. In August 1966, Tao was elected member of the Politburo of the Eighth CCP Central Committee.

Despite his rapid rise to the center, Tao was branded as a counterrevolutionary revisionist and purged in 1967. He died on November 30, 1969.

REFERENCES

Wolfgang Bartke, *Who's Who in the People's Republic of China* (Armonk, NY: M. E. Sharpe, 1981); William W. Whitson with Chen-Hsia Huang, *The Chinese High Command* (New York: Praeger, 1973).

TEN MARSHALS OF THE PEOPLE'S LIBERATION ARMY. See MILITARY RANKS

THAGLA RIDGE. See SINO-INDIAN BORDER WAR

THIRD FIELD ARMY When the Third Field Army formed in 1949, many of its leaders, about 45 percent, were from the Jiangxi area and were part of the **Jiangxi Soviet. Chen Yi, Xiang Ying**, who led the **New Fourth Army**, **Su Yu**, and Tan Zhenlin are most closely associated with the lineage of the Third Field Army. Some **Fourth Front Army** units were merged with the **First Front**

Army units to form the Third Field Army, and the leaders of the army tended to dominate the leadership posts in coastal China, including Shanlong, Jiangsu, Anhui, Zhejiang, and Fujian Provinces.

Chen Yi and Su Yu were designated to lead the Third Field Army in 1949, and their families are still bonded together by the marriage of Chen's son, Chen Xiaolu, to Su Yu's daughter, Su Huining, both of whom served as military officers. Tan Zhenlin, who was active in the Fujian area leading guerrilla forces, also was one of the important leaders of the army, which, when formed, included units from the First Front Army and the Fourth Front Army.

The units that eventually formed the Third Field Army were caught up in the **New Fourth Army Incident**, when **Ye Ting** was captured by the Nationalist forces, and Xiang Ying was killed. In Shandong Province, such leaders as **Xu Xiangqian** and Han Xianchu dominated the army. Zhang Aiping, a former minister of national defense, is also a cadre of the Third Field Army.

After the **Civil War**, Chen Yi was appointed mayor of Shanghai, which allowed him to consolidate power in East China and ensured that Third Field Army cadre got good positions in that area. Su Yu went into Nanjing to organize a government, and Tan Zhenlin went to Hangzhou.

During the **Korean War**, about half of the forces in the Third Field Army participated in the war. The Ninth Army, led by Song Shilien, was badly mauled by the U.S. First Marine Division in the Korean War during the Chosin Reservoir Campaign. About 9,000 Communist troops reportedly froze to death during the campaign. As might be expected from an organization whose political and administrative center was Shanghai, from which many of the leaders of the Cultural Revolution originated, Third Field Army leaders were affected by the turmoil. Chen Yi stayed foreign minister during the Cultural Revolution, despite having been criticized heavily. Su Yu remained a manager of the National Defense Industries Office. He died in 1985.

REFERENCES

William W. Whitson with Chen-Hsia Huang, *The Chinese High Command* (New York: Praeger, 1973); ZGDBKQS.

THIRD LINE (INDUSTRIES) Known as the "Third Line" or the "Third Front," in the 1960s and early 1970s much of China's national defense industrial base was moved from coastal and urban areas to more remote locations in southern and western China. **Mao Zedong**'s intention, along with the National Defense Industry Office, was to ensure the survival of China's military industrial base in the event of a major conventional or nuclear attack on China by the Soviet Union or the United States. As much as 50 percent of China's defense expenditures during the 1960s may have been devoted to this strategic relocation of industry, which included the development of the requisite rail, road, and housing infrastructure to support the industries. Although the program was designed to ensure the survivability of the industrial base, it created tremendous coordination and distribution problems for Chinese defense industries, creating disastrous inefficiencies and piecemeal and inadequate production. By the 1990s, a general effort of "defense conversion" civilianized some of the production lines of some of these industries, leading to joint venture investment in China's interior and modernizing some factories. Nonetheless, the majority of Third Line industries are a drag on China's economy today.

REFERENCES

Paul H. Folta, *From Swords into Plowshares? Defense Industry Reform in the PRC* (Boulder, CO: Westview Press, 1992); John Frankenstein and Bates Gill, "Current and Future Challenges Facing Chinese Defense Industries," *The China Quarterly*, no. 146 (June 1996); Bates Gill and Taeho Kim, *China's Arms Acquisitions from Abroad: A Quest for Superb and Secret Weapons* (Oxford: Oxford University Press, 1995); Mel Gurtov, "Swords into Market Shares: China's Conversion of Military Industry to Civilian Production," *The China Quarterly*, no. 132 (December 1992).

THIRD REVOLUTIONARY WAR. See CIVIL WAR

"THREE-ALL CAMPAIGN" In response to the successful August–December 1940, Communist-led **People's Liberation Army** (PLA) **Hundred Regiments Campaign**, the Japanese went on the offensive to eradicate the Communists. The Japanese response was a major effort during 1941–1942 in the North China Plain area. In March 1941, 10 Japanese divisions began a counteroffensive in response to the PLA's Hundred Regiments Campaign. The Japanese attacked the PLA forces under the command of **Liu Bocheng** in the Taihang Mountains with a corps of three divisions (the 36th, 37th, and 41st); the Japanese 12th Corps attacked PLA forces in the area northwest of Beijing in the Hebei–Inner Mongolia border region. The "Three-All Campaign," intended to "kill all, burn all, destroy all," was a means to punish the entire Chinese populace because of peasant support for the Communist guerrilla forces.

This 18-month Japanese effort, according to some historians, "may have surpassed even the **Rape of Nanjing** in sheer brutality." In addition to killing all of the Communist political activists they could identify, the Japanese reorganized Chinese local administration in areas under their own control. The Japanese also instituted a system of group responsibility (*baojia*), punishing entire villages or production units for attacks on Japanese forces. A system of blockhouses and ditches was built with conscripted labor to hamper Communist resupply operations.

Under the command of Japanese general Okamura, the "Three-All Campaign" was a calculated effort to strangle economic support for Communist forces and to terrify the local populace into submission. It forced the Communist forces to rely on mobile guerrilla warfare for the next few years of war.

REFERENCES

Chalmers Johnson, *Peasant Nationalism and Communist Power* (Stanford, CA: Stanford University Press, 1962); William W. Whitson with Chen-Hsia Huang, *The Chinese High Command* (New York: Praeger, 1973); ZGRMJFJZS, vol. 2.

TIANANMEN INCIDENT. See ZHOU ENLAI

TIANANMEN SQUARE MASSACRE (June 4, 1989) On June 4, 1989, after a period of two weeks of martial law declared to stop popular demonstrations in Beijing against Communist Party corruption, nepotism, and authoritarian control, the **People's Liberation Army** (PLA) moved into the city with deadly force, killing between 2,500 and 3,700 demonstrators and bystanders, mostly students and workers. In addition, in reaction to news gathering by foreign press correspondents and the presence of foreign diplomatic observers, the PLA staged an incident at the foreign diplomatic compound at Jianguomenwai, in eastern Beijing on June 7. Claiming to have been fired on by a sniper from the diplomatic compound, PLA troops opened fire with automatic weapons into the apartments of foreign diplomats, press correspondents, and military attachés in an attempt to drive foreign eyes out of the capital.

The demonstrations began in response to the death of Hu Yaobang, a popular reformer who was general secretary of the Communist Party when dismissed in 1987 by strongman **Deng Xiaoping**. Hu was replaced by another reformer from Deng's own province, Sichuan, Zhao Ziyang. When Hu Yaobang died of a heart attack on April 15, 1989, students in Beijing's universities began to mourn his death. The mourners demonstrated against the Communist Party, noting that Hu was an honest official of great integrity who had been mistreated by the party. While the leaders of the Chinese Communist Party (CCP) formed a funeral committee, mourners began to flood Tiananmen Square.

Throughout late April and early May, the demonstrations in Beijing escalated as news of open attacks on the Communist Party and its leadership was published in the press and in underground journals by students and intellectuals. There were direct attacks in print on Deng Xiaoping and his family complaining about instances of corruption and nepotism. Within the Communist Party, orthodox Marxist hard-liners and conservative veterans of the **Long March** lined up together against the reform programs, which Zhao Ziyang represented. As Deng Xiaoping came under increasing popular attack, it became clear that he, too, must cut his ties to Zhao and purge him. Meanwhile, Zhao Ziyang, in an attempt to save himself and his reform programs, started an inner-party fight to retain his position. He also attempted to use the demonstrations to his advantage. Using the local party branches as his power base, Zhao, as CCP general secretary, mobilized more popular support, creating larger demonstrations and marches in Beijing and other major cities around China. These demonstrations involved not only students but state workers urged by their CCP work units to participate and the general populace as well. Meanwhile, Soviet president and party chairman Mikhail Gorbachev arrived in China on a scheduled state visit. By mid-May, the demonstrations and marches in Beijing had grown to such an intensity that Gorbachev's motorcade could not move around the capital. The students occupying Tiananmen Square ignored government pleas to leave the square, causing an embarrassment to the state and the party when Gorbachev's formal welcome ceremony had to be canceled. As the protesters began a hunger strike calling for reform, international support began to pour in, including financial support from Hong Kong. Chinese students and activists around the world flooded Beijing with facsimile (fax) messages encouraging the demonstrators. Zhao Ziyang, in perhaps an effort to absolve himself of guilt in the impending crackdown, told Gorbachev in front of reporters that Deng Xiaoping, not Zhao, was in charge of the party and the country. On May 17, over 2 million marchers demonstrated in the streets of Beijing in support of reforms.

In a strong conservative effort to maintain party rule, five old, Long March veterans and Communist Party organizers lined up with Deng in favor of a crackdown: Chen Yun (84 years old at the time), **Wang Zhen** (81), **Yang Shangkun** (82), **Peng Zhen** (87), and **Li Xiannian** (80). These veterans called the demonstrations a ''turmoil,'' implying that the situation had degenerated to the level of the disorder of the Cultural Revolution. On May 20, martial law was declared in Beijing by Premier Li Peng and Mayor Chen Xitong. Troops from the PLA attempted to enter the city to enforce the martial law order, but they were blocked by thousands of protesters. The PLA was paralyzed by its own refusal to act against the populace and by the continuing struggle within the party leadership. As military and party leaders lined up behind the conservatives, retired senior PLA generals called for restraint and asked that the PLA not be used against the general populace; these included Ye Fei, former defense minister Zhang Aiping, **Xiao Ke**, former chief of the General Staff Yang Dezhi, Chen Zaidao, Li Juqiu, **Xu Xiangqian**, and **Nie Rongzhen**. Zhao Ziyang was dismissed from office on May 26, and the city of Beijing was surrounded by elements of as many as 14 of China's 24 group armies (each group army is the size of a U.S. corps, with about 40,000–50,000 soldiers). Elements from the following group armies were identified in the vicinity of Beijing: 12th, 14th, 16th, 21st, 24th, 28th, 38th, 39th, 40th, 47th, 54th, 64th, 65th, and the 15th Airborne Army. In addition, the Sixth Tank Division and the First Tank Division gathered around the city. The PLA attempted to enter the city and reach Tiananmen Square unarmed and peacefully on the morning of June 3, but the soldiers were repulsed by crowds.

Early in the morning on June 4, 1989, troops advanced into Beijing from all sides of the city. They ran into stubborn popular resistance organized around roadblocks constructed from city buses and trucks. Using the tanks to break the barricades, the PLA punched through the roadblocks, but troops and vehicles were pelted with rocks, gasoline bombs, and Molotov cocktails. Road barriers were used by protesters to break the tracks on tanks and armored personnel carriers, and gasoline was poured into engine intakes, incinerating the vehicles. Escaping crews were attacked by the angry protesters. The PLA opened fire on people in the streets, killing innocent bystanders as well as demonstrators. The Chinese government claimed that only 23 students had been killed, and none of those in Tiananmen Square. Although the *New York Times* gave an estimate of 900 people killed, U.S. Embassy sources in the Chinese government, by June 5, put the death toll at 2,500. In spring 1996, a PLA defector in Hong Kong said that the death toll among the civilian populace was actually over 3,700 people, but these figures are not verified. As for the claim by the PLA that there were no deaths among the demonstrators on Tiananmen Square, one officer of the U.S. Embassy on the square when the PLA initiated its assault witnessed several Chinese demonstrators being shot in the head and killed by the PLA.

REFERENCES

Timothy Brook, *Quelling the People: The Military Suppression of the Beijing Democracy Movement* (New York: Oxford University Press, 1992); Michael Fathers and Andrew Higgins, *Tiananmen: The Rape of Peking* (Toronto: Doubleday Canada, 1989); Theodore Han and John Li, *Tiananmen Square Spring 1989* (Berkeley: Institute of East Asian Studies, University of California, 1992); Jane's Information Group, *China in Crisis: The Role of the Military* (Surrey, U.K.: Jane's Information Group, 1989); Jin Jiang and Qin Zhou, trans., *June Four: A Chronicle of the Chinese Democratic Uprising* (Fayetteville: University of Arkansas Press, 1989).

TIANJIN, TREATY OF (1858) After the search and seizure of a Hong Kong-registered, British ship, the *Arrow*, by Chinese authorities, the British seized Guangzhou (Canton) and sailed a fleet north, attacking Chinese ports and naval vessels on the route. In May 1958, as they had in 1840–1842, British forces again closed on Tianjin and took control of the **Dagu forts** protecting approaches to the city. In the face of this threat, the Qing emperor signed a treaty allowing the British to place an ambassador in Beijing, opened travel anywhere in China to British nationals with passports, and protected the preaching of the Christian religion in China. The Treaty of Tianjin also opened up a total of six more treaty ports in Manchuria, Taiwan, Guangdong, and Hainan Island to British trade.

Chinese authorities later refused to permit foreign ambassadors to live in Beijing, despite this treaty clause. The British again assaulted but failed to seize the Dagu forts near Tianjin, even though supported by ships under the command of Commodore Joseph Tatnall, an American naval officer. The British chief negotiator responded by marching British troops to Beijing and burning the emperor's Summer Palace, the **Yuanmingyuan** (now called the "old Summer Palace") on October 16, 1860. The Chinese emperor agreed to add a "Convention of Peking" to the Treaty of Tianjin, expressing regret for harassing British representatives, opening Tianjin and other places as treaty ports, and ceding part of the Kowloon Peninsula to Britain. (See **Peking Convention** and the **Arrow War**.)

REFERENCES

John K. Fairbank, *Trade and Diplomacy on the China Coast* (Cambridge: Harvard University Press, 1953); Gerald Graham, *The China Station: War and Diplomacy, 1830–1860* (New York: Oxford University Press, 1978); Jonathan D. Spence, *The Search for Modern China* (New York: W. W. Norton, 1990).

U

UNIT 731 This is the identifier for a secret Japanese biological warfare experimentation unit that was established in Harbin, China, during the period of Japanese occupation there. In 1932, a Japanese physician, Ishii Shiro, established an ''Epidemic Prevention Laboratory'' in a military medical school in Tokyo. Concurrently, in Manchuria, with military assistance, Ishii established the ''Togo Unit'' in a village 100 kilometers southeast of Harbin. The unit in Tokyo experimented with biological warfare defenses, while the Manchurian unit conducted research on offensive methods and produced biological agents. In 1936, the Togo Unit was redesignated the Epidemic Prevention Unit of the Kwantung Army (it was also known as the Ishii Unit). After **World War II** broke out in China, Japan also built a production facility for biological weapons in Pingfan, 25 kilometers southeast of Harbin, which was completed in August 1940. By 1941, the Ishii Unit had been renamed again, as the ''Epidemic Prevention and Water Purification Department'' of the Kwantung Army. However, internally it was referred to as the **Manchukuo** Unit 731. Ishii became a colonel in the Japanese army and established branch units in Beijing, Nanjing, Guangzhou, and Singapore.

In 1939, Unit 731 employed a typhoid fever pathogen against Russian forces on the Mongolian border. The Japanese also experimented by spreading bubonic plague in Chengdu and Ningbo in 1941, killing about 100 people in Ningbo. In addition to using Chinese citizens for biological experiments, the Japanese conducted frostbite experiments on Chinese prisoners to determine how to best treat cold injuries. Ishii was promoted to general by the end of the war.

REFERENCES

Michael Lindsay, *The Unknown War: North China 1937–1945* (London: Bergstrom and Boyle, 1975); Yuki Tanaka, *Hidden Horrors: Japanese War Crimes in World War II* (Boulder, CO: Westview Press, 1996).

UNITED FRONT This is a term that has two meanings. In the context of Chinese Communist theory, it means to "unite with secondary enemies while attacking the major ones." The 12th Congress of the Executive Committee of the Communist International (ECCI, or Comintern) in Moscow in October 1932 asked all national Communist Parties to establish committees and depots to initiate mass movements when required at some point in the struggle against imperialism. The Chinese Communist Party (CCP) used the slogan of "uniting with secondary enemies" on April 10, 1934, in a denunciation of the Japanese occupation of Manchuria.

In the context of Nationalist Party (Guomindang, or KMT)-CCP politics, the United Front meant the agreement by forces of each side to combine their efforts to build the country. Comintern agents in China in 1923 convinced leaders of the two parties to combine forces to "end warlordism and foreign humiliation." The first United Front leadership met at **Whampoa Military Academy** in May 1924 to create a military organization designed to defeat **warlord** forces. The United Front conducted a moderately successful **Northern Expedition** but broke down in 1927 after the Shanghai uprisings in April, the **Canton Coup**, the **Nanchang Uprising**, and the **Autumn Harvest Uprisings**.

The second United Front was formed in order to fight the Japanese, as a result of the **Xi'an Incident** in 1936. The second United Front lasted from 1937 to 1945, but the **New Fourth Army Incident**, where Nationalist units attacked Communist military forces in January 1941, is an example of how much mutual distrust and enmity existed between CCP and KMT forces.

REFERENCES

Edmund O. Clubb, *Twentieth Century China* (New York: Columbia University Press, 1964); Mao Zedong, "On the International United Front against Fascism," *Selected Works*, vol. 3 (Beijing: Foreign Languages Press, 1967); Benjamin Schwartz, *Chinese Communism and the Rise of Mao* (Cambridge: Harvard University Press, 1951).

U.S. ARMY OBSERVER GROUP, YAN'AN. See DIXIE MISSION

U.S. MARINES IN NORTH CHINA (1945–1949) Units of the U.S. Marine Corps were assigned to secure northern Chinese ports to facilitate the Japanese surrender, extract Allied internees, and establish efficient communications between Lieutenant General **Albert Wedemeyer**'s command and American forces in the Pacific. The marines soon found themselves guarding railroads and otherwise becoming entangled in an ever-widening web of political as well as military responsibilities. It took the involvement of General **George C. Marshall** to repatriate the Japanese and extract the combat veterans of the Third Amphibious Corps.

The Third Amphibious Corps, led by Major General Roy Geiger, comprised the First Marine Division (led by Major General Pedro del Valle) and the Sixth Marine Division (led by Major General Lemuel Shepherd, Jr.). The First Marine

Air Wing and various service units brought the corps to approximately 80,000 officers and men. These were experienced combat troops recuperating from the Okinawa campaign and preparing for Operation Olympic, the November 1945 invasion of Japan.

With the sudden end of **World War II**, General MacArthur, supreme commander of the Allied powers (SCAP), issued General Order Number One, which required all Japanese troops to surrender to designated Allied Theater commanders. Thus, the 1.5 million Japanese troops in the China Theater were Generalissimo **Chiang Kai-shek**'s responsibility. In cable exchanges with the Joint Chiefs of Staff (JCS), General Wedemeyer, commanding general, China Theater, made clear that Chinese government forces were incapable of moving from the Chinese interior to Japanese-occupied China in sufficient strength to accept the Japanese surrender and provide minimum security. The expeditious surrender and disarming of Japanese troops to Chinese authorities required U.S. transportation and ground assets.

North China held a major portion of Japanese forces and prisoner of war (POW) camps for Allied internees and was the farthest distance for Nationalist government forces to travel. On September 30, 1945, the First Marine Division came ashore in the **Dagu**-Tianjin area, the gateway Western and Japanese troops used 45 years earlier to move across the flat North China Plain to suppress the **Boxer Rebellion**. Units of the First Marines moved inland and cleared a road to, and garrisoned, Beijing. On October 6, the First Marines and a general from the Nationalist government in Nanjing took the surrender of 50,000 Japanese troops in the Tianjin area. On October 11, the Sixth Marine Division assumed its duties at Qingdao, a deepwater anchorage on the southern coast of the Shandong Peninsula that had been first developed by the Imperial German Navy. The Japanese garrison at Qingdao formally surrendered on October 25.

Shortly thereafter, General George Stratemeyer, Wedemeyer's deputy, cabled the JCS that the Third Amphibious Corps had completed its mission of repositioning sufficient Nationalist government troops to facilitate Japanese repatriation. He recommended that the marines be withdrawn in mid-November 1945. Instead, the political and military situation in North China deteriorated as China teetered close to **Civil War**. Wedemeyer's command reported that the Nanjing government gave little priority to removing the Japanese from North China. On the contrary, it was using these forces to guard urban areas and the road and rail infrastructure. The burgeoning Civil War drew both the marines and the Japanese into guarding the railroad lines that brought crucial coal from northern Chinese mines to Central China's cities and industries and keeping a local peace.

The marines were stuck. During November, seven resolutions calling for the withdrawal of the marines from North China were introduced in Congress. Marine outposts and fliers became targets of harassing fire at the instigation of both the Nationalist and Communist Chinese factions. Ambassador **Patrick J. Hurley**, in China for almost a year, suddenly resigned on November 27, clouding the policy issues. The afternoon of Hurley's resignation, President Truman designated chief of staff general George C. Marshall as a presidential envoy to head a special U.S. mission to conciliate the Chinese factions.

General Marshall arrived in China on December 20, 1945, and by January 10 had negotiated a cease-fire in North China. Four weeks later, the Nanjing government agreed to a plan to accelerate the repatriation of Japanese troops and civilians. Major shipping assets arrived, and North China was cleared of Japanese by April (**Japanese Repatriation from China**). Simultaneously, the marines began withdrawing. The Sixth Marine Division was reduced to the point where it was redesignated the Third Marine Brigade on April 1, 1946. The brigade disbanded on June 10, the day on which the Third Amphibious Corps was redesignated Marine Forces, China. At this point the marines came under General Marshall's operational command and were primarily used to support Tripartite Executive Headquarters, which oversaw the cease-fire. General Marshall continued to reduce the marine presence in China, at times not informing the Nanjing government of his moves. The parallel reduction of Japanese and marines continued. General Marshall departed China in late January 1947. With the end of the Executive Headquarters, the main duty tying the remaining 13,000 marines to China was completed. A small marine presence remained in Tianjin and Qingdao until the end of the Civil War in 1949, but the combat veterans of World War II who had detoured through North China had long departed.

REFERENCES

Bevan G. Cass, *History of the Sixth Marine Division* (Washington, DC: Infantry Journal Press, 1948); Warren I. Cohen, ed., *New Frontiers in American–East Asian Relations* (New York: Columbia University Press, 1983); Herbert Feis, *The China Tangle: The American Effort in China from Pearl Harbor to the Marshall Mission* (Princeton: Princeton University Press, 1953); Foreign Relations of the United States, 1945, vol. 7: *The Far East: China* (Washington, DC: Government Printing Office, 1969); Foreign Relations of the United States, 1946, vol. 10: *The Far East: China* (Washington, DC: Government Printing Office, 1972); Foreign Relations of the United States, 1947, vol. 7: *The Far East: China* (Washington, DC: Government Printing Office, 1972); Benis Frank and Henry Shaw, Jr., *History of U.S. Marine Corps Operations in World War II: Victory and Occupation* (Historical Branch, Headquarters, U.S. Marine Corps, 1968); Harry Harding and Yuan Ming, eds., *Sino–American Relations, 1945–1955: A Joint Assessment of a Critical Decade* (Wilmington, DE: Scholarly Resources Imprints, 1989); Charles F. Romanus and Riley Sunderland, *United States Army in World War II: China–Burma–India Theater: Time Runs Out in CBI* (Washington, DC: Government Printing Office, 1959); U.S. Department of State, *United States Relations with China: With Special Reference to the Period 1944–1949* (Washington, DC: Government Printing Office, 1949).

David E. Reuther

V

VIETNAM, SELF-DEFENSIVE COUNTERATTACK AGAINST (February 17–March 17, 1979) Between 1978 and early 1979, Vietnamese military forces began operations in Cambodia designed to drive the Chinese-supported Pol Pot regime from power. By January 1979, Vietnamese forces had seized Phnom Penh and were soon poised on the Cambodian–Thai border, where they threatened Thailand. Beijing, a major supporter of Pol Pot, had good relations with the Thai government. In response to Vietnam's military moves, the Chinese government complained about a series of violations along the Sino–Vietnamese border, threatening punitive action over alleged Vietnamese incursions into China. During January and into mid-February the **People's Liberation Army** (PLA) moved main-force field armies and divisions from around China to staging areas north of the Vietnam border. Between 30 and 40 divisions of soldiers were eventually assembled. At the same time, fearing a Soviet response in the north because Hanoi and Moscow had signed a Treaty of Friendship and Cooperation in November 1978, China made preparations for combat in the north, for a Soviet counterattack.

In the south, Chinese forces aligned on the Vietnamese border in one front, while in the north another front was formed. The Northern Front comprised Shenyang, Beijing, Jinan, Lanzhou, and Xinjiang Military Regions (MR). **Li Desheng**, commander of the Shenyang MR, was appointed frontal commander. Preparations in the north were mainly defensive, including the evacuation of civilians along the Sino–Soviet border and raising readiness levels of Chinese forces.

The Southern Front was divided into two theaters of war or zones of operation (*Zhanqu*), the Eastern in Guangxi Province opposite Langson and the Western in Yunnan Province opposite Laokai. The Southern Front was given to **Xu Shiyou**, who, as Guangzhou MR Commander, commanded the Eastern Theater of Operations. **Yang Dezhi**, the Kunming MR commander, was in control of the Western Theater of Operations. In Beijing, **Deng Xiaoping** controlled forces on both fronts for the Central Military Commission, with Marshals **Xu Xiangqian** and **Nie Rongzhen** as deputies. **Geng Biao** was chief of the General Staff for the operation.

On February 17, 1979, Chinese forces attacked across the Vietnamese border, advancing along five main axes: in the Eastern Theater of operations against Lang Son and Cao Bang; in the West against Ha Giang, Lao Cai, and Lai Chao. Logistics and service support for Chinese forces came from local force and militia units in the two military regions.

The Vietnamese responded to the preattack Chinese buildup by moving forces north from around Hanoi and by withdrawing some main force units from Cambodia. Vietnam also relied heavily on local force and militia units. China had announced its intention only to "punish Vietnam" for border incursions and said that the attack would be limited to an advance no more than 50 kilometers into Vietnam. After initial advances against very heavy resistance, Chinese forces halted their advance, concentrating on consolidation around major cities, and began an orderly withdrawal, which was completed by March 17, 1979. In the process of the campaign the Vietnamese claimed to have killed or wounded 42,000 Chinese, while Beijing claimed it had inflicted 50,000 casualties on Vietnam with a loss of only 20,000 people. The Vietnamese estimates are probably closer to the actual outcome.

In the conflict, the PLA units suffered from poor command and control, poor logistics, and an inability to coordinate large formations on the battlefield. After the attack on Vietnam the PLA began to discuss reinstituting a rank structure to facilitate battlefield command and control, restructure its Group Armies, train for combined operations, develop Rapid Reaction Forces, and reorganize its military logistics structure.

REFERENCES

Kenneth W. Allen et al., *China's Air Force Enters the 21st Century* (Santa Monica, CA: Rand Corporation, 1995); King C. Chen, *China's War with Vietnam, 1979: Issues, Decisions and Implications.* (Stanford, CA: Hoover Institution Press, 1987); Harlan W. Jencks, "China's Punitive War on Vietnam, a Military Assessment," *Asian Survey* 19 (August 1979): 806–815.

W

WANG DONGXING (1916–   ) Throughout his entire career, Wang Dongxing was associated with the Central Guards, the 8341 Unit of the **People's Liberation Army** (PLA). He was **Mao Zedong**'s personal bodyguard. After Mao's death, Wang helped to arrest the **Gang of Four**, consisting of Mao's wife, Jiang Qing, Wang Hongwen, Zhang Qunqiao, and Yao Wenyuan. Wang was later removed from all of his party and state posts after **Deng Xiaoping**'s accession to power. Wang Dongxing was born in 1916 in Jiangxi Province. Orphaned as a youth, he joined the **Jiangxi Soviet**, where he served in a guard unit in 1933. During and after the **Long March** Wang was part of the personal bodyguard of Mao Zedong. Once the Communist Army reached **Yan'an**, Wang stayed with Mao and commanded Mao's bodyguards through 1949. With the establishment of the People's Republic of China, Wang headed up the elite and secretive "8431 Unit," a guard force that protected China's top leaders. Wang accompanied Mao to Moscow on Mao's only overseas trip in December 1949 and was appointed deputy director of the Eighth Bureau, Ministry of Public Security. Wang also served as vice minister of public security (1955–1959), vice governor of Jiangxi Province (1958), and vice minister of public security in 1960. During the **Cultural Revolution**, Wang's unit was probably responsible for the arrests of the major victims of the turmoil, including **Peng Zhen, Liu Shaoqi**, and Deng Xiaoping.

In 1967, Wang served as director of the General Office of the Chinese Communist Party (CCP) Central Committee, a position that allowed him to clear all information that went to Mao.

In October 1976, as commander of the 8431 Security Forces guarding the Politburo members living in Beijing, he played a key role in the arrest and purge of the Gang of Four. Wang believed that Mao had designated Hua Guofeng as his successor. In 1980, once Deng Xiaoping returned to power, Wang was removed from all party and state posts at the Fifth Plenum of the CCP 11th Central Committee.

REFERENCE

Wolfgang Bartke, *Who's Who in the People's Republic of China* (Armonk, NY: M. E. Sharpe, 1981).

WANGHIA, TREATY OF (1844) An American ambassador from Massachusetts, Caleb Cushing, was sent to China as ambasssador to represent President John Tyler in 1843. Cushing arrived in China in February 1844 and negotiated a treaty that was concluded at Wanghia, a small village near Macao. The American treaty paralleled the British **Treaty of Nanjing** in its provisions granting rights to foreigners in China and opening trade ports. The Treaty of Wanghia permitted Americans to establish hospitals, churches, and cemeteries at five trade ports. It also contained an extraterritoriality clause, giving U.S. officials permission to hold trials under U.S. law and to punish Americans who committed crimes in China. Chinese law had no jurisdiction over Americans. The Chinese also granted the United States the right to employ scholars and to study the Chinese language. One feature of the U.S. treaty that differed significantly from the Treaty of Nanjing is that the United States agreed to turn over for punishment under Chinese law anyone engaged in the opium trade. A French treaty with provisions very close to those of the American treaty was negotiated in October 1844.

REFERENCES

Gerald Graham, *The China Station: War and Diplomacy, 1830–1860* (New York: Oxford University Press, 1978); Jonathan D. Spence, *The Search for Modern China* (New York: W. W. Norton, 1990).

WANG JINGWEI (1853–1944) Wang Jingwei was born Wang Zhaoming on May 4, 1853, in Guangzhou. His family, however, was from Shaoxing, Zhejiang Province. Wang was close to the leader of the **Republican Revolution** (1911), Sun Yat-sen and was himself a senior Guomindang leader. In 1940, Wang became the head of the puppet Chinese regime sponsored by Japanese forces in Nanjing.

In an ironic coincidence, Wang Jingwei's ancestral home in Shaoxing, Zhejiang Province, was also the home of **Chiang Kai-shek**. Wang and Chiang competed against each other for years for leadership of the Guomindang. In fact, Wang was probably closer to Sun Yat-sen than was Chiang. Wang attended Tokyo Law College on a government scholarship, where he earned a degree in 1906, having focused on constitutional law and political theory. He had joined the Tung Meng-hui, which was formed in Japan in 1905 by Sun Yat-sen and other anti-Manchu Chinese. Wang also wrote for the Sun Yat-sen newspaper *Min Pao*.

In 1907, when Sun Yat-sen was forced to leave Japan, Wang Jingwei accompanied Sun to Southeast Asia, where he worked to gather financial support for the Tung Meng-hui. Wang went to Beijing in 1910, where he attempted to assassinate Prince Ts'ai Feng (Cai Feng) by blowing up a bridge across which the prince was to pass. Wang was arrested in April 1910 and was under sentence of execution when the **Wuchang Revolt** in October 1911 brought down the Qing dynasty. From 1912 to 1917, Wang was in France. He returned to China in 1917, where he worked with Sun Yat-sen to build support for the military.

Prior to the **Northern Expedition**, Wang Jingwei went with Sun to Beijing and met with the **warlords Zhang Zuolin, Feng Yuxiang**, and Duan Qirui in an effort to bring about national unity. After Sun Yat-sen's death in 1925, Wang was one of the senior members of the Guomindang. He led the left-wing faction, which advocated cooperation with the Communists (he may well have met some of them in France). Wang continued to exercise leadership in the Guomindang through 1940, after competing with Chiang Kai-shek. In fact, on March 21, 1939, when Wang was in Hanoi, an assassination attempt by persons believed to be Nationalist agents under Chiang's control wounded Wang and killed his companion.

Wang traveled to Shanghai after this incident, where he talked to Chinese who were active in the Japanese-controlled puppet government. He also went to Tokyo in May and October 1939, where he secretly signed an agreement to cooperate with Japan. Finally, on November 30, 1940, Tokyo signed a treaty with the puppet government in Nanjing. Wang signed another, broader treaty of alliance between his Nanjing government and Tokyo on October 30, 1943.

In 1944, Wang Jingwei traveled to Japan for medical treatment for the bullet wounds he suffered in Hanoi in 1939. He died on November 10, 1944.

REFERENCE

Howard L. Boorman, ed., *Biographical Dictionary of Republican China*, 4 vols. (New York: Columbia University Press, 1970).

WANG SHUSHING (1905–1974) Wang Shushing was born in Making County, Hubei Province, the site of the **Huangan-Macheng Uprisings**. He spent a great deal of his career in that area, including service with the **New Fourth Army**. Wang was born into a landlord family and received a high school education, although he did not attend a university. He was one of the officers promoted to general when the **People's Liberation Army** (PLA) instituted a rank system in 1955. Wang started out as a guerrilla leader in the Dabie Mountain (Dabieshan) area of the **Eyuwan Soviet**. When **Zhang Guotao** formed the **Fourth Front Army**, Wang Shushing served as one of its deputy commanders. In fact, when the Fourth Front Army split with the First Front Army during the **Long March**, after the **Maoergai Conference**, Wang led the right column of the Fourth Front Army into Xinjiang in an attempt to establish a secure base there.

Later in 1937, when the Fourth Front Army rejoined Communist forces, forming the **Eighth Route Army**, Wang Shushing was sent back into the Dabie Mountain area of the old Eyuwan Soviet under the command of **Ye Ting**, with the forces that formed the New Fourth Army. During this period, Wang developed a close working relationship with **Li Xiannian**. Wang stayed in the Hubei region after **World War II** and the **Civil War**, serving as commander of the Henan–Hubei Military Region and the Hubei Military District. When the South-Central Military Region formed, Wang became a deputy commander. He was moved to the center in 1956 and, after arriving in Beijing, was appointed director of the PLA General Ordnance (General Logistics) Department. From 1960 to 1974, Wang was a political commissar of the Academy of Military Science, a position he held until the time of his death. From 1969 to 1974, Wang also served as a member of the Communist Party Central Military Commission. He died on January 7, 1974.

REFERENCES

Wolfgang Bartke, *Biographical Dictionary and Analysis of China's Party Leadership, 1922–1988* (New York: K. G. Saur, 1990); ZGDBKQS.

WANG ZHEN (1908–) Wang Zhen was born in 1908 in Liuyang County, Hunan Province. He attended only three years of primary school. At the age of 12, Wang Zhen left his home (his father was a poor peasant) to work and live in Changsha, the provincial capital. He served in the Hunan Provincial Army as a private and also worked as a fireman and boiler tender on a steam locomotive on the Canton-Hank'ou (Guangzhou-Hankou) Railway. Wang Zhen actually joined the Nationalist Party (Guomindang, or GMD) in 1925, perhaps at the urging of the Communist movement. He attended the third class of the **Whampoa Military Academy** and with the Nationalist Army took part in the **Northern Expedition**. Then, for a time, he worked to organize railway workers to join the labor movement. Wang led a women's detachment in the **Nanchang Uprising** in August 1927 and became a full Communist Party member in that year.

In 1928 and 1929, Wang Zhen studied in Moscow, probably at the Frunze Academy with **Liu Bocheng**. During 1930–1931, he worked with **Zhang Yunyi** in Jiangxi Province helping to organize guerrillas. When that guerrilla unit was designated the Red Sixth Corps of the **Chinese Workers and Peasants Red Army**, Wang Zhen served as the political commissar of the corps. In 1933, when the Second Front Army was formed, the Sixth Red Corps was incorporated into the Second Front Army, and Wang stayed on as a division, and later corps, political commissar, serving in that position during the **Long March**, from 1934 to 1936. After the **People's Liberation Army** forces reached **Yan'an** and set up a base area there, the Front Armies converted into divisions of the **Eighth**

Route Army and the **New Fourth Army**. Wang Zhen was made commander of the 359th Brigade, 120th Division, Eighth Route Army. The division was formed from the forces of the Second Front Army. The 359th Brigade was assigned an operational area in the Shaanxi–Gansu–Ningxia area, which contains some of the least productive agricultural land in China and is one of the poorest areas in the country. Wang Zhen is famous for having conducted the **Nanniwan Experiment**, which achieved agricultural self-sufficiency for his brigade and still serves as the model for the People's Liberation Army logistics system today. After 1949, Wang was sent into Xinjiang Province, in western China, where he served as deputy commander, political commissar, and commander of the Xinjiang Military District between 1951 and 1953, pacifying the local Muslim minority populace and establishing Communist Party control over the region.

Because of his experience as a youth working on steam locomotives as a fireman, Wang Zhen was made the commander and political commissar of the People's Liberation Army Railway Engineering Corps in 1954. This military organization worked to build and maintain the railways as strategic lines of communication for the People's Liberation Army. From 1956 on, Wang was a Communist Party Central Committee member and also served on the Politburo and the Central Military Commission. He was untouched by the **Cultural Revolution** and was later a vice premier from 1975 to 1980 and president of the Chinese Communist Party Central Committee's Party School from 1982 to 1987. In 1988, he was elected vice president of China by the National People's Congress.

REFERENCES

Wolfgang Bartke, *Biographical Dictionary and Analysis of China's Party Leadership, 1922–1988* (New York: K. G. Saur, 1990); William W. Whitson with Chen-Hsia Huang, *The Chinese High Command* (New York: Praeger, 1973).

WAR OF LIBERATION. See CIVIL WAR

WAR TO OPPOSE THE UNITED STATES AND ASSIST KOREA (Korean Conflict or Korean War) The "War to Oppose the United States and Assist Korea" (Kangmei Yuanchao) is the official Chinese name for the **Korean War**. The phrase encapsulates Chinese propaganda at the time, which blamed the United States and the Republic of South Korea (ROK) for their aggression against Kim II Sung and the North Korean people. The Chinese have consistently blamed the United States and ROK forces for initiating the war and have stressed their own role as one of giving spontaneous and voluntary aid to a neighbor in its time of need in the noble fight against imperialism.

Telegrams between **Mao Zedong** and Joseph Stalin and other documents and memoirs that were released in the 1980s reveal that the leadership was also concerned with China's pragmatic self-interests when it entered the Korean War. There was a perception among Chinese leaders, Mao, in particular, that war between China and the United States was inevitable. Therefore, they reasoned, it was better to fight the United States early, before China had rebuilt and, therefore, had less to lose. It was also to China's advantage to fight the United States away from Chinese soil, but still close enough to rely on its own logistical support.

Anti-Americanism built up in China prior to the start of the Korean War, beginning in late 1946, when the American navy transported Nationalist (Guomindang) forces to Manchuria to block Communist forces from entering the region. U.S. support for **Chiang Kai-shek** on Taiwan further angered Beijing. Mao believed the United States would soon attack China by air or sea or even by land through Vietnam or the Korean Peninsula to help reinstate the Nationalists to power and to expand U.S. power throughout East Asia. To counter this threat, Beijing took measures to strengthen the organization and disposition of forces along its eastern coast and establish strategic reserves, even before all remnants of the Nationalist forces had been defeated. The fear of American attack also served as the catalyst for hasty efforts to establish a modern air force and navy to defend China.

During the War to Oppose the United States and Assist Korea, anti-American propaganda was widespread. Movements, such as the "Hate America Campaign," and repeated accusations that the United States used bacterial warfare against the Korean and Chinese people helped sustain support for the war effort in China during the early years of the founding of the People's Republic of China. Out of a sense of patriotism, many Chinese parents even named their children after slogans of the day, such as "Support Korea" (Yuanchao).

REFERENCES

Jürgen Domes, *Peng Te-huai: The Man and the Image* (Stanford, CA: Stanford University Press, 1985); Alexander L. George, *The Chinese Communist Army in Action: The Korean War and Its Aftermath* (New York: Columbia University Press, 1967); Samuel B. Griffith II, *The Chinese People's Liberation Army* (New York: McGraw-Hill, 1967); Nie Rongzhen, *Inside the Red Star: The Memoirs of Marshal Nie Rongzhen*, trans. Zhong Renyi (Beijing: New World Press, 1988); Peng Dehuai, *Memoirs of a Field Marshal*, trans. Zheng Longpu (Beijing: Foreign Languages Press, 1984); Shu Guang Zhang, *Mao's Military Romanticism: China and the Korean War, 1950–1953* (Lawrence: University Press of Kansas, 1995).

Susan M. Puska

WARD, FREDERICK TOWNSEND (1831–1862) Frederick Townsend Ward, the commander of the **Ever-Victorious Army**, which helped to suppress the **Taiping Rebellion**, was born on November 29, 1831, in Salem, Massachusetts. His father was involved in the shipping trade as both a ship's master and a ship broker. In 1847, when a company of Massachusetts Volunteers going to fight in the Mexican War paraded through Salem, Ward sought to join them. However, he was returned to his home as underage. Ward was then sent to sea by his father on the merchant ship *Hamilton*, bound for Hong Kong, captained by a relative, William Henry Allen. Ward returned to Salem in 1848 and enrolled in the American Literary, Scientific, and Military Academy in Vermont (this evolved into Norwich University, which is still a military institution today). On December 16, 1849, Ward again went to sea aboard the *Russell Glover*, captained by his father, bound for San Francisco. Ward also seems to have traveled to South America in 1850 and 1851, where he claimed to have met Guiseppe Garibaldi. In 1851, Ward sailed from San Francisco to Shanghai, where he engaged in the coastal trade. He signed on as a first officer on the ship *Gold Hunter*, which carried Chinese workers to Mexico in 1852. In Mexico, Ward signed on with William Walker, who had declared himself president of the "Republic of Lower California," in Sonora, Mexico. Ward deserted from the Walker group and attempted to start a business shipping scrap metal to New York but failed and returned to California. In February 1854, Ward embarked on the ship *Westward Ho!* for Hong Kong. Later in 1854, Ward returned to Salem. He soon left again as a volunteer lieutenant with the French army to serve in the Crimean War. He left French service in 1857 and returned to China, where he was first mate on a coastal steamer, returning to New York and then back to Shanghai in 1859 in the company of **Henry Burgevine**. In June 1860, Ward gained the trust of Shanghai banker Yang Fang and organized the **Foreign Arms Corps**, with Burgevine, to defend against the Taipings. This evolved into "Ward's Corps of Disciplined Chinese." Ward was later made a colonel in the Imperial Qing "Army of the Green Standard." As he prepared to marry Yang Fang's daughter, the "Foreign Arms Corps" was renamed the "Ever-Victorious Army" by the local governor, Xue Huan.

Ward married Yang Fang's daughter in Shanghai in 1862 and was promoted to brigadier general in the Qing Army. He was the only foreigner authorized to command Chinese troops, and the Ever-Victorious Army was the only Sino–foreign force authorized by the Qing throne.

In the attack on Yuyao, near Ningbo, Ward was shot in the abdomen by a Taiping musket ball. He was evacuated to Ningbo on the British gunboat *Hardy*, but he died on September 22, 1862, at the age of 30.

REFERENCES

Caleb Carr, *The Devil Soldier: The Story of Frederick Townsend Ward* (New York: Random House, 1992); Michael Franz, *The Taiping Rebellion: History and Documents*, 3 vols. (Seattle: University of Washington Press, 1966); Andrew Wilson, *The "Ever-Victorious Army"* (London: William Blackwood and Sons, 1868; reprinted Novato, CA: Presidio Press, 1991).

WARLORDS (Early 1900s) After the establishment of Nationalist China (the Republic of China) in 1911 and the death of the first president, Sun Yat-sen, in 1925, power on the Chinese mainland devolved from the central government to regional military governors who, with private or provincial armies, were local military dictators. Most of these military leaders, or warlords, received sponsorship and financial support from foreign powers, either commercial entities or foreign countries, and China was divided up into their spheres of influence. They supervised regional military affairs, sought legitimacy from charisma, and dominated civil affairs.

The north-central area of China and the North China Plain was controlled at some times by Wu Peifu and at other times by Feng Yuxiang; the coastal developed areas of Shandong, Jiangsu, Anhui, Zhejiang, and Fujian by Feng Xi; Manchuria by **Zhang Zuolin**; Shanxi by **Yan Xishan**; Sichuan by Liu Xiang and Yang Sen; Guizhou by Yuan Cuming; Yunnan by Dang Zhiyao; and Hubei by Wu Peifu. The central government was unable to control the warlords or to compel their obedience to its edicts. The warlords were engaged, among themselves, in what has been described as "a European-like model of balance of power politics." Only after the **Northern Expedition** (Beifa, 1926–1928), led by **Chiang Kai-shek**, did the central government begin to gain control over the regional feifdoms, and even then, there was trouble with the growing Communist movement. The basis of warlordism was the personal nature of allegiance and patron–client relationships inherent in regional politics. This tendency was by the weakening of the Qing dynasty and the need to raise local military forces to combat such uprisings as the **Taiping Rebellion** and the **Nian Rebellion**. Some attribute the tendency to Qing leaders such as **Zeng Guofan** and **Zuo Zongtang**, who fought the Taiping and Nian rebels with forces they raised themselves. Today, the rotation of military region commanders by the Central Military Commission of the Chinese Communist Party after the establishment of the People's Republic of China in 1949 is a method designed to prevent local **People's Liberation Army** (PLA) commanders from developing power structures independent of the party center and the central government and turning into "new warlords."

REFERENCES

Lucian W. Pye, *Warlord Politics: Conflict and Coalition in the Modernization of Republican China* (New York: Praeger, 1971); James E. Sheridan, *Chinese Warlord: The Career of Feng Yu-Hsiang* (Stanford, CA: Stanford University Press, 1966); Arthur Waldron, "The Warlord: Twentieth-Century Chinese Understandings of Violence, Militarism, and Imperialism," *The American Historical Review*, Vol. 26, No. 4 (October 1991), pp. 1073–1100; William W. Whitson with Chen-Hsia Huang, *The Chinese High Command: A History of Communist Military Politics, 1927–71* (New York: Praeger, 1973).

WASHINGTON CONFERENCE (1921–1922) The representatives of the United States, Great Britain, Japan, France, and Italy met in Washington, D.C., between November 1921 and February 1922 to decide on the limitation of naval armaments among the Great Powers and the status of Pacific area holdings and claims at the invitation of U.S. president Warren G. Harding. Three treaties were agreed at the conference. The first was a Four-Power Treaty among the United

States, Japan, Great Britain, and France on the status of China and the Pacific Islands, which also sought to avert a potential war between Great Britain and the United States. Great Britain was committed to enter a war in support of Japan by the Anglo–Japanese Treaty of 1905 (renewed in 1911), and tension in the Asia-Pacific made war between Japan and the United States look possible. The second treaty was the Five-Power Treaty on the question of arms limitations, where the four powers were joined by Italy. The Five-Power Treaty attempted to reach an equilibrium in naval construction measured in tonnage of capital combat ships of 5:5:3 among Great Britain, the United States, and Japan. The five powers, which discussed arms, also participated in a third set of talks with China, Belgium, the Netherlands, and Portugal to discuss the general situation in the Asia-Pacific region, concluding in the Nine-Power Treaty. While the main focus of the conference was to avoid a naval arms race among Japan, the United States, and Great Britain (the five-power conference), the meeting of the nine powers principally focused on the future of the claims of each of these powers on China, which had been carved into concession areas. The Nine-Power Treaty decided by the Washington Conference dealt with issues such as tariff rates, taxation of foreign ventures in China, cotton exports, leased territories and extraterritoriality in concession areas, open-door clauses in international agreements with China, and the maintenance of foreign post offices. Most important in the context of China's military history, however, may have been the negotiations over the question of China's **warlords**, police and railway guards on foreign-operated railroads in China, arms trafficking, and the Chinese Eastern Railroad. Japan's settlements in Manchuria and on the Shandong Peninsula, a fact of life since the **Sino–Japanese War** (1894–1895) and the Russo–Japanese War (1904–1905), were a major question addressed at the conference, as were British claims to Hong Kong and the port of Weihaiwei on the Shandong Peninsula. The conference was held between November 19, 1921, and January 31, 1922. Other issues discussed were Great Britain's naval base in Singapore and Japan's presence in the Soviet Far East. For China, however, the conference was significant in that the presence of foreign troops in concession areas was ratified, leaving Japanese forces in control of the Korean Peninsula and garrisoned in parts of Manchuria and the Shandong Peninsula. These same troops were later involved in the murder of the warlord **Zhang Zuolin** in Manchuria, the **Jinan Incident** in 1928, the **Nine One Eight Incident** (Mukden Incident) in Manchuria in 1931, and the **Marco Polo Bridge Incident** beginning **World War II** in China in 1937.

REFERENCES

Erik Goldstein and John Maurer, eds., *The Washington Conference, 1921–1922: Naval Rivalry, East Asian Stability and the Road to Pearl Harbor* (Essex, U.K.: Frank Cass, 1994); Arthur Waldron, *How the Peace Was Lost* (Stanford, CA: Stanford University Press, 1992); Wun-sze King, *China at the Washington Conference, 1921–1922* (New York: St. John's University Press, 1963).

WEDEMEYER, ALBERT C. (1897–1989) General Albert C. Wedemeyer was born in Omaha, Nebraska in July 1897. He is most closely identified as an anti-Communist who was a strong supporter of **Chiang Kai-shek**. He was first assigned to Asia in 1943, after serving out the early years of the war in the War Department as a war plans and strategy staff officer, with close ties to General **Marshall**. At the time of the U.S. entry into the war, in December 1941, General Wedemeyer had been newly promoted to lieutenant colonel. In June 1942, Wedemeyer was promoted to colonel; to brigadier general in July 1942; to major general in September 1943; and to lieutenant general in January 1945. Wedemeyer was once the subject of a major Federal Bureau of Investigation inquiry into the disclosure of classified war plans to the press. A *Chicago Tribune* correspondent, Chester Manly, published the content of what was purported to be a summary of U.S. war plans, drafted by Wedemeyer, on December 5, 1943. He was also suspected of having a pro-Nazi leaning, an allegation that he strongly denied. He was a graduate of the U.S. Military Academy (in 1919), and he had overseas service in the Philippines, China, and Germany. Wedemeyer was a graduate of the German War College and lived in Germany between June 1936 and fall 1938, which was the basis for the charge that he sympathized with the Nazis.

In his own memoirs, General Wedemeyer assessed Chiang Kai-shek as a "most loyal and least demanding ally," a view that was at odds with General **Stilwell**'s assessment of Chiang and his character. This disagreement was reflective of Wedemeyer's deep, long-standing personal conflict with Stilwell after Wedemeyer's assignment to the Far Eastern Theater.

After the Casablanca Conference on January 24, 1943, General Wedemeyer flew to New Delhi, India, and to China to brief Allied war commanders on the impact of the decision at Casablanca on the China Theater. He spent the early part of February 1943 in China and met Stilwell as well as then-major general **Claire L. Chennault**. Wedemeyer also met with Chiang Kai-shek to discuss the results of the Casablanca Conference as well as the provision of arms and ammunition to China along the **Burma Road** and by air (the "Hump" flights).

In April 1943, as relations between General Stilwell and his air commander, General Chennault, worsened, Wedemeyer attended meetings in the office of General **George Marshall** in Washington. Wedemeyer's own account of these meetings records that Stilwell had an extremely low opinion of Chiang Kai-shek and did not trust Chennault. Wedemeyer, in contrast, seems to have liked Chennault immensely.

In October 1943, General Wedemeyer was promoted to major general and sent to India as a member of British admiral Mountbatten's Southeast Asia command staff. Wedemeyer's responsibilities were for logistics and campaign planning, a position similar to the one he had occupied in the War Department in Washington. He was deputy chief of staff, Southeast Asia Command. Wedemeyer attended the December 1943 "SEXTANT" war planning conference in Cairo, Egypt, which was also attended by Chiang Kai-shek. Wedemeyer opposed

General Stilwell's concept of a ground campaign through Burma, and opposed Stilwell on the construction of the Burma Road as an all-weather highway. Instead, agreeing with Chennault's concept, Wedemeyer believed that resupply to China was sounder from a tactical standpoint.

On October 27, 1944, General Wedemeyer was notified that he was to be ordered to China by General Marshall to assume command of the China Theater of Operations (separated from the Burma–India Theater), with concurrent duties as chief of staff to Chiang Kai-shek. His command comprised all of the Chinese mainland, including Manchuria, Indochina, and the offshore islands of China with the exception of Taiwan, which was held by the Japanese. The mission assigned Wedemeyer was (1) to advise Chinese forces on the conduct of the ground war; (2) to assist China in ground operations, training, and logistics support and to carry out air operations using U.S. combat forces; (3) to withhold the use of U.S. forces from any mission involving the suppression of internal civil strife; and (4) to act as Chiang Kai-shek's chief of staff. Stilwell had been recalled to Washington after a long, open, and sometimes bitter conflict with Chiang Kai-shek over the conduct of the war against Japan. Chennault, who was close to Chiang, remained in China as the commander of the China Air Task Force, which evolved out of the **Flying Tigers** and later became the 14th Air Force. Wedemeyer also inherited a number of staff officers from Stilwell, many of whom advised Wedemeyer that "it was impossible to learn anything, or to secure any cooperation, from Chiang Kai-shek's military headquarters." Wedemeyer concentrated his efforts on selecting about 350 of the best divisions in the Chinese army and training 39 of them as a combat force (see **ALPHA Force**). Wedemeyer also pressed Generalissimo Chiang Kai-shek continuously to improve the low pay and poor food supplies provided to Nationalist military forces. Wedemeyer believed that the poor food and allowances were a significant factor leading to corruption among Nationalist officers.

After the surrender of the Japanese, Wedemeyer remained in China as commander of American forces in China until April 1946, when he returned to the United States for the second time since the Japanese surrender, for consultations. Wedemeyer had been considered for appointment as ambassador to China but, because of strong opposition to his appointment from the Chinese Communist Party leaders, did not receive that appointment. After a short period of recuperation in the United States, General Wedemeyer was appointed commander of the Second Army, one of the U.S. Continental Armies, headquartered at Fort Meade, Maryland. From that post he continued to comment on U.S. China policy in personal correspondence to the secretary of defense (Forrestal), urging that the United States support the Nationalist government and Chiang Kai-shek against the Communists. Wedemeyer returned to China between July 16 and September 18, 1947, to assess "the political, economic psychological, and military situations" on the orders of President Truman. He advised that the United States extend military aid and advice to Nationalist forces, which were about to lose control of Manchuria, the Shandong Peninsula, and Hebei. Wedemeyer concluded that "a China dominated by Chinese Communists would be inimical to the interests of the United States." General Wedemeyer died on December 17, 1989. He was buried in Arlington National Cemetery.

REFERENCES

Keith E. Eiler, ed., *Wedemeyer on War and Peace* (Stanford, CA: Hoover Institution Press, 1987); Barbara W. Tuchman, *Stilwell and the American Experience in China, 1911–45* (New York: Macmillan, 1971); Albert C. Wedemeyer, *Wedemeyer Reports!* (New York: Henry Holt, 1958); U.S. Department of State (Far Eastern Series 30), *United States Relations with China: With Special Reference to the Period 1944–1949* (Washington, DC: Government Printing Office, 1949).

WHAMPOA MILITARY ACADEMY (Huangpu Military Academy) The Nationalist Party (Guomindang) First National Congress organized the Nationalist Army along Russian lines in January 1924. The congress also decided to establish a military academy at the naval base of Whampoa, in Canton (Guangzhou).

The National Cadet Corps at the Whampoa Military Academy was formed in May 1924 of about 500 men culled from a pool of 3,000 applicants. General **Chiang Kai-shek** was the president of the academy, and Liao Zhongkai was appointed its civilian administrator. Among the political advisers at the school was **Zhou Enlai**, from the Chinese Communist Party (CCP). In addition to a staff drawn from the Nationalist and Communist Parties, a group of Soviet military and political advisers assisted in the establishment and administration of the Whampoa Military Academy. Chiang had visited Moscow in October 1923, on behalf of Chinese Nationalist president Sun Yat-sen (Sun Zhongshan). On Chiang's return, in December 1923, he met Vasily K. Bluecher, who later became chief military consultant from the Soviet Comintern to the Guomindang. Bluecher, using the nom de guerre General Galen, became one of the faculty members of the academy.

Chinese faculty members of the Whampoa Military Academy were drawn from the graduates of the Japanese Military Academy (Shikan Gakko) and China's Baoding and Yunnan Military Schools. The Whampoa Military Academy had both political and military goals, with a curriculum intended to produce ideologically sound officers able to advance the ideals of the Chinese Nationalist Revolution and to defeat the **warlord** armies. The initial course of instruction at the Whampoa Military Academy was six months long, but it was later increased to two years. The academy trained the officer Corps that later led both the Communist and Nationalist Armies. In addition, the Stalinist Soviet political and military party infrastructure that Chiang learned from Bluecher formed the basic structure of political and propaganda work in both the Nationalist and Communist Armies, a form that survives intact today. See **Canton Coup**.

REFERENCES

JGDBKQS, vol. 2; Frederick Fu (F. F.) Liu, *A Military History of Modern China: 1924–1949* (Princeton: Princeton University Press, 1956); William W. Whitson with Chen-Hsia Huang, *The Chinese High Command* (New York: Praeger, 1973); ZGRMJFJDSD.

WORLD WAR II (ANTI-JAPANESE WAR) By early 1937, Japanese troops were firmly in control of Manchuria, having created the Independent Republic of **Manchukuo**. Around Shandong, Tianjin, and Beijing, Japanese forces were garrisoned in strength. They conducted regular field exercises on the North China Plain around Hebei, which they were entitled to do by the Boxer Protocol of 1900. When the **Marco Polo Bridge Incident** took place on July 7, 1937, Japanese troops were conducting night maneuvers about 10 miles southwest of Beijing. In reaction to the clash with Nationalist forces, by July 27, Japanese troops seized the bridge and further reinforced their presence, dominating the region around Beijing and Tianjin.

Although no war was officially declared in 1937, **Chiang Kai-shek** attempted to bomb the Japanese fleet in Shanghai, missing the fleet but killing hundreds of civilians. This act prompted Japan to send 15 new divisions to North and Central China. Through August and September 1937, Japanese forces pressed their attack on Chinese forces in Hebei along the northern rail lines, capturing Baoding in September, Shijiazhuang in October, and moving west along the main rail line, bringing Taiyuan, Shanxi Province, under attack in November 1937. In the central coastal region, the Japanese conducted an amphibious landing in Hangzhou Bay, threatening Shanghai, which they seized that month. Advancing from Hangzhou Bay and Shanghai, Japanese forces attacked Nanjing in late 1937.

On December 12, 1937, Nationalist forces began a disorganized retreat, abandoning Nanjing to the Japanese forces. Here, in the infamous **Rape of Nanjing**, Japanese forces are reported to have slaughtered as many as 300,000 Chinese, including young children, gang-raped some 20,000 Chinese women, killed 30,000 fleeing soldiers, and murdered 12,000 other civilians. The Nationalist armies simply abandoned the region, retreating into Central China during 1938.

In May 1938, Chinese forces managed to hold the Japanese in a defensive battle near Xuzhou, in southern Shandong, killing some 30,000 Japanese attackers. Nonetheless, the city of Xuzhou and its strategic rail junction fell to the Japanese in May 1938. In another delaying action, Nationalist forces destroyed the dikes on the Yellow River, holding the Japanese advances but destroying hundreds of small Chinese villages. By October 25, 1938, the Japanese took control of the island river port city of Wuhan, consolidating control of most of northern and eastern China. The Nationalist government fled into Sichuan, establishing a capital in Chongqing, Sichuan Province. Japanese forces firmly controlled Manchuria, the entire east coast of China, Taiwan, Guangdong, and Indochina. In northwest China, the Communist Party and its forces controlled parts of Shaanxi, Shanxi, Inner Mongolia, Ningxia, and Lanzhou. The Nationalists controlled southwest and South China. China's main link to the outside world was the **Burma Road**, running from Kunming in Yunnan Province into Burma, where it reached the port of Rangoon. The Communists and Nationalists organized a National Army, with a goal of cooperating in many areas against the Japanese. The Communist forces in **Yan'an** combined into the **Eighth Route Army**, which operated in northern China, while the forces of the old Communist Second and **Fourth Front Armies**, organized into the **New Fourth Army** in South-Central China.

In 1940, the Communist forces launched the **Hundred Regiments Campaign** as a means to take pressure off Nationalist forces in northern China. The Nationalist–Communist **United Front** cooperation broke down, however, as Nationalist forces grew resentful of Communist successes in the Yangtze River delta area, east of Shanghai. Nationalist generals insisted that Communist forces shift their operations farther west, across the Yangtze River. The Communist New Fourth Army ignored ultimatums from the Nationalists, made no attempt to cross the Yangtze, and between January 7 and 13, 1941, came under attack by the Nationalists, killing at least 3,000 Communist troops. This was known as the **New Fourth Army Incident**.

U.S. personnel came to China's aid in the war. As early as 1937, an adviser to Chiang Kai-shek, **Claire Lee Chennault**, who was a former U.S. Army Air Corps pilot, organized a group of volunteer pilots to fly for China. By 1941, President Roosevelt had shipped 100 P-40 fighters to China as part of a Lend-Lease program designed not to violate U.S. neutrality in the war against Japan. Chennault was also permitted to recruit Army Air Corps pilots for service in China as ''volunteers,'' where they became known as the **Flying Tigers**.

After the United States entered the war on December 7, 1941, it began to grant loans and more Lend-Lease funds to China, bolstering Nationalist coffers. President Roosevelt sent Lieutenant General (later, General) **Joseph W. Stilwell** to China to command the China–Burma–India Theater of war. Known as ''Vinegar Joe,'' Stilwell also served as the liaison to Chiang Kai-shek. In an internal argument within the U.S. policy community, precedence was given to Chennault and airpower in the prosecution of the war, while building the ground forces and conducting a ground offensive against the Japanese were put on hold. Communist guerrilla forces and Nationalist units, through 1942 and 1943, tied down hundreds of thousands of Japanese troops in a holding action, while Allied forces fought the Japanese across the islands of the South Pacific.

At the Cairo Conference, in December 1943, Chiang Kai-shek met with Roosevelt and Churchill, securing an agreement that after the war Manchukuo (Manchuria) and Taiwan would be returned to Chinese control. By 1944, Stilwell's forces, with the British and the Chinese, had opened a new offensive through Burma. Air forces, operating out of Chongqing, meanwhile struck at Japan and Thailand.

In summer 1944, in reaction to the air raids against Japanese forces, Japanese troops initiated **Operation Ichigo**. This was a major ground and air offensive out of Shanxi, Henan, Anhui, and Guangdong intended to cripple Chinese resistance. By this time, the United States tried to encourage better cooperation between the Communists and the Nationalists, placing a **U.S. Army Observer Group at Yan'an**. The United States also wanted to send in underground forces of the Office of Strategic Services (OSS) to work with Communist militia units against the Japanese. The initiative in the war stalled, but as the atomic bomb dropped on Japan in 1945, Russian forces attacked into Manchuria supported by Communist military forces, drawing the **Anti-Japanese War** to a close and setting the stage for the **Civil War**.

REFERENCES

Iris Chang, *The Rape of Nanjing: The Forgotten Holocaust of World War II* (New York: Basic Books, 1997); Herbert Feis, *The China Tangle: The American Effort in China from Pearl Harbor to the Marshall Mission* (Princeton: Princeton University Press, 1953); Frederick Fu (F. F.) Liu, *A Military History of Modern China: 1924–1949* (Princeton: Princeton University Press, 1956); JGRMJFJZS, vol. 2; Barbara W. Tuchman, *Stilwell and the American Experience in China: 1911–45* (New York: Macmillan, 1971).

WUCHANG UPRISING (1911) The Wuchang Uprising was a revolt against the Qing (Manchu) dynasty that took place on October 10, 1911, in the city of Hankou. The date October 10 (''Double Ten Day'') now marks the day associated with the establishment of the Republic of China. It is celebrated on Taiwan as National Day. Because the day marks the rebellion against dynastic rule, it is also celebrated in the People's Republic of China (PRC) as having unleashed the forces that led to the establishment of the PRC.

The cities of Wuchang, Hanyang, and Hankou, on the Yangtze River, have linked today into Wuhan. From about 1904 onward, groups of radical Chinese students, many of whom had studied abroad, linked with disaffected military officers and workers to form antidynastic revolutionary cells in Hankou and Wuchang. In Sichuan Province, the ''Railway Protection Movement,'' a protest against Qing attempts to nationalize locally owned railroads, threw the province of Sichuan into turmoil in September 1911. This violence incited the rebels in Wuchang, which was also a major rail and transportation hub, to plot a rebellion. The revolutionary cells in Hankou and Wuchang had attracted followers from secret societies as well as about one-third of the troops in the Hubei New Army, totaling between 5,000 and 6,000 men.

An accidental explosion of a bomb on October 9, 1911, in the revolutionary headquarters prompted an investigation by local police. As police began to arrest revolutionaries, the rebels advanced their plans. On October 10, army units that had taken up the revolutionary cause seized the ammunition depot of the city and attacked Wuchang's main forts. The first unit to act was the Wuchang Eighth Engineer Battalion, but other military units joined the rebellion, which spread to Hankou. By October 22, despite reactions from the Qing government, the rebellion spread to army units in Changsha, Hunan Province, and Xi'an, Shaanxi Province. Northern generals ignored Qing dynasty orders to move troops against the rebels, and by November 1911, the Qing dynasty had complied with demands to promulgate a constitution, establish a parliament, and elect a premier. **Yuan Shih-kai** was elected premier on November 8, 1911. Sun Yat-sen returned to China from France in December, and on January 1, 1912, he was inaugurated as "provisional president" of a new republic of China, with the capital in Nanjing. The Manchu Court announced the abdication of the emperor Puyi (then six years old) on February 12, 1912.

REFERENCES

Jonathan D. Spence, *The Search for Modern China* (New York: W. W. Norton, 1990); Mary C. Wright, ed., *China in Revolution: The First Phase* (New Haven, CT: Yale University Press, 1968).

WUHAN INCIDENT (July 1967) After being powerless to cope with student uprisings during the **Cultural Revolution** in Wuhan in mid-July 1967, Wuhan Military Region commander Chen Zaidao permitted students supporting the **People's Liberation Army** (PLA) to kidnap and hold two Beijing officials. In the ensuing struggle to restore order, PLA units killed well over 1,000 **Red Guards** and students.

Although the Wuhan Incident was not the only instance of the use of military force to restore order when the Red Guards went too far, it was perhaps the most dramatic event, where PLA leaders took a clear stand against Beijing's leftist **Gang of Four**, led by Jiang Qing.

After serious incidents in Sichuan and Yunnan, where the PLA killed or wounded thousands of Red Guard radicals, spurred to action by the Cultural Revolution's leaders in Beijing, two officials, Wang Li and Xie Fuzhe, were returning to Beijing via Wuhan. The two were arrested by Red Guard factions, possibly with the help or at least tacit agreement of Chen Zaidao. In a protest against the Cultural Revolution Group in Beijing, Chen declared himself unable to effect the release of the officials. Eventually, only intervention by airborne forces of the PLA's 15th Army and other main-force PLA combat units was able to restore order and ensure the release of the Cultural Revolution Group officials. The incident triggered a reaction in the PLA, with **Lin Biao**, on August 9, 1967, denouncing political commissars and purging the leadership of the General Political Department. PLA forces worked to restore order, often violently, and the Maoist Cultural Revolution Group lost power.

REFERENCES

Parris Chang, *Radicals and Radical Ideology in China's Cultural Revolution* (New York: Columbia University Press, 1973); Lynn White, *Policies of Chaos: The Organizational Causes of Violence in China's Cultural Revolution* (Princeton: Princeton University Press, 1989).

X

XI'AN INCIDENT (December 12, 1936) Two Nationalist generals, **Zhang Xueliang** and Yang Hucheng, in cooperation with Chinese Communist forces, arranged for the capture of **Chiang Kai-shek** at Lintong, outside Xi'an on December 12, 1936. They turned Chiang over to a Chinese Communist Party (CCP) delegation and forced Chiang to agree to work with the Communists in a **United Front** effort of national resistance against Japan.

After the arrival of Communist forces in Yan'an in October 1935, Yung Hucheng and Zhang Xueliang began to cooperate with Communist forces against the Japanese. Zhang's father, **Zhang Zuolin**, the Manchurian **warlord** leader, had been blown up on a train in Manchuria in 1928, and Zhang Xueliang had been forced out of Manchuria by the Japanese in 1931. Zhang was supposed to coordinate attacks on the **Yan'an** Soviet area for Chiang Kai-shek in another "bandit suppression campaign" (or **Encirclement Campaign**) designed to destroy the Communist forces. Instead, Zhang entered into negotiations with **Zhou Enlai** on coordinated anti-Japanese actions. In December 1936, Chiang Kai-shek flew to Xi'an, pushing Zhang Xueliang to step up the Encirclement Campaign and to crush the Yan'an Communist Base. When Chiang's forces opened fire on student demonstrations in the city of Xi'an, who were protesting Japanese occupation of China, Zhang Xueliang sought to force Chiang Kai-shek to focus on Japanese aggression instead of the Chinese Communist forces. Talks between Zhang and Chiang broke down on December 11, 1936.

On December 12, 1936, forces of Zhang Xueliang, aided by Yang Huchang, attacked Chiang's headquarters at a resort in Lintong. They killed Chiang's guards and stormed his villa, but Chiang escaped in pajamas into the hills above the resort. Within hours, by daylight on the twelfth, Zhang's men captured Chiang Kai-shek. A former adviser to Zhang Xueliang, who also served as Chiang's adviser, W. H. Donald, traveled to Xi'an with Chiang's brother-in-law, **T. V. Soong**, and Nationalist intelligence chief **Dai Li** to negotiate Chiang's release. Zhang Xueliang and Yang Hucheng put forth an eight-point program calling for an end to the **Civil War** in China and for a period of United Front cooperation against Japan. For the Communist side, Zhou Enlai reached Xi'an in a plane sent by Zhang. Zhou proposed a national United Front against Japan under Chiang Kai-shek's leadership, that a line of demarcation be established between Communist and Nationalist forces, and that a national conference on the "salvation" of China be held that included Communist representatives. **Ye Jianying** was part of the CCP delegation to the talks. The talks ended on December 24, 1936, with Chiang agreeing to end suppression campaigns against the Communists, to reorganize the national government, to release political prisoners, and to permit Zhang Xuechang and Yang Hucheng to direct forces in northwest China. Chiang returned to Nanjing on December 25, reaching the city on the twenty-sixth, accompanied by Zhang Xueliang. However, Zhang was soon arrested by Chiang's forces and, on December 31, 1936, was sentenced to 10 years in prison for "organizing followers to coerce one's superior with brute force." Zhang was later given amnesty and put under house arrest. The agreements, however, generally held up through the war against Japan, until the outbreak of the **Civil War** after the Japanese surrender. Yang Hucheng was murdered on September 17, 1949 in Chongqing, allegedly by Nationalist agents.

REFERENCES

James Bertram, *First Act in China: The Story of the Sian Mutiny* (New York: Viking Press, 1938; reprinted 1973); Mi Zhanchen, *The Life of General Yang Hucheng* (Hong Kong: Joint, 1981); Wu Tien-wei, *The Sian Incident: A Pivotal Point in Modern Chinese History* (Ann Arbor: University of Michigan Press, 1976).

XIANG ARMY The Xiang Army was a local, private defense force created by a Qing official from Hunan Province, **Zeng Guofan**. In 1852, Zeng and his brothers organized a force of local peasants to protect family property against Taiping rebels. The army was named for the Xiang River, which flows through Hunan Province. The Xiang Army proved to be very effective against the rebel forces during the **Taiping Rebellion**. It was led by local gentry from Hunan, and its soldiers were all locally conscripted peasants.

The Xiang Army is only one of a number of locally raised and financed forces that operated against rebels and bandits in China. These local armies, really the forerunners of the **warlord** armies of the early 1900s, were accepted and even encouraged by the Qing dynasty, first because they proved to be cheaper than using imperial forces against rebels and bandits, and second because they responded to localized threats more quickly and vigorously because the leaders were tied to the area through property holdings. However, need for local forces like the Xiang Army and the **Huai Army**, which was organized later to suppress bandit forces in the **Nian Rebellion**, demonstrated the difficulty the Qing rulers were having keeping China unified under their rule. After the fall of the city of Nanjing, the Taiping capital, in 1864, the Xiang Army was disbanded.

REFERENCES

Caleb Carr, *The Devil Soldier: The Story of Frederick Townsend Ward* (New York: Random House, 1992); Franz Michael, *The Taiping Rebellion: History and Documents*, 3 vols. (Seattle: University of Washington Press, 1966); Ssu-yu Teng, *The Taiping Rebellion and the Western Powers: A Comprehensive Survey* (Oxford: Clarendon Press, 1971).

XIANG YING (1898–1941) Xiang Ying was born in Huangpo County, Hubei Province, into the family of a scholar who was living in poverty. Xiang received only four years of formal, primary school education. Xiang Ying started out as an early and influential Communist Party member, joining the party in 1922. He was deeply involved in party organizational work during the early part of his career, particularly in organizing labor movements in China. In 1928, he was secretary of the Jiangsu Province Communist Party Bureau. He was also a Communist International (Comintern) supervisory member and was the Chinese representative to the Moscow-controlled Pan-Pacific Trade Union, serving as the organization's secretary.

In 1931, as the **Jiangxi Soviet** formed, Xiang Ying was the secretary of the Red Army Central Revolutionary Committee. He also served as political commissar for the Red Army from 1934, during the **Long March**. He was concurrently a Communist Party Central Committee member and a Politburo member continuously from 1928 to the time of his death in 1941. Xiang Ying differed often with **Mao Zedong** and considered Mao an elitist who relied on pedantic references to ancient Chinese novels, a man out of touch with current reality. In 1937, when the **New Fourth Army** was formed, Xiang Ying was assigned as the deputy commander under **Ye Ting**. Because of Ye Ting's early service with the Nationalists and his departure from the Communist cause after the failed **Autumn Harvest Uprisings**, Xiang's assignment was probably as much to ensure that the Communist Party had control over the New Fourth Army as to take advantage of Xiang's skills at party organization work. Xiang Ying was killed in the **New Fourth Army Incident**, at Maolin in southern Anhui Province, when elements of the Nationalist Army attacked the **People's Liberation Army**. Ye Ting was captured in the same incident.

REFERENCES

Wolfgang Bartke, *Biographical Dictionary and Analysis of China's Party Leadership, 1922–1988* (New York: K. G. Saur, 1990); William W. Whitson with Chen-Hsia Huang, *The Chinese High Command* (New York: Praeger, 1973).

XIAO JINGGUANG (1903–1989) Xiao Jingguang was one of the officers appointed to the rank of general in 1955, when the **People's Liberation Army** (PLA) established a system of **military ranks**. His career is noteworthy because at one point, in the 1930s during the **Encirclement Campaigns**, Xiao was tried by the Chinese Communist Party, expelled from the party, and then imprisoned.

He was later restored to his party positions by **Mao Zedong**, with whom Xiao Jingguang had particularly close relations.

Xiao Jingguang was born in Shashi County, Hunan Province, in 1903, into a wealthy family of landlords. He attended the First Normal School in Changsha, Hunan Province, where, in 1920, he began a long association with Mao Zedong and **Liu Shaoqi**. In fact, while in Changsha he joined the Soviet Russia Research Association, established at the First Normal School by Mao and Liu. Xiao joined the Socialist Youth Corps in Shanghai in 1921 and later, in 1923, traveled to the Soviet Union, where he attended Sun Yat-sen University (the University of the Toilers of the East) and joined the Communist Party. He also studied military science there. After his return to China in 1924, Xiao Jingguang attended the **Whampoa Military Academy** and graduated from its first class. He stayed at the Whampoa Academy and taught the fourth class of Whampoa graduates in 1925. Xiao Jingguang took part in the **Northern Expedition** as a Communist Party representative in the Fifth Division of the Second Corps. He fought at the **Nanchang Uprising** and later in Nanjing as part of the Communist uprising.

After the Nanchang Uprising, when the Communist and Nationalist Parties split, Xiao Jingguang returned to the Soviet Union, where he studied at the Military and Political College (Red Army Academy) in Leningrad for three years. Among his classmates were **Ye Jianying** and **Liu Bocheng**. Another classmate of Xiao's at the Military and Political College was Chiang Ching-kuo (Jiang Jingguo), the son of Guomindang leader **Chiang Kai-shek**. Chiang Ching-kuo went on to become the minister of defense and then president of the Republic of China, on Taiwan.

When the Red Army Academy of China was started in Ruijin, in the vicinity of the **Jinggangshan Revolutionary Base Area**, Xiao Jingguang was the commandant and the commander of the cadet regiment. Xiao did not take part in the first three of the Encirclement Campaigns but worked closely, instead, with the Moscow-controlled Communist Internationale (Comintern) adviser, Otto Braun (Li De), who was based in Ruijin as Comintern representative. When the Fifth Army formed with a base of 20,000 Nationalist soldiers who defected to the Communist side in December 1931, Xiao became political commissar of the Fifth Red Army Corps. He was also commander of the Seventh Red Army during the Fourth Encirclement Campaign. In 1932–1933, as Mao Zedong's support began to wane, Mao was removed from his posts at the **Ningdu Conference**. Xiao Jingguang was accused by **Zhou Enlai** of being a rightist, probably because of Xiao's close association with Otto Braun, among other things. Xiao was tried by the Communist Party with Zhou in control of the trial, expelled from the Communist Party, and given a five-year prison sentence. Reportedly, Xiao attempted suicide in disgrace. His case was later revisited, and he was assigned to the Red Army University to work, with his jail sentence suspended. After the **Zunyi Conference**, during the **Long March**, when Mao's primacy over the party and the PLA was reempowered, Xiao was restored to full Communist Party membership by Mao. During the Long March, Xiao served as the chief of staff of the Cadre's Regiment of the **First Front Army** and as commander of the Seventh Red Army (Corps).

In 1936, after the **Eighth Route Army** formed, and the Communist forces were headquarters in **Yan'an**, Xiao Jingguang was the commander of the 29th Independent Division. He also served as chief of staff of the Eighth Route Army and commander of rear services for the Shaan–Gan–Ning (Shaanxi–Gansu–Ningxia) border region, in which Yan'an was located. As the **Civil War** began in 1946, with the movement of Communist forces into place for the **Manchurian Campaign**, Xiao served as a deputy commander of the **Northeast Democratic United Army** under the leadership of **Lin Biao**. He was later commander of the North Manchurian Military Region. Xiao Jingguang moved south to command the First Army and later the 12th Army, moving the First Army into Hunan and Guangxi Provinces, where he accepted the surrender of Nationalist general Chen Mingren. Xiao later served as deputy commander of the **Fourth Field Army** and commander of the Wuhan Garrison and the Hunan Military District.

In 1950, Xiao Jingguang was appointed the commander of the People's Liberation Army Navy, a position that he held until 1979, for over 28 years. He was a member of the National Defense Council and was appointed a general when ranks were given to the PLA in 1955. He retired from active service in 1979, without having been harmed by the **Cultural Revolution**. Even after his retirement, however, he remained on the Communist Party Committee of the PLA Navy. When the Central Advisory Commission for the Party and the Military was formed by **Deng Xiaoping** to facilitate the leadership transition in the military by a younger generation of leaders, Xiao was appointed to the Central Advisory Commission. From 1956 through 1982, Xiao was successively a member of the Eighth, Ninth, 10th, and 11th Central Committees of the Chinese Communist Party. He died on March 29, 1989.

REFERENCES

Wolfgang Bartke, *Biographical Dictionary and Analysis of China's Party Leadership, 1922–1988* (New York: K. G. Saur, 1990); William W. Whitson with Chen-Hsia Huang, *The Chinese High Command* (New York: Praeger, 1973).

XIAO KE (1908–) Xiao Ke was born in 1908 in Jiahe County, Hunan Province. His father was a poor, but educated, member of the gentry class, and Xiao Ke received a classical education at his hands. Xiao later entered middle school and attended a teacher's college (normal school). He left Hunan and attended the **Whampoa Military Academy**, after which he joined a police regiment in Canton (Guangzhou), where the academy is located. In 1927, Xiao Ke joined the 24th Division, led by **Ye Ting**, and began what was to become a long association with Ye. Xiao acted as a political instructor in the division, for which he was trained at the Whampoa Military Academy, and joined the Communist Party in 1927.

Xiao Ke took part in the **Nanchang Uprising** and then returned to the south, in the Guangzhou area, taking part in the **Autumn Harvest Uprisings**. He moved around South China in late 1927, leading a small guerrilla band, which he turned over to the control of **Zhu De** and **Chen Yi** when he joined them in south Hunan in January 1928. The group then moved into the **Jinggangshan Revolutionary Base Area**, and Xiao became a regimental commander in the Fourth Corps. Between 1930 and 1932, Xiao Ke commanded a division in the Sixth Corps, where he was associated with **He Long** and **Wang Zhen** as part of the group of officers who led the **Second Front Army**. Xiao Ke fought in the **Encirclement Campaigns** around the Jinggangshan Base Area in the Jiangxi Soviet. Then, during the Fifth Encirclement Campaign, as the **Long March** started, Xiao left the base area on August 1, 1934, to link up with He Long, in the area of the Central Soviet in Jiangxi, marching out with a force of 9,000 men. Xiao and He Long began a romance with a pair of sisters, whom they married at about the same time, strengthening the bonds between the two leaders.

In 1937, Xiao Ke became commander of the 31st Army of the **Fourth Front Army**. As the **Eighth Route Army** formed after the Long Marchers reached **Yan'an**, Xiao commanded the 358th Brigade of the 120th Division. He also served as deputy commander of the 120th Division. Between 1937 and 1946, Xiao Ke served in the Hebei area, at one time as commander of the West Beijing (Xiping) Military District and later as commander of the Hebei-Rehe-Liaoning Region. During the **Manchurian Campaign** of the **Civil War**, Xiao was director of the Rear Services (Logistics) Department of the **Northeast Democratic United Army**. In 1949, Xiao became the vice president of the North China Military Academy of the **People's Liberation Army** (PLA). After serving as chief of staff of the **Fourth Field Army** in 1949–1950, Xiao began to serve in positions responsible for PLA education and training, as director of the People's Revolutionary Military Training Council from 1951–1954, as deputy director of the Training Department of the PLA General Staff Department from 1954 to 1957, and as director of the Training Department from 1957 through 1959. In 1959, Xiao Ke was sent to be the vice minister of State Farms and Land Reclamation, probably because of his close association with **Peng Dehuai**, who as chief of the General Staff Department was purged by **Mao Zedong**. With the **Cultural Revolution** heating up in 1966, Xiao Ke was removed from all posts, including his membership on the Eighth Communist Party Central Committee, and he was purged of all party and government positions. In 1972 or 1973, Xiao Ke was reactivated, indicating that his problems with the party were as much related to problems with **Lin Biao** as problems with Mao, since the rehabilitation came not long after Lin's death.

Xiao Ke resumed his military career in 1973 as commandant of the PLA Military and Political Academy. In 1978, when that institution was converted into the Academy of Military Science of the PLA, Xiao remained as its commandant. He retired from active duty in 1985 and joined the Central Advisory Commission of the Communist Party Central Committee. At the commemoration of the 50th anniversary of the end of **World War II**, in August 1995, Xiao Ke gave the keynote speech in the Great Hall of the People in Beijing.

REFERENCES

Wolfgang Bartke, *Biographical Dictionary and Analysis of China's Party Leadership, 1922–1988* (New York: K. G. Saur, 1990); William W. Whitson with Chen-Hsia Huang, *The Chinese High Command* (New York: Praeger, 1973).

XISHA ISLANDS DEFENSIVE CAMPAIGN (Paracel Islands) (1974) A neomaritime spirit arose in China with the reappearance of **Deng Xiaoping** in 1973 as the vice chairman of the Chinese Communist Party (CCP) **Central Military Commission** (CMC). Soon after that, the PRC initiated its campaign to seize the Xisha (Paracel) Islands from the South Vietnamese. Deng had close contacts with the navy leader, **Su Zhenhua**, and was said to be a leading voice supporting a more maritime-oriented approach for the **People's Liberation Army** (PLA) Navy, which had been principally a coastal, ''brown water'' force. (''Blue water'' navies sail the open seas. ''Brown water'' navies restrict their operations to the littoral or continental shelf.) Thus, Deng appears to be closely linked with Xisha events as a means of developing a more active maritime role for China.

On January 11, 1974, the PRC Ministry of Foreign Affairs released a public warning that the PRC had indisputable sovereignty over the **Paracels**, adding that the ''sea areas around them also belong to China.'' Four days later, the PLA Navy began moving forces into position. According to South Vietnamese survivors, China first began to move fishing vessels and Chinese fishermen into the area. On January 15, South Vietnamese ships allegedly attempted to displace the Chinese fishermen and two Chinese patrol boats from the islands. By that time, the PLA had assembled a naval force consisting of about 11 warships carrying more than 600 assault troops. Fighter aircraft based on Hainan Island, about 130 miles northeast of the Xishas, were available for air cover. On January 17, several Chinese patrol vessels arrived, followed by 10 other warships. The South Vietnamese had a former U.S. Coast Guard cutter with a five-inch gun and four destroyer escorts, each armed with two three-inch guns. There was no air support available to South Vietnam. In an ensuing naval engagement, one of the destroyer escorts was reported to have been hit by a missile from a Chinese patrol craft. Within two days, the Chinese had landed and taken control of the Xishas. Chinese press releases on the incident emphasized that the participation of ''people's-militia fishermen'' played a role in the seizure of the islands. The Chinese claim that South Vietnamese forces suffered 300 casualties in the engagement and that the People's Republic of China (PRC) captured 49 Vietnamese military personnel, who were later returned to Vietnam.

This action marked the first time China used military force after the United States had improved relations with China (January 1972), and Washington took no action. Hanoi was not in a position to react either. The Xisha operation of 1974 marked the only Chinese amphibious operation of any distance, and it involved only 11 ships and some 600 ground troops.

REFERENCES

Frederica M. Bunge, ed., *China, a Country Study* (Washington, DC: Government Printing Office, 1981); Gerald Segal, *Defending China* (London: Oxford University Press, 1985); Sun Zhen, ed., *PLA Forces* (Hong Kong: CONMILIT Press, 1986); Bruce Swanson, *Eighth Voyage of the Dragon—A History of China's Quest for Seapower* (Annapolis, MD: Naval Institute Press, 1982); ZGDBKQS.

Vance H. Morrison

XU GUANGDA (1908–between 1966 and 1979) Xu Guangda was a Moscow Frunze Academy-trained officer who was associated with **He Long** throughout most of his military career. Xu was generally opposed to **Mao Zedong**'s **People's War** strategy and favored, instead, a professional, regularized military. He died after having been ''persecuted'' by the **Gang of Four** during the **Cultural Revolution**. Xu Guangda was one of 10 officers appointed to the rank of general in 1955.

Xu Guangda was born in 1908 in Changsha, Hunan Province. His family was well-to-do enough to pay for an education, and Xu later graduated from the **Whampoa Military Academy**. Xu participated in the **Nanchang Uprising** in 1927 and was an early organizer of the **People's Liberation Army** (PLA). When the Red Sixth Army (Corps) was established in Hubei in 1929, Xu helped to organize the unit under He Long and served as its chief of staff. Xu was sent to Moscow for training at the Lenin University and the Frunze Academy in the mid-1930s. He was dean of education at the *Kang-Da*, or **Anti-Japanese Military and Political College**, in **Yan'an**, after his return to China in 1936 or 1938. In 1942, Xu Guangda commanded the Second Brigade of the 120th Division, **Eighth Route Army**. He stayed with the division until 1945.

During the **Civil War**, Xu Guangda continued to serve under He Long. He commanded the Third Column of the Northwest Field Army and participated in the capture of Lanzhou, the capital of Gansu Province, in the Xining Campaign of 1949. After the **Field Army System** formed, Xu commanded the Second Army (Corps) of the First Field Army. Based on his formal Soviet training in tactics at Frunze, Xu was put in command of PLA armored forces in 1951, a position he continued to hold until his purge in 1967. During the **Korean War** in 1952, Xu led the second of the armored divisions to move into Korea as part of the **Chinese People's Volunteers**. He had organized and trained the division himself. Xu was appointed a vice minister of national defense in 1959, concurrent with his being commander of armored forces, and he held that position until he was purged as an anti-Communist rightist in the midst of the Cultural Revolution. He was posthumously rehabilitated by **Deng Xiaoping** in 1979. A *People's Daily (Renmin Ribao)* article on July 9, 1979, said that Xu Guangda had "died heroically after frantic persecution by **Lin Biao** and the 'Gang of Four.' "

REFERENCES

Wolfgang Bartke, *Biographical Dictionary and Analysis of China's Party Leadership, 1922–1988* (New York: K. G. Saur, 1990); William W. Whitson with Chen-Hsia Huang, *The Chinese High Command* (New York: Praeger, 1973).

XU HAIDONG (1900–1970) Xu Haidong was one of the 10 officers appointed as generals when ranks were established in the **People's Liberation Army** (PLA) in 1955. His life is noteworthy for his dedication to the army and the way he organized his troops for battle. His death, however, may be more noteworthy. Despite having been wounded in combat nine times and having been partially bedridden for almost 18 years, Xu Haidong was persecuted during the **Cultural Revolution**, sent out of Beijing to Henan Province, and allegedly killed by loyalists of **Mao Zedong, Lin Biao**, and the **"Gang of Four"** for his opposition to Mao's policies.

Xu Haidong was born in Dawu County, Hubei Province, on June 17, 1900. He was one of 10 children. His father made pottery. Xu attended only four years of primary school and then became an apprentice potter, raised ducks, and worked in a factory. In 1925, Xu Haidong joined the Communist Party. He spent a total of about six years in the military forces of **warlords** and the Nationalist Army, but after the **Northern Expedition** he began to organize guerrilla resistance units in Hubei. During the Guomindong campaign of "White Terror" to rid Hubei of leftists and Communists, Xu slowly organized a small guerrilla band operating in the area of the **Macheng Uprising**, southwest of the **Dabieshan** Mountains into the Red 31st Division. This unit was the forerunner of the Red 11th Corps. By 1931, Xu Haidong was commander of the 25th Red Army (Corps), which was part of the **Fourth Front Army**.

When the Dabieshan Revolutionary Base area containing the **Eyuwan Soviet** was abandoned by the main body of the Fourth Front Army in 1932, during the fourth **Encirclement Campaign**, Xu Haidong commanded the stay-behind forces. In 1934, when the stay-behind force of the Fourth Front Army made its own **Long March** to join the rest of the force, Xu led it from the Eyuwan Soviet area in September 1934 into the area of the Wei River near the city of Xi'an, Shaanxi Province. He finally arrived in the Xi'an area with his force in June 1935. He also led the "Eastern Campaign" of the Fourth Front Army in Shanxi. When the **Eighth Route Army** was organized in 1937, Xu commanded the 344th Brigade, 115th Division. He moved south with his brigade to join **Zhang Yunyi** and the **New Fourth Army**, where he was deputy commander of New Fourth Army forces north of the Yangtze River and secretary of the Shandong Province Communist Party Committee.

In 1947, Xu Haidong was sent to organize logistics forces but entered the hospital in Dabian. He had been in and out of hospitals because of combat wounds since 1939. Xu was appointed to the Eighth Central Committee of the Chinese Communist Party in 1956 but was hospitalized periodically. Then, at the **Lushan Conference**, Xu sided with **Peng Dehuai**, attacking Mao Zedong's economic policies in the "Great Leap Forward." In 1966, he again disagreed with Mao over the policies of the Cultural Revolution, particularly the practice of attacking career cadres with a long history of working for the party and the army. He was, nonetheless, still appointed to the Communist Party Ninth Central Committee as a full member in 1969. However, on October 25, 1969, Xu was sent to Zhengzhou, Henan Province, with his family, as an "anti-party element," allegedly by Lin Biao and the "Gang of Four." He died on March 25, 1970, in Zhengzhou. Xu Haidong was posthumously rehabilitated by **Deng Xiaoping** on January 25, 1979, as one of eight deceased revolutionary cadre who had been persecuted to death by Lin Biao and the Gang of Four during the Cultural Revolution.

REFERENCES

Wolfgang Bartke, *Biographical Dictionary and Analysis of China's Communist Party Leadership, 1922–1988* (New York: K. G. Saur, 1990); He Bian, "Xu Haidong," in *Jiefangjun Jiang Ling Zhuan*, vol. 1, pp. 471–508 (Beijing: Jiefangjun Chubanshe, 1986).

XU XIANGQIAN (1901–) Xu Xiangqian is one of the marshals of the **People's Liberation Army** (PLA) named in 1955. He was born in Wutai County, Shanxi Province, and graduated from Taiyuan Normal College in 1923. After graduation, he enrolled in the **Whampoa Military Academy** in June 1924 and took part in the **Northern Expedition**. He moved to Wuchang (one of the cities forming Wuhan) and taught at the military academy there, where he joined the Communist Party. Xu Xiangqian was doing underground organizational work in Guangzhou during the **Nanchang Uprising** and did not take part in that event. However, he did take part in the **Guangzhou Uprising** in 1927. Xu was a commander in the **Fourth Front Army** and was a leader of the **Eyuwan Soviet**. Naturally, he was closer to **Zhang Guotao** in military style than **Mao Zedong**, rejecting peasant-based guerrilla warfare. Xu was subjected to heavy party criticism after the purge of Zhang Guotao by Mao in 1935 but was named a deputy commander of the 129th Division of the **Eighth Route Army**. Throughout his career, Xu was associated with **Ye Jianying, Liu Bocheng, Peng Dehuai**, and **Nie Rongzhen**. He was a **Cultural Revolution** survivor but was criticized heavily for a time in 1967. Nonetheless, Xu was a Politburo and Cultural Revolution Group member in 1967. He was also a Central Committee member in 1969. In 1978, Xu was appointed minister of national defense.

REFERENCES

Wolfgang Bartke, *Who's Who in the People's Republic of China* (Armonk, NY: M. E. Sharpe, 1981); William W. Whitson with Chen-Hsia Huang, *The Chinese High Command* (New York: Praeger, 1973).

Y

Y-FORCE The "Y-Force" was the designation for 15 Nationalist divisions stationed in Yunnan Province during the early days of the U.S. involvement in **World War II**. American plans called for a two-axis invasion of Burma to wrest it from Japanese control. The acronym for the plan among American staff officers was the X-Y Plan. The Western axis of the attack was to come from India and was called the "X-Force," while the eastern axis was from China, moving south from Yunnan Province. This was called the "Y-Force." General **Stilwell** sought to train 30 Chinese Nationalist divisions to be used against the Japanese in Burma, but Generalissimo **Chiang Kai-shek** objected, arguing that he could not spare the people and could not supply the troops. Stilwell wanted to train the cadres of the divisions, hoping that these trained cadres would, in turn, train the troops of the divisions. Eventually, Stilwell got agreement from Chiang for a force of only about 20 divisions, 5 of which were trained in Ramargh, India. Some of these troops were airlifted to India from China, whereas others had marched out of Burma under pursuit after a defeat by the Japanese. The other 15 divisions were located in Yunnan Province, where they were based and trained, and made up the Y-Force of the American X-Y Plan.

Stilwell never got complete cooperation from Chiang on this plan. Although Chiang eventually acceded to giving Stilwell control of the Y-Force, it became a point of bitter contention between the two men. At one point Stilwell wrote home in a letter describing his dealings with Chiang that working with the generalissimo was difficult and involved "trying to guide and influence a stubborn, ignorant, prejudiced, conceited despot who never hears the truth except from me and finds it exceedingly hard to believe" (Tuchman, *The American Experience in China*, p. 317). Beginning in August 1942, as Stilwell worked to build the X-Force in India, the Chinese fed in soldiers who were of poor quality and in poor physical condition. Many were rejected for the training in India by the Americans, but they were funneled directly into the Y-Force by the Nationalist leadership. Eventually, as late as November 3, 1942, Chiang finally agreed to give Stilwell command of the divisions trained at Ramargh and to put 15 divisions, comprising the Y-Force, into Yunnan Province but under Chinese command. Stilwell was allowed to pick the divisions and the commanders. This force was still part of the American plan for a well-trained force of 30 Chinese divisions. Finally, a school was established in Kunming, Yunnan Province, with the agreement of the governor there, Long Yun, and 30 divisions were designated for inclusion in the U.S. plans for China's army, which included the 15 divisions designated for the attack into Burma in the Y-Force. The force was finally committed into action in the Salween Offensive along the **Burma Road** in May and June 1944. The Y-Force succeeded in driving the Japanese back from a few areas in China along the Burma Road, but the offensive was halted against stiff Japanese resistance short of the border with Burma.

REFERENCES

Center for Military History, *China Defensive* (Washington, DC: U.S. Army Military History Institute, 1996); Charles F. Romanus and Riley Sunderland, *The United States Army in World War Two. The China-Burma-India Theater: Stilwell's Mission to China* (Washington, DC: Historical Division, Department of the Army, 1953); Joseph W. Stilwell, *The Stilwell Papers*, ed. Theodore H. White (New York: Sloan Publishers, 1948); Barbara W. Tuchman, *Stilwell and the American Experience in China, 1911–45* (New York: Macmillan, 1971).

YALTA AGREEMENT (1945) The Yalta Agreement was signed by the United States, the United Kingdom, and the Soviet Union on February 11, 1945, setting the conditions under which the Soviet Union would enter the war against Japan. Although the agreement was signed secretly, without the participation or consent of the Chinese government or Generalissimo **Chiang Kai-shek**, it dealt with matters concerning the sovereignty and territorial integrity of China. In the agreement, the Soviet Union acknowledged that China had full sovereignty over Manchuria but was granted by the United States rights to the port at Dalian (**Port Arthur**) and the China Eastern Railway (the Manchurian Railway). This agreement, for practical purposes, restored rights that were the Soviet Union's based on older, czarist agreements with the Qing dynasty. Stalin, in return for these concessions and others involving the Polish border and Europe, agreed to join **World War II** in the Pacific Theater against Japan. The United States and Great Britain also gave the Kurile Islands to the Soviet Union.

In response both to a deterioration in relations between the Soviet Union and the Nationalist government in China and to Nationalist Chinese unwillingness to enter into any agreements of cooperation and trust with the Chinese Communist Party or its forces in the fight against Japan, President Roosevelt sent Vice President Wallace to China as his emissary in June 1944. One goal of the Wallace mission was to smooth Sino–Soviet relations, and on his way to China, Wallace met with the U.S. ambassador to the Soviet Union, W. Averill Harriman, in Tashkent, on June 15–16, 1944. Harriman also accompanied Wallace to Alma Ata on June 17, continuing to brief Wallace on the Sino-Soviet situation. During these meetings, Harriman briefed Wallace on a conversation between Harriman, Marshal Stalin, and Soviet foreign minister Molotov that had taken place on June 10, 1945. Stalin agreed with the United States that "under the circumstances" Chiang Kai-shek must be supported by the nations allied against the Germans and the Japanese. Stalin also encouraged the United States to push harder for Chiang's military cooperation with the Chinese Communists against the Japanese. In response to a cable that Harriman sent to Washington about the conversation, which emphasized the need for military cooperation in China between the Communists and the Nationalists against the Japanese, the State Department sent instructions to the U.S. embassies in Moscow and Chongqing that set out the U.S. approach to Sino–Soviet relations and, in substance, foreshadowed the agreements at Yalta. U.S. ambassador to China, Clarence E. Gauss, also gave these instructions to Vice President Wallace in preparation for Wallace's talks with Chiang when Wallace arrived in China.

The U.S. ambassadors were told to take the position that (1) Manchuria and Xinjiang Provinces were part of China and should be treated by Moscow as such; (2) the status of Mongolia (Outer Mongolia, as it was known) should be resolved practically, and since that region was autonomous, the Chinese government should not attempt to assert control over the region; (3) Korea should be recognized as an independent state, and its people should ultimately determine their form of government; (4) the U.S. government was not concerned with "doctrinal questions" between contending Chinese groups and recognized the Nationalist government as the primary political authority in China (the Japanese had formed their own "Quisling" government under **Wang Jingwei**, in China and Pu Yi in Manchuria, or **Manchukuo**); (5) the U.S. government would like to see all Chinese forces—not just part of them—take part in the United Nations fight against Japan; and (6) the U.S. government wanted the Chinese and the Soviet governments to remain friendly allies. During their talks, Chiang Kai-shek indicated to Wallace that he was eager to come to terms with the Soviet Union and to improve relations. Chiang also asked that the United States act to find ways to help any future Sino-Soviet discussions turn out well. In a memorandum to Vice President Wallace of June 23, 1944, Chiang also offered to "go more than halfway" in reaching an understanding with the Soviet Union if the United States could bring about a meeting between Chinese and Soviet representatives. In effect, the United States took these comments by Chiang as its "passport" for negotiating an agreement in Yalta with the Soviet Union that was concluded in secret and only later briefed to the Chinese by Ambassador **Hurley** in April 1945.

The main motivation behind signing the Yalta Agreement for the United States and Great Britain was military. Japan had maintained its Kwantung Army in Manchuria intact throughout the war, and the United States had no assurance that the atom bomb, secretly under development, would work. For the United

States, therefore, it was necessary to open another campaign in North China to tie down Japanese forces in the event that the United States had to carry out its planned invasion of Kyushu and the subsequent campaign to take, one by one, the islands making up Japan. To have had to carry out the invasion of Japan would have been very costly, perhaps disastrous if the troops of the Kwantung Army were available to reinforce the Japanese homeland. The invasion of Japan, which was planned to start with the assault on Kyushu, was called "Operation Olympic" and was scheduled to begin in November 1945. Because they were concerned that the information about the agreement would be leaked to Japan by elements in China, Roosevelt and Stalin decided to conclude the Yalta Agreement in secret and brief Chiang Kai-shek only after the fact. When Ambassador Hurley finally informed Chiang Kai-shek of the terms of the agreement on June 15, 1945, Chiang did not react strongly; his reaction indicated that the Soviets may have already discussed the terms of the Yalta Agreement with Chiang.

Concluding the Yalta Agreement in secret still remains a source of controversy and tension in Sino–U.S. relations. In a 1989 meeting in Beijing, **Deng Xiaoping** told President George Bush that "one of the results of the Yalta Conference held by the Soviet Union, Great Britain, and the United States was to Divide up China. . . . Yalta not only severed Outer Mongolia from China, but also brought the northeastern part of China into the Soviet sphere. . . . I hope you will look at the map to see what happened after the Soviet Union severed Outer Mongolia from China . . . a huge chunk of the north cut away" (Bush and Scowcroft, *The World Transformed*, pp. 94–95).

REFERENCES

John Robinson Beal, *Marshall in China* (New York: Doubleday, 1970); George Bush and Brent Scowcroft, *A World Transformed: The Collapse of the Soviet Empire, the Unification of Germany, Tiananmen Square, the Gulf War* (New York: Alfred A. Knopf, 1998); Herbert Feis, *The China Tangle: The American Effort in China from Pearl Harbor to the Marshall Mission* (Princeton: Princeton University Press, 1953); U.S. Department of State (Far Eastern Series 30), *United States Relations with China: With Special Reference to the Period 1944–1949* (Washington, DC: Government Printing Office, 1949).

YAN XISHAN (Yen Hsi-shan) (1883–1960) A **warlord** in Shanxi Province from the time of independence (1911) through the Revolutionary War period, Yan Xishan developed his own ideology combining elements of socialism, capitalism, and Confucian authoritarianism. At different times Yan allied with **Chiang Kai-shek** and arrived at a modus vivendi with the Japanese. However, he remained staunchly anti-Communist.

Yan Xishan was born in Hebian, outside Taiyuan, in 1883. He enrolled in the National Military College and accompanied his class to Japan in 1904. Yan studied in Japan for two years and then entered the Imperial Military Academy, graduating in 1909. His period of study in Japan impressed him with Japanese nationalism and militarism and influenced his rule in Shanxi. On his return to

China, Yan was appointed the director of a military school in Shanxi. He later commanded the Second Brigade of the New Shanxi Army. He became ''military governor'' of Shanxi in the 1911 revolution but later withdrew into northwest Shanxi until he allied with **Yuan Shih-kai**. From 1917 on, he controlled Shanxi Province.

Yan organized the ''Patriotic Self-Sacrifice League'' in Shanxi. He emulated many of the unsuccessful reforms of the *Tong Zhi* restoration of the 1860s that were instituted by **Zeng Guofan**. During the **Northern Expedition** in 1928, Yan occupied Beijing and placed units of his army in Tianjin by agreement with the Guomindang. He is known for having industrialized the province and having created the Taiyuan Arsenal with a staff of Chinese technicians working with German and Swedish engineers. He organized the Shanxi Military Technical School, and by 1926 the arsenal employed about 8,000 workers. During the 1930s Yan broke cooperation with the Guomindang and formed his own regime in Shanxi. For some time he gave Communist organizers such as **Bo Yibo** a free hand to develop a militia-like organization and guerrilla forces. In October 1937, Shanxi fell into Japanese hands, and Yan withdrew his forces to Linfen. He later consolidated his rule in far southwest Shanxi. For a short time, Yan allied with the Communist forces, but during 1941 and 1942, he entered into negotiations with the Japanese to end fighting in Shanxi in return for a Japanese withdrawal. By 1944, he was fighting the Communists again.

Yan accompanied the Guomindang government to Taiwan in 1949 and served as premier until March 1950. He was later made a ''senior adviser'' to Chiang Kai-shek, but remained nearly a prisoner on Taiwan until his death on May 24, 1960.

REFERENCES

Donald G. Gillin, *Warlord: Yen Hsi-shan in Shansi Province, 1911–1949* (Princeton: Princeton University Press, 1967); Lincoln Li, *The Japanese Army in North China* (London: Oxford University Press, 1975); Shanxi Renmin and Zhengzhi Xieshang Hui, *Wo Suo Zhidao de Yan Xishan* [The Yan Xishan I Knew] (Taiyuan: Taiyuan Yinshuachang, 1986).

YAN'AN This city in northeast Shaanxi Province was the seat of power for the Chinese Communist Party from the end of the **Long March** in 1937 to the breakdown of talks with the Guomindang (KMT) in 1947 during the **Civil War**. The Nationalist government recognized Yan'an as the seat of government for the Shaanxi–Ningxia–Gansu Border Region, which was essentially under Communist control during the **Anti-Japanese War. Mao Zedong** was the dominant figure at the base area, where he developed an entire body of ideology addressing Communist Party development, administration, self-criticism, and political persuasion. The Central Communist Party School was located at Yan'an, along with the Anti-Japanese Military and Political College, known as *Kang-Da*. In 1944, the United States sent a **Military Observer Mission**, the **Dixie Mission**, to Yan'an to assess the strength of the Communist movement. Today, the term "Yan'an Spirit" still symbolizes for the Communist Party the ideal of simple, honest administration under arduous conditions.

REFERENCES

Mark Selden, *The Yenan Way in Revolutionary China* (Cambridge: Harvard University Press, 1971); ZGDBKQS; ZGRMJFJZS, vol. 2.

YANG QIANG DUI. See FOREIGN ARMS (RIFLE) CORPS

YANG SHANGKUN (1907–1998) Yang Shangkun was a central figure in the **People's Liberation Army**, the Chinese Communist Party (CCP), and civil-military relations in China, especially during the **Tiananmen Square Massacre**. He was born in Sichuan Province, like **Deng Xiaoping**, to a family of well-to-do peasants. Yang was an early revolutionary who joined the Communist Youth League in 1925 and became a CCP member in 1926. He was among the Party members sent to the Soviet Union in 1926, where he attended Sun Yat-sen University in Moscow from 1927 to 1931. After his return to China, Yang immediately went to work in political organs as the chief of the Propaganda Department of the All-China Federation of Trade Unions in the area of Jiangsu Province and Shanghai. By 1933, as the Nationalists began their suppression campaigns against the CCP, Yang moved to the area of Ruijin in the Jiangxi Soviet of the **Jinggangshan Revolutionary Base Area**.

From 1933 to 1935, Yang was the director of the Political Department of the **First Front Army**. He was a strong supporter of **Mao Zedong** at the 1935 **Zunyi Conference** during the **Long March**. During the **Anti-Japanese War**, Yang worked closely with **Zhu De** and **Peng Dehuai** in directing guerrilla operations. By 1948, Yang moved into the position of director of the General Office of the Central Committee of the Chinese Communist Party, an extremely powerful position that he continued to occupy until he was purged in the **Cultural Revolution** in 1966. Interestingly, in May 1966, Yang was head of the Secret Police. He was named in a "May 16 Circular" as one of the group of five people involved in "counterrevolutionary activities," along with the head of the CCP in Beijing, **Peng Zhen**; PLA chief of the **General Staff Department, Luo Ruiqing**; Minister of Culture Lu Dingyi and Lu's wife. A speech by **Liu Shaoqi** on May 18, 1966, denounced Yang and the other four senior CCP officials at a Politburo meeting. Ironically, by August 1966, Liu was purged along with Deng Xiaoping. Yang was not rehabilitated and restored to power in the party until 1978 (by Deng Xiaoping), and he was made political commissar of the Guangzhou Military Region and vice-governor of Guangdong Province. Yang returned to Beijing in 1980 as a member of the 11th Central Committee and served also as executive vice chairman and secretary general of the **Central Military Commission** (CMC). He remained on the CMC even after being elected president of China in 1988. Yang was married to a playwright, Li Bozhao, who died in 1985.

Yang was a strong supporter of the crackdown on demonstrators during the Tiananmen Square Massacre, as were many of the other former victims of the Cultural Revolution. By the early 1990s, Yang's half-brother, Yang Baibing, was made director of the **General Political Department** of the PLA. Yang was still relatively healthy compared to the retired, but behind-the-scenes strongman, Deng Xiaoping, and in 1992, in Yang's effort to challenge Deng's hand-picked successor Jiang Zemin, Deng forced Yang and his half-brother to retire. Many of the Yang brothers' protégés were also demoted or forced into retirement from the PLA at the same time. Yang Shangkun died on September 14, 1998, in Beijing.

REFERENCES

Wolfgang Bartke, *Who's Who in the People's Republic of China* (Armonk, NY: M. E. Sharpe, 1981); Lowell Dittmer, *Liu Shaoqi and the Chinese Cultural Revolution: The Politics of Mass Criticism* (Berkeley: University of California Press, 1974); Martin Ebon, *Lin Biao: The Life and Writings of China's New Ruler* (New York: Stein and Day, 1970); *The Washington Post*, September 15, 1998.

YANMEN GUAN AMBUSH **He Long**, leading the **Eighth Route Army**'s 120th Division, operated in the Shanxi–Ningxia–Gansu border region. In September 1937, He Long led the division across the Yellow River from Shaanxi into Shanxi Province. Seeking a decisive engagement under conditions favorable to Communist forces, the 120th Division harassed Japanese supply lines. He Long initially struck at the Japanese forces that had attacked and overrun Datong in northern Shanxi. As the Japanese moved south, toward Taiyuan, the 358th Brigade's 716th Regiment kept pressure on the Japanese columns with screening and reconnaissance operations. On October 18, 1937, the regiment executed an ambush, attacking a Japanese supply column at Yanmenguan (Yanmen Pass), killing 300 Japanese soldiers and destroying 20 vehicles. Continuing the effort to sever Japanese lines of communication in support of **Yan Xishan**'s Nationalist Army's defense of Taiyuan against the Japanese, the 120th Division's forces stayed in the area.

After a night attack on October 20, the division's 716th Regiment again carried out a successful ambush of a Japanese supply column in the Yanmenguan area on the twenty-first. This effort caught hundreds of Japanese vehicles and about 200 soldiers in the ambush. The Japanese responded with air attacks against the 120th Division's forces. In ground action, the Japanese committed the ''Ushiromiya Division'' to an offensive in northwest Shanxi Province, which eventually forced He Long to withdraw and move his division farther north. The effort against the Japanese supply lines permitted Yan Yishan's defense of Taiyuan more time and significantly delayed the Japanese effort. Yan eventually came to some accommodation with the Japanese, although the city eventually fell to their forces.

REFERENCES

ZGDBKQS, vol. 2; ZGRMJFJZS, vol. 2.

YE JIANYING (1897–1986) One of the marshals of the **People's Liberation Army** (PLA), Ye served as minister of defense from 1975 to 1978, a position from which he provided some cohesiveness and stability for the PLA through the years after **Mao Zedong**'s death. In spite of his high military rank, during his address at the 60th anniversary of the founding of the Communist Party, in 1981, he argued against the restitution of rank in the army.

Ye Jianying was born on May 14, 1897, in Guangdong Province. He was a Hakka minority with strong family ties to Guangdong. Ye attended the Yunnan Military Academy in Guangzhou. He commanded the 21st Division of the National Army during the **Northern Expedition**. Ye took part in the **Guangzhou Uprising** (part of the **Autumn Harvest Uprisings**) on the Communist side. He subsequently studied military science in Moscow with the Comintern from 1928 to 1931, returning to the Jiangxi Soviet in the **Jinggangshan Revolutionary Base Area**.

During the **Long March**, Ye sided with **Zhang Guotao** against Mao Zedong, arguing for creation of a regular conventional force in a safe area to be used on the battlefield against Japan and the Nationalists. Ye was the chief of staff of the **Eighth Route Army** during the **Anti-Japanese War**. Later in the war he served as the Communist forces' liaison officer in the joint headquarters of the Nationalist, Communist, and U.S. armed forces. During the **Civil War**, Ye was a deputy chief of the General Staff of the communist forces. Later he was mayor of Beijing and chairman of its military committee. Ye also served as the commander of the Guangdong Military District and mayor of Guangzhou.

In 1955, Ye Jianying was promoted to the rank of marshal. He was appointed commandant of the Academy of Military Science in 1958, holding that position for a few years and minister of national defense from 1975 until 1978. Ye Jianying died in 1986.

REFERENCES

Wolfgang Bartke, *Who's Who in the People's Republic of China* (Armonk, NY: M. E. Sharpe, 1981); Tong Bai, ed., *Junren Shouce* (Beijing: Changzheng Chubanshe, 1987).

YE TING (1896–1946) Ye Ting was born on September 10, 1896, in Guangdong Province. He spent most of his adolescent and adult life in the military as a student at military academies, in **warlord** armies, as a leader of Nationalist forces, in the **Northern Expedition**, and as one of the foremost guerrilla leaders in the **People's Liberation Army** (PLA) in the Dabie Mountains around the

Eyuwan Soviet. He was captured by the Nationalists in the **New Fourth Army Incident** and spent five years in a Nationalist prison. After his release from prison, he was killed in a plane crash while returning to the Communist headquarters in **Yan'an** from Chongqing.

Ye Ting attended the Guangdong Province Army School for his primary education and then went to the Wuchang Number Two Reserve Military School for his middle school education. He graduated from the Baoding Military Academy outside Beijing and took part in Sun Yat-sen's October 10, 1911, revolution. He was a member of the Nationalist Army in 1919 and became a member of the Nationalist Party (Guomindang). In 1924, Ye was sent to the University of the Toilers of the East in Moscow. He also attended the Chinese class of the Red Army College there. While in Moscow, Ye Ting joined the Communist Party. When he returned to China, Ye was made chief of staff of the Nationalist Fourth Army, in Guangzhou. He took part in the Northern Expedition and was named as one of the outstanding generals of the campaign, after which he served as the deputy commander of the 25th Division of the Nationalist Fourth Army (Corps) and commander of the Nationalist 24th Division. When the **Nanchang Uprising** took place, he was the commander of the 11th Corps of the Nationalist Army, and he brought this corps, including its 24th and 25th Divisions, into the Communist force, joining the Red Army. **He Long** also moved to the Red Army at the same time, bringing with him the 20th Corps. These five divisions, led by He and Ye, formed the core of the force that attacked the arsenal in the Nanchang Uprising and later dispersed to conduct the **Autumn Harvest Uprisings** in an attempt to foment a general, Marxist revolution in China.

Ye Ting commanded forces in the Chinese Workers and Peasants Revolutionary Army during the 1927 uprisings but by 1928 had serious trouble with the Communist Party. He spent a period of about 10 years in Germany, France, Singapore, and, finally, Macao. The base of his problem seemed to have been serious disagreements with **Mao Zedong**, for whom Ye Ting seemed to have little respect (although by 1937, when he was given command of the New Fourth Army, Ye acknowledged Mao's leadership of the party). Ye also was criticized by **Zhou Enlai** for having lost faith in the Communist revolutionary cause and for advancing his own self-interest over that of the party. When Ye was in Macao, **Chiang Kai-shek** offered him a government position, which Ye refused, stating that he was a soldier, not a bureaucrat. After **World War II** broke out, Ye Ting traveled to Yan'an and on September 29, 1937, accepted command of the New Fourth Army, which was to have fought in cooperation with the Nationalists, from Mao Zedong. **Xiang Ying** was sent to act as Ye Ting's deputy commander. Both Ye and Xiang Ying had little respect for Mao and considered Mao to be pedant and elitist.

On January 4, 1941, at Maolin, during the New Fourth Army Incident, when Nationalist forces attacked the Communist New Fourth Army, Xiang Ying was killed and Ye Ting was captured by the Nationalists. Ye spent the next five years in prison and was released only in 1946, in Chongqing. After his release, Ye petitioned Mao Zedong to renew his Communist Party membership, and Mao granted this request, calling Ye to Yan'an for a party assignment. However, the aircraft on which Ye Ting flew to Yan'an crashed en route, killing Ye Ting on April 6, 1946.

REFERENCES

William W. Whitson with Chen-Hsia Huang, *The Chinese High Command* (New York: Praeger, 1973); Xiao Zhaoran, *Zhonggong Dangshi Jianming Cidian* [A Concise Dictionary of the Chinese Communist Party's History] (Beijing: Jiefangjun Chubanshe, 1986); Xing Huo Ranliao Bianjibu, ed., *Jiefangjun Jianglingzhuan* [Chronicle of the General Officer Leadership of the People's Liberation Army], vol. 3 (Beijing: Jiefangjun Chubanshe, 1986).

YI HE TUAN. See BOXER REBELLION

YIJIANGSHAN CAMPAIGN (1954) The **People's Liberation Army** (PLA) began to shell the islands of Quemoy and Mazu, just off the mainland coast, in spring 1954. At President Eisenhower's direction, the United States moved naval and air forces into the area and provided logistical support to the Nationalist garrisons on Quemoy and **Mazu**. The Nationalists had also stationed about a regiment of troops on the Yijiangshan Islands, part of the Dachen Island Group, 10 miles off the mainland coast north of Mazu, between Fuzhou and Shanghai.

The PLA began a measured buildup of forces off Taiwan in September 1954. Over a two-month period, the PLA conducted air reconnaissance sorties over the Dachen Island group, while ground and air forces rehearsed an amphibious assault in assembly areas. Then on November 1, 1954, the PLA Air Force initiated what was to be a 78-day air campaign to blockade the islands and to reduce nationalist defenses. The PLA Navy also assisted in the blockade and shelled the islands. On January 18, 1955, the PLA mounted an amphibious assault against Yijiangshan Island supported by air attacks and naval gunfire. The PLA succeeded in capturing Yijiangshan Island and the Nationalist general in charge of the Dachen Island Group. In follow-up operations between February 2 and February 9, 1955, PLA forces attacked and seized a series of four other islands along the coast off Zhejiang Province.

REFERENCES

Kenneth W. Allen et al., *China's Air Force Enters the 21st Century* (Santa Monica, CA: Rand Corporation, 1995); John W. Garver, *Foreign Relations of the People's Republic of China* (Englewood Cliffs, NJ: Prentice-Hall, 1993); Tong Bai, ed., *Junren Shouce* (Beijing: Changzheng Chubanshe, 1987).

YOKE FORCE. See Y-FORCE

YUAN SHIH-KAI (1859–1916) Yuan Shih-kai was born in Henan Province in 1859. He came from a family of Qing dynasty court officials, some of whom were civil officials and others military. However, Yuan failed the Imperial civil service examination twice and was posted to Korea, then a Manchu tributary state. After the **Sino-Japanese War** (1894–1895), Yuan Shih-kai was sent to Tianjin, where he served as governor. Yuan managed to build a military force for the Qing Court, which he kept intact through the **Boxer Rebellion**. As the Qing dynasty began to rebuild its military forces after the Boxer Rebellion, Yuan Shih-kai was appointed the Beiyang (northern) commissioner. He instituted a series of military reforms as the Qing "New Army" grew in strength, until he eventually had personal control over six divisions, which formed the **Beiyang Army**.

Yuan Shih-kai was a favorite general of Qing empress Cixi, and his personal power was weakened after her death in 1908. Nonetheless, Yuan's base of power in Beijing and his ties to the Beiyang Army and New Army forces were instrumental in the formation of a government for the Republic of China. Yuan negotiated the abdication of the Qing dynasty in 1911. He had been elected premier by the National Assembly, a Qing Court-organized parliament. Yuan was elected provisional president of China on February 14, 1912, by the national council that formed in Nanjing by the republican government. Yuan's position was weakened and his prestige lowered considerably when Japan presented the Twenty-One Demands to China. Toward the end of 1915, Yuan tried to restore the position of emperor. A special assembly convened by Yuan voted that Yuan become emperor, and Yuan accepted on December 12, 1915. Yuan Shih-kai initiated a new regime on January 1, 1916, calling himself the Hongxian emperor. However, Yunnan Province declared independence in late December 1915, followed by a similar declaration from Guizhou and Guangxi. Having lost control of much of China, Yuan Shih-kai issued a declaration in March 1916 that he would cancel the monarchy. By the time he died of uremia on June 16, 1916, the central government had lost control of most of South China, opening up the **warlord** period.

REFERENCES

Hsi-sheng Ch'i, *Warlord Politics in China, 1916–1928* (Stanford, CA: Stanford University Press, 1977); James E. Sheridan, *Chinese Warlord: The Career of Feng Yu-hsiang* (Stanford, CA: Stanford University Press, 1966); Ernest Young, *The Presidency of Yuan Shi-K'ai* (Ann Arbor: University of Michigan Press, 1977).

Z

Z-FORCE An American plan for a 30-division force of Nationalist troops to be used against the Japanese in **World War II** was conceived by General **Stilwell**. Although the force never really grew to that size, some Nationalist forces were trained and equipped by the United States according to Stilwell's plan. One of the elements of the Allied plan for the attack on Burma, which was to advance on an axis along the **Burma Road** from Yunnan Province, China, was designated the **Y-Force**, or Yoke Force (''yoke'' was the U.S. military phonetic for the letter Y). Y-Force eventually consisted of 15 divisions trained in Kunming, the provincial capital of Yunnan. Stilwell outlined a second ''30-division plan,'' to train another 30 divisions of Nationalist troops with American advisers and to equip them through the U.S. Lend-Lease program. This was dubbed the Z-Force, or Zebra Force. To train the Z-Force personnel, Stilwell envisioned the establishment of a second infantry training center to complement the one in Kunming, at Guilin, Guangxi Province. The objective of the force was to first secure the U.S. and Chinese airfields in East China, and then, in a broad offensive, to attack Japanese forces in the Yangtze River valley. The U.S. War Department only reluctantly acted on this plan, since the Y-Force had not yet been fully equipped or trained. By 1944, the training center at Guilin was established, but before it could begin to push through an effective 30-division-sized force, it came under attack by **Operation Ichigo**, the Japanese offensive launched on April 19, 1944. What divisions were trained at the Guilin Training Center did not fare well against the Japanese, and the Z-Force plan never reached fruition.

REFERENCES

Center for Military History, *China Defensive* (Washington, DC: U.S. Army Military History Institute, 1996); Charles F. Romanus and Riley Sunderland, *The United States Army in World War Two. The China-Burma-India Theater: Stilwell's Mission to China* (Washington, DC: Historical Division, Department of the Army, 1953); Barbara W.

Tuchman, *Stilwell and the American Experience in China, 1911–1945* (New York: Macmillan, 1971).

ZEBRA FORCE. See Z-FORCE

ZENG GUOFAN (Tseng Kuo-fan) (1811–1872) Zeng Guofan was a scholar and official forced by the crisis of the **Taiping Rebellion** to turn to military matters in an effort to defeat the Taipings. Zeng was born in Xiangxiang, Hunan Province, on November 28, 1811. He obtained a Jinshi degree by 1838 and entered the prestigious Hanlin Academy. By 1849, he was a vice president of its Board of Rites. During the Taiping Rebellion, Zeng was sent to Hunan to organize a regional militia to fight the Taipings. He studied military strategy for a time and then worked in Hunan to build a force that met the high moral standards of the Confucian system he had overseen on the Board of Rites. Zeng sought to recruit an army that was built on personal loyalty, along Confucian lines. He sought officers of character loyal to himself and had each commander select his subordinate leaders. Lower-level commanders personally recruited their own enlisted soldiers. By doing so, Zeng created a patron–client network of leaders and followers who were linked by loyalty to each other and shared a common ethic. However, these same traits and methods are often criticized as having set the stage for the warlordism that plagued China 50 to 70 years later.

Zeng Guofan trained the Hunan Army against the Taiping rebels and also organized a supporting naval and riverine force. He also worked with the Anglo–French and foreign armies raised by the Shanghai–Ningbo area, the **Ever-Triumphant Army** and the **Ever-Victorious Army**. Zeng's brother took part in the final suppression of the Taipings in Nanjing in 1864. Zeng also was the official who was most involved in the agreement to provide the **Lay–Osborn Fleet** for China, which it appeared he sought to keep under his own control. As the **self-strengthening movement** gained strength, Zeng Guofan sent a subordinate to the West to purchase machine tools for the purpose of establishing an arsenal. Zeng was instrumental in setting up the **Jiangnnan Arsenal**, later, the Jiangnan Shipyard.

REFERENCES

Arthur W. Hummel, ed., *Eminent Chinese of the Ch'ing Period (1644–1912)* (Washington, DC: Government Printing Office, 1943); John L. Rawlinson, *China's Struggle for Naval Development, 1839–1895* (Cambridge: Harvard University Press, 1967).

ZHANG GUOTAO (1897–1979) Zhang Guotao established the military Department of the Chinese Communist Party (CCP) in 1925. He was an advocate for the military "expert," who was a proficient small unit leader well drilled in tactics and the employment of weaponry. Whereas **Mao Zedong** advocated the use of maneuver and deception, avoiding battle except on terms favorable to the army, Zhang sought traditional battles between organized military units on conventional military terms. In 1932, after successive **Encirclement Campaigns** by the Nationalist Army, Zhang set out with his Fourth Front Army from the

Eyuwan Soviet Base Area on his own **Long March** into the Sichuan–Shaanxi area. He abandoned his base area in Sichuan–Shaanxi in spring 1935, moving south and east to link up with the **First Front Army** on its own Long March. Zhang seized an area in western Sichuan, near Mougong, in June 1935, which the First Front Army used to rest and regroup during the Long March. But this time, Mao Zedong and Zhang, who had always disagreed over military style, split again at the **Maoergai Conference**, with Mao arguing for a move to the north toward Inner Mongolia, while Zhang wanted to move farther west. Zhang's idea was to use a secure area in western China to develop a tactically proficient, professionally led main force capable of meeting the Japanese or the Nationalists on the battlefield. At the Maoergai Conference, the Fourth Front Army was broken up, and Zhang was made the political commissar of the Red Army.

Zhang Guotao was born in Jiangxi Province in 1897. He attended Beijing University, where he was involved in student activism during the May 4 Movement of 1919. In October 1920, he joined a Communist study group and participated in the organization of the CCP in Shanghai in July 1921. He was named head of the Military Department of the CCP in 1925. Zhang went to Moscow for training in 1928 with the Communist International. He returned to China in 1931, where he led the Eyuwan Soviet. For the rest of this military career he opposed Mao Zedong. He was accused by Mao of cooperating with the KMT from 1938 and functioning as a KMT agent. In 1949, Zhang fled to Hong Kong and subsequently to Canada, where he lived until his death in Toronto in 1979.

REFERENCES

William W. Whitson with Chen-Hsia Huang, *The Chinese High Command* (New York: Praeger, 1973); ZGDBKQS, vol. 2.

ZHANG WANNIAN (1928–) General Zhang Wannian was appointed the senior uniformed military officer in China at the 15th Congress of the Chinese Communist Party (CCP) in September 1997, when he was made first vice chairman of the Communist Party's **Central Military Commission** (CMC). In that position, Zhang became the direct subordinate of CCP chairman, China's president, and chairman of the CMC Jiang Zemin. Zhang Wannian was also appointed a member of the CCP Political Bureau (Politburo) along with General **Chi Haotian** at the 15th Party Congress. However, unlike their predecessors, Generals **Liu Huaqing** and **Zhang Zhen**, neither Zhang Wannian nor Chi Haotian was appointed to the Politburo Standing Committee.

Zhang Wannian was born in Huangxian County, Shandong Province, in August 1928. He enlisted in the **People's Liberation Army** in August 1944 and served initially in an independent battalion in Shandong Province. From 1945 to 1950, Zhang served in the 12th Division, Fourth Column, of the **Northeast Democratic United Army (Northeast China Field Army)**. He started out in the 35th Regiment and later was a platoon leader and deputy political instructor in the 36th Regiment. During the time he was assigned to the northeast, Zhang also was a communications section chief and a regimental operations officer. He took part in the campaign to secure the western portions of Liaoning Province and Shenyang during the **Manchurian Civil War Campaign** and the **Siping offensive and defensive campaigns**. As the **Civil War** fighting moved south out of Manchuria, Zhang Wannian fought in the **Beiping–Tianjin Campaign** (Ping-Jin Campaign). From 1950 to 1956, Zhang was assigned to South China. He was one of the defenders of Dongshan Island, off Shantou, Guangdong Province, during the **Dongshan Island "defensive" battle** in 1953. During the 1996 exercises off China's coast in the Taiwan Strait, Zhang was sent back to observe the exercises, probably as much for psychological warfare value as because he was then a vice chairman of the CMC. Zhang was appointed the director of the headquarters for operations against Taiwan during those exercises, during which he commanded the forces of the Guangzhou and Nanjing Military Regions. Zhang Wannian also served with the **Fourth Field Army** and is associated with the discredited PLA leader **Lin Biao** because of that service. Zhang is a graduate of the PLA Nanjing Military Academy, which he attended from 1956 to 1958.

During the 1979 **Self-Defensive Counterattack against Vietnam**, Zhang Wannian was the commander of the 127th Division, 43d Group Army, and concurrently served as the deputy army commander. From 1982 to 1985, Zhang was deputy commander of the Wuhan Military region, which at the time functioned as China's strategic reserve. In 1985, Zhang was appointed deputy commander of Guangzhou Military Region (MR) and was made commander of the military region in 1987. After commanding Guangzhou MR until 1990, Zhang was transferred to Jinan Military Region by Jiang Zemin, CCP general secretary. In 1992, Zhang was made chief of the General Staff Department, a position he held until 1995, when he was elevated to the Central Military Commission as a vice chairman.

REFERENCES

Jiefang Junbao, Beijing, December 18, 1997; *Kuang Chiao Ching*, no. 304, Hong Kong, January 16, 1998, pp. 18–22; *Sing Tao Jih Pao*, Hong Kong, May 12, 1996; *Who's Who in China* (Beijing: Foreign Languages Press, 1995).

ZHANG XUELIANG (1901–) Zhang Xueliang was born in Haicheng District, Liaoning Province, in 1901, although some biographies place his date of birth in 1898. He was the eldest son of Marshal Zhang Zuolin, **warlord** of Manchuria who headed the **Fengtian Army**. Zhang Xueliang enrolled at the National University in Beijing, but he was recalled by his father to attend the Fengtian Military Academy in Mukden (Shenyang) to study military science. He was promoted directly to colonel by his father in 1919 and given command of his father's personal bodyguard force. When the Anhui–Zhili War broke out (see **Anfu Clique**), Zhang Xueliang was given command of the Third Mixed Brigade of the Fengtian Army. He was promoted to brigadier general in November 1920 and was sent by his father to observe the autumn military maneuvers by the Japanese army in Japan. He returned with strongly anti-Japanese attitudes but was sufficiently impressed with the Japanese armed forces to recommend sweeping reforms of the Fengtian Army. He participated in the first **Fengtian–Zhili War** (Zhili later became Hebei Province) in 1922. At the war's end he was given command of the First Fengtian Army and was concurrently made commandant of the Fengtian Military Academy.

Zhang Xueliang won a significant battle at the Great Wall during the second Fengtian–Zhili War in 1924. He continued to move forces south to Shanghai, pacifying the lower Yangtze Valley in 1925. Later, he was appointed director of the Beijing War College. When Guo Songling, a former Zhang Zuolin subordinate, conspired with **Feng Yuxiang**, of the Zhili Clique, to overthrow Zhang Zuolin's control of the Manchurian region, Zhang Xueliang was almost executed by his father for having failed to detect the conspiracy. During the **Northern Expedition**, Zhang Xueliang fought as part of the National Pacification Army.

Zhang Zuolin, Zhang Xueliang's father, was killed on June 3, 1928, when a train carrying him from Beijing to Shenyang (then called Mukden) was blown up by Japanese agents. At the time, Zhang Xueliang was in Beijing. The younger Zhang immediately went to the pass between the Great Wall and the sea, Shanhaiguan, in northeast Zhili. He left Yang Yuting, Zhang Zuolin's chief of staff, in Shanhaiguan as a rear guard and, disguised as a private soldier, sneaked into Mukden undetected by the Japanese or his own enemies. On July 4, 1928, Zhang was made commander in chief of the ''Manchurian Peace Preservation Force,'' giving him effective control of the northeast. In a July 4, 1928, meeting with **Chiang Kai-shek**, Zhang accepted him as leader of the Nationalist government and accepted his leadership of the Kuomintang political organization in Manchuria but negotiated concessions from Chiang for joining the National government. Chiang agreed not to send southern-based troops into Manchuria, agreed not to establish a Kuomintang political branch in Zhang's region, and ceded the Jehol District to Manchuria as a fourth province. The agreement was to be effective on July 22, when the Nationalist flag was to fly over Manchuria.

The Japanese, however, were not happy about the incorporation of Manchuria into the Nationalist government. The Sino–Japanese Commercial Treaty of 1896 was set to expire on July 20, 1928, and Chiang Kai-shek did not intend to renew the treaty. On August 10, two days after a funeral service for Zhang Zuolin, Japanese consul-general in Mukden Hayashe delivered a message from Prime Minister Tanaka of Japan warning against joining the Nationalist government and seeking to develop close Manchurian–Japanese business. Zhang agreed to protect Japanese business rights in Manchuria but announced his intent to join the Nationalist government. Finally, on December 29, 1928, Zhang Xueliang raised the flag of the Nationalist government in Mukden. In return, Chiang Kai-shek appointed him commander in chief of the Manchuria Border Defense Army.

When Feng Yuxiang and **Yan Xishan**, respectively, warlords in Hebei and

Shanxi, opposed Chiang Kai-shek in 1930, Zhang refused to join that coalition. Feng had been Zhang's father's enemy. Zhang's active support for Chiang Kai-shek eventually forced Yan Xishan to withdraw from his coalition with Feng.

Chiang Kai-shek urged Zhang Xueliang not to react to the Mukden Incident of September 18, 1931 (**918 Incident**), seeking to avoid war with Japan. Under pressure from the Council of the League of Nations, which met at China's request in response to the incident, Japan offered to negotiate with China but ignored the League of Nations request to withdraw troops from Manchuria. By March 1, 1932, Henry Pu-yi accepted the position of regent of the Japanese puppet republic of **Manchukuo**. Zhang later resigned all his posts and on March 10, 1933, at Baoding, south of Beijing, handed over control of his army to Chiang Kai-shek. Zhang also relinquished his position as chairman of the Peking branch of the National Military Council. He was replaced by Nationalist minister of war General Ho Ying-chin (He Yingqin), who also took command of the entire Manchurian Army, which had retreated south. Zhang went to Europe with his family and political adviser, William Donald. He met Mussolini in Italy and Hitler and Goering in Germany and spent about six months in Europe.

On January 8, 1934, Zhang returned to Shanghai. He was appointed a deputy commander of "Bandit Suppression" in Hunan, Hebei, and Anhui by Chiang on February 7, 1934, where he was in charge of Communist extermination campaigns, the **Encirclement Campaigns**. His forces were among those that pressured the Communists to begin the **Long March**.

After the Communist forces captured Zhang Xueliang's 109th division, in 1936, they indoctrinated the troops and returned them to Zhang's control. In May 1936, Zhang entered into a secret agreement with the Communists to cooperate in the fight against Japan. He had met **Zhou Enlai** on April 9, 1936, in **Yan'an**, where they discussed Guomindang–Communist cooperation. In October 1936, in Loyang, Zhang raised with Chiang the question of a **United Front** with the Communists against Japan, but Chiang refused. Meanwhile, Zhang continued secret meetings with Zhou Enlai. After again raising the subject of a United Front with Chiang in a meeting in the city of Luoyang on December 3, 1936, he returned to Xi'an. Chiang Kai-shek arrived in Xi'an on December 4, 1936. Over an eight-day period, Zhang and Chiang continued to discuss the subject of a United Front, but Chiang refused to budge and instead was prepared to order another Bandit Suppression Campaign. At midnight on December 12, Zhang and General Yang Hucheng acted in the **Xi'an Incident**, capturing Chiang Kai-shek and holding him hostage for talks with the Communists.

After the United Front was agreed to in Xi'an, Zhang Xueliang traveled to Nanjing. He was court-martialed by the Nationalist government and put under house arrest. In 1946, as the Communists began to flee the mainland, he was transferred to Taipei, Taiwan, where he was placed under house arrest. He is said to have remained under house arrest until 1961 but never left his Taipei home after that date. Biographic accounts published as late as 1997 still do not give a date of death for Zhang Xueliang.

REFERENCES

Keiji Furuya, *Chiang Kai-shek, His Life and Times* (New York: St. John's University Press, 1981); Robert Payne, *Chiang Kai-shek* (New York: Weybright and Talley, 1969); Tong Te-kong and Li Tsung-jen, *The Memoirs of Li Tsung-jen* (Boulder, CO: Westview Press, 1979); Susan Fu Tsu, *A Study of Chang Hsuel-liang's Role in Modern Chinese History* (Ann Arbor, MI: University Microfilms International, 1984).

ZHANG YUNYI (1892–1974) Zhang Yunyi was one of 10 officers appointed as generals in 1955. He was born in 1892 in Wencheng County, Hainan Island, Guangdong Province. His entire education from primary school through college, a total of nine years, was in military schools. After attending military primary and middle (secondary) schools, Zhang Yunyi graduated from the sixth class at the Baoding Military Academy in 1919. As one of the officers who were sent to the **Whampoa Military Academy** to help establish that institution, Zhang met **Zhou Enlai** and joined the Communist Party. He also participated in the **Northern Expedition** as a division chief of staff. He was one of the participants in the **Nanchang Uprising** in 1927. After a short stint with the Chinese Communist Party (CCP) Center in Shanghai, in 1928, he was sent into Guangxi Province, where he worked with **Deng Xiaoping** to establish the Seventh Corps (Army) of the **Chinese Workers and Peasants Red Army**. Zhang moved north to Jiangxi Province with two divisions during the **Encirclement Campaigns** and took part in the **Long March** with the **First Front Army**.

In 1936, when the First Front Army reached **Yan'an**, Zhang Yunyi was made chief of staff of the Military Affairs Committee of the Communist Party. This organization is now the **Central Military Commission**, and the function of the chief of staff is analogous to that of the head of the general office of the Central Military Commission. As the **New Fourth Army** was formed in South China, Zhang was sent to the south to command the Third Detachment of the New Fourth Army. He stayed with the New Fourth Army throughout the remainder of **World War II**, commanding the forces north of the Yangtze River and the Second Division of the army. After the defeat of Japan, Zhang Yunyi was made deputy commander of the Shandong Military District, an area that corresponds to the Jinan Military Region in contemporary times. He organized the campaign on the Shandong Peninsula to control the Jinan–Qingdao railway line, and in a series of engagements along the rail line, Zhang's forces, along with those of **Xu Shiyou**, were defeated by Nationalist troops in a battle at Gaomi on October 8, 1946. About 30,000 Communist troops were reportedly killed in two weeks of this campaign in November 1946. Zhang continued to operate in the Shandong area with the East China Field Army and during 1947–1948 was the deputy commander and chief of staff of the army. When it became the **Third Field Army** in 1949, Chen Yi was commander, and Zhang Yunyi was chief of staff.

After liberation from the Nationalists in 1949, Zhang Yunyi was sent back into Guangxi, where he had operated in 1929–1931. He concurrently served as deputy commander and political commissar of the Guangxi Military District and the secretary of the Guangxi Province Communist Party Committee from 1949 to 1955. He was recalled to Beijing in 1954–1955, where he was a member of the National Defense Council. From 1963 through 1969, Zhang Yunyi was also a deputy secretary of the Chinese Communist Party Central Military Commission. He was essentially untouched by the **Central Revolution** and died on November 19, 1974.

REFERENCES

Wolfgang Bartke, *Biographical Dictionary and Analysis of China's Party Leadership, 1922–1988* (New York: K. G. Saur, 1990); Chalmers Johnson, *Peasant Nationalism and Communist Power* (Stanford, CA: Stanford University Press, 1962); William W. Whitson with Chen-Hsia Huang, *The Chinese High Command* (New York: Praeger, 1973).

ZHANG ZUOLIN. See WARLORDS; ZHANG XUELIANG

ZHAO NANQI. See CHO NAM GI

ZHENBAO ISLAND CLASH (March 2, 1969) After about a year of confrontations between the **People's Liberation Army** (PLA) and the Russian army over border demarcation on the eastern border, a major clash broke out at Zhenbao Island (called Damansky by the Russians) on March 2, 1969. Zhenbao Island is located on the Ussuri River between the cities of Khabarovsk and Vladivostok. The Chinese maintained that the border between the two countries followed the "Thalweg Principle," or the central line of the main channel, putting the island on the Chinese side; Moscow claimed that the Chinese banks of the Amur and Ussuri Rivers were the border, putting some 600 disputed islands on the Russian side.

As each side patrolled, physical clashes became more common, until on March 2, 1969, a Chinese patrol crossing the frozen river to the island was challenged by Russian forces. Automatic weapons fire from the Chinese bank hit the Russian forces, killing seven Russians and wounding 23 men. The Chinese, who said their patrol was fired on first by the Russians, also claim to have had several casualties. It is not clear from accounts by either side, however, whether the PLA instigated the incident or the Soviets initiated fire. On March 4 and 12, the Russians reinforced the island and flew reconnaissance aircraft along the border. Then, on March 15, 200 Russian infantry men supported by 30 armored vehicles again tried to seize the island. Clashes continued through March 17, when both sides de-escalated the conflict. Tensions continued, however, for several years.

REFERENCES

Harold C. Hinton, "Conflict on the Ussuri, A Clash of Nationalisms," *Problems of Communism* 20, no. 1–2 (January–April 1971): 48–59; Neville Maxwell, "The Chinese

Account of the 1969 Fighting at Chen Bao,'' *China Quarterly*, no. 56 (Fall 1973): 730–739; Tong Bai, ed., *Junren Shouce* (Beijing: Changzheng Chubanshe, 1987).

ZHILI CLIQUE The Zhili Clique was a grouping of **warlords** and military leaders in northern China centered around the **Beiyang Army**. The other major group of North China warlords was the **Anfu Clique**, centered in Anhui Province, led by Duan Qirui (Tuan Ch'i-rui). The Zhili Clique was initially led by the Beiyang warlord Feng Guozhang (Feng Kuo-chang). After Feng's death in 1919, Cao Kun (Ts'ao K'un), who for a time served as president of the Republic of China, led the Zhili Clique. The warlord Wu Peifu was also a powerful clique leader, as was the brother of Cao Kun, Cao Rui, who was based in Tianjin. The two major North China warlord cliques fought several local wars for control of the area, the first of which was the Zhili-Anfu War of July 1920. The Zhili Clique allied with the **Fengtian Army**, controlled by the warlord Zhang Zuolin of Manchuria (the father of **Zhang Xueliang**, one of the generals who captured **Chiang Kai-shek** in the **Xi'an Incident**). Between 1922 and 1924, the Zhili Clique fought two wars with the Fengtian Army of Zhang Zuolin, the **Fengtian-Zhili Wars**. These local wars were fought primarily for nominal control of the government of the newly formed Chinese Republic and continued until the **Northern Expedition** unified the warlords in support of the republic.

REFERENCES

Donald G. Gillin, *Warlord: Yen Hsi-shan in Shanxi Province, 1911–1949* (Princeton: Princeton University Press, 1967); Diana Lary, *Warlord Soldiers: Chinese Common Soldiers, 1911–1937* (New York: Cambridge University Press, 1985); James E. Sheridan, *Chinese Warlord: The Career of Feng Yu-hsiang* (Stanford, CA: Stanford University Press, 1966).

ZHILI-FENGTIAN WAR. See FENGTIAN-ZHILI WAR

ZHOU ENLAI (1899–1976) Zhou Enlai was one of the most interesting and influential figures in the Chinese Communist Party (CCP). He directed the Party Central Committee's Military Affairs Department from April 1927 and was extremely influential in military policy until he was removed from the post during the **Long March** by a Central Committee decision at the **Zunyi Conference**. Zhou participated in the planning of the **Nanchang Uprising** (August 1, 1927) and was a key figure in the birth and development of the Red Army. Zhou, at one point, sided with **Li Lisan**'s strategy of concentrating on armed uprisings of rich peasants and workers, which called for seizing key cities. This probably belied his own upbringing as the son of a wealthy, middle-class Zhejiang family. He also seemed sympathetic to Li Lisan's strategy of concentrating the army for a general attack on the Nationalists, set forth in Central Committee Circular 70 of February 26, 1930. From 1931 until his removal from the Central Military Commission at the Zunyi Conference in 1935, Zhou was probably the foremost opponent of the Maoist strategy of guerrilla warfare. Zhou supported the Russian model of a trained, professional Red Army and agreed with the emphasis on the control of cities, railroads, rivers, and strategic lines of communication.

Zhou Enlai was born in 1899 in Shaoxing, Zhejiang Province. His father was a well-educated man. He lived for some time in his grandfather's house in Huaiyun, Jiangsu Province. At the age of 12, in 1911, Zhou moved to an uncle's home in Shenyang. Zhou then attended Nankai middle school in Tianjin in 1913. He also studied at Waseda University and Japan University in Tokyo from 1917 to 1919, moving to Nankai University in Tianjin in 1919. Zhou Enlai was imprisoned for a short time by the Nationalist government for his participation in the May 4 Movement protests in Beijing in 1919. After six months in jail, he went to France on a work-study plan, where he founded the Chinese Communist Youth League with Li Lisan in 1922, as the Paris Branch of the CCP. Zhou also studied for a year in Germany. He returned to China in 1924, where he directed the Guangdong Province CCP Committee Military Affairs Department. Zhou took part in the Guomindang (KMT)-CCP **United Front** and was director of the Political Department of the **Whampoa Military Academy** under **Chiang Kai-shek**. In underground work, Zhou helped to investigate and organize workers uprisings in Shanghai in 1925, 1926, and 1927. He fled from Chiang's forces in Shanghai in April 1927 to Hankow (one of the three cities making up what is now the city of Wuhan) where he was elected to the Politburo of the CCP and made head of the Military Department. After the failure of the Nanchang Uprising, Zhou Enlai attended the Sixth CCP Congress, held in Moscow in 1928. He returned to Shanghai later in 1928 to work with Li Lisan to organize armed uprisings in the cities. Zhou joined the **Jiangxi Soviet** in 1931, where he continued to oppose **Mao Zedong**'s ideas on guerrilla warfare. After the Long March, Zhou reached an accommodation with Mao and continued in the party leadership as Mao's deputy or alter ego. He was the main negotiator with Chiang Kai-shek after the **Xi'an Incident** in 1936, helping to form the United Front between the Communist and Nationalist parties during the **Anti-Japanese War**. Zhou also served as the main liaison officer between the CCP and the Nationalists in Chongqing. From 1945 to 1947, Zhou was involved in attempting to negotiate a coalition government with the Nationalists. He later served as a Politburo member and premier.

Zhou Enlai died on January 8, 1976. Zhou's death was marked on April 5, 1976 (Qingming Day, for the honoring of the dead), by mass demonstrations on Tiananmen Square. Called the "**Tiananmen Incident**," these demonstrations were interpreted by the CCP as mass criticism of Mao Zedong and the **Cultural Revolution**.

REFERENCES

Wolfgang Bartke, *Who's Who in the People's Republic of China* (Armonk, NY: M. E. Sharpe, 1981); *Chinese Communist Who's Who* (Taipei: Institute for International Relations, 1967); Jonathan D. Spence, *The Search for Modern China* (New York: W. W. Norton, 1990).

ZHU DE (1886–1976) One of the most influential and experienced military leaders of the **People's Liberation Army** (PLA), Zhu De was also one of the few leaders who had received a professional military education.

Zhu De was born in Yilong County, Sichuan Province, in 1886. Zhu entered the Yunnan Military Academy in Kunming in 1908 and joined the Tongmenghui, the revolutionary society led by Sun Yat-sen. After graduating from the academy, Zhu at times taught there but also participated in military campaigns with the armies of Yunnan **warlords**. He commanded border units along the Sino–Lao and Sino–Vietnamese border in 1913–1915. Between 1917 and 1922, Zhu De is reported to have developed a strong opium habit, from which he recovered only in 1922 at a hospital in Shanghai. From 1922 to 1924, Zhu studied military science, politics, and history at the University of Göttingen, Germany. He reportedly met **Zhou Enlai** in Germany, and, after organizing a Guomindang (KMT) Party branch in Berlin, he worked closely with German Communists. Zhu was arrested by the German police in 1926 and subsequently returned to China.

Zhu De in 1927 was one of the oldest and most senior of the Communist military leaders. He was one of the planners of the **Nanchang Uprising**. On August 1, 1927, Zhu moved into Fujian Province with about 900 troops that he organized into the "Ninth Revolutionary Army." In comparison to **Mao Zedong**, who preferred guerrilla units and tactics, by virtue of his early military training, Zhu concentrated on conventional forces that had undergone regular military training. Nonetheless, by 1928, Zhu joined forces with Mao at the **Jinggangshan Revolutionary Base**. Zhu was a major figure on the **Long March**, where his tactics and use of maneuver helped to get the Communist forces to **Yan'an**. He was made commander in chief of the **Eighth Route Army** in 1937.

By the time of the establishment of the People's Republic of China (PRC) in 1949, Zhu De, along with Mao Zedong, **Liu Shaoqi**, Zhou Enlai, and **Chen Yun**, was one of the most prominent figures in China. By 1954, the position of PLA commander in chief was abolished, and Zhu De held no more military posts.

Zhu De had one wife die in childbirth, one killed by warlords, and a third executed by Nationalist leaders. His fourth wife, Kang Keqing, participated in the Long March. Zhu De died on July 6, 1976, two months before Mao Zedong's death and six months after the death of Zhou Enlai.

REFERENCES

Wolfgang Bartke, *Who's Who in the People's Republic of China* (Armonk, NY: M. E. Sharpe, 1981); William W. Whitson with Chen-Hsia Huang, *The Chinese High Command* (New York: Praeger, 1973).

ZUNYI CONFERENCE (1935) At the Zunyi Conference, which took place during the **Long March, Mao Zedong** emerged as the leader of the Chinese Communist Party (CCP) **Central Military Commission**. The conference took place January 6–8, 1935, in Zunyi, Guizhou Province. It was attended by about 20 Central Committee members, including Politburo members, as well as the Communist International's (Comintern) military adviser to the CCP, Li De (Otto Braun). After heated argument within the party over the loss of the **Jiangxi Soviet** area in the **Jinggangshan Revolutionary Base Area** because of Nationalist pressure during the fifth **Encirclement Campaign, Zhou Enlai** and Li De were severely criticized. Zhou was relieved of his position as chairman of the Central Military Commission and replaced by Mao Zedong. **Peng Dehuai** also gained a stronger position as a result of the meeting. The political commissar system in the Red Army was strengthened, while the army was reorganized into columns, instead of divisions, for the Long March. The army sought to avoid a static defense strategy, relying instead on mobility to engage Nationalist forces at time advantageous to the Red Army; in other words, guerrilla warfare. In meeting notes, the strongest criticism was reserved for ''Comrade XXX'' for the decision to adopt static defenses. That was probably a reference to Zhou Enlai. Plans were also made for a coordinated effort to link forces with **Zhang Guotao's Fourth Front Army**, which had moved into Sichuan earlier in the face of heavy Nationalist military pressure against **Eyuwan Soviet**. Although Mao's leadership and strategy were challenged again at the **Maoergai** and **Lianghekou Conferences** during the Long March, the Zunyi Conference set the tone for the future **People's Liberation Army** (PLA) and is probably the most important of the military policy meetings.

REFERENCES

Warren Kuo, *Analytical History of the Chinese Communist Party*, Vol. 3: (Taipei: Institute of International Relations, 1968); ZGRMJFJZS, vol. 1.

ZUO ZONGTANG (Tso Tsung-tang) (1812–1885) Zuo Zongtang was a civilian leader pressed into service and charged with raising military forces for the Qing dynasty. He was instrumental in fighting several peasant rebellions in China, including the **Taiping Rebellion**, the **Nian Rebellion**, operations against Muslim rebels in northwest China, and the **Franco–Chinese War**. Zuo was born on November 10, 1812, in Xiangyin, Hunan Province. His early studies focused on China's geography and military classics, but he failed the metropolitan-level Jinshi examination three times. Nonetheless, he maintained an interest in scholarship and compiled an atlas of Chinese military strategy from a historical perspective. In 1852, Zuo Zongtang was put in charge of military affairs in the Hunan–Hubei area, where he organized operations against the Taiping rebels in the Wuchang (now, Wuhan) area. After he came to the attention of **Zeng Guofan**, Zuo supervised military affairs in Hunan.

Zuo was charged with raising a force of 5,000 men for service against the

Taiping rebels in 1860. His troops routed the Taipings from Changsha and from Wuyuan, Anhui Province, by December 1861, after which he was appointed governor-general of Zhejiang Province (January 23, 1862). After a series of battles, he recovered control of the province from the Taipings and was concurrently appointed governor of Fujian in May 1863. In 1864, in Hangzhou, Zhejiang Province, Zuo experimented with the use of steamboats on West Lake (Xihu) in Hangzhou. He later established the **Fuzhou Dockyard** at Mawei in Fuzhou, Fujian Province. This was to become the Fuzhou Navy Yard operated by Shen Baozhen. In 1866, Zuo Zongtang was appointed governor-general of Shaanxi and Gansu and in a series of campaigns against Muslim rebels secured Xi'an and Lanzhou by 1872. He was then sent to Xinjiang to manage military affairs. Through a combination of military campaigns and agreements with rebels, Zuo secured Xinjiang and Chinese Turkestan by 1878. He later was called upon to quell rebellions in Shandong and to return to Fujian in the **Franco–Chinese War** in December 1884. He died on September 25, 1885, just after a peace agreement was signed with the French.

REFERENCES

Arthur W. Hummel, *Eminent Chinese of the Ch'ing Period, 1644–1912* (Washington, DC: Government Printing Office, 1943); John L. Rawlinson, *China's Struggle for Naval Development, 1839–1895* (Cambridge: Harvard University Press, 1967).

SELECTED BIBLIOGRAPHY

Academia Sinica. *Collected Papers from the Conference on the Self-Strengthening Movement in Late Ch'ing China, 1860–1894*. 2 vols. Taipei: Institute of Modern History, 1987.

Allen, Kenneth W. et al. *China's Air Force Enters the 21st Century*. Santa Monica, CA: Rand Corporation, 1995.

Ames, Roger, trans. *Sun-Tzu: The Art of Warfare*. New York: Ballantine Books, 1993.

An Tai-sung. *The Sino–Soviet Territorial Dispute*. Philadelphia: Westminister Press, 1973.

Angin, W. A. ''Compiled Records of the Commander in Chief, United States Asiatic Fleet,'' Shanghai, December 1938. Milton E. Miles Papers. Hoover Institution Archives. Hoover Institution on War, Revolution, and Peace, Stanford, CA.

Appleman, Roy E. *U.S. Army in the Korean War: South to the Naktong, North to the Yalu (June–November 1950)*. Washington, DC: Center of Military History, U.S. Army, 1961.

Austin, Greg. *China's Ocean Frontier: International Law, Military Force, and National Development*. Sydney: Allen and Unwin, 1998.

———, ed. *Missile Diplomacy and Taiwan's Future: Innovations in Politics and Military Power*. Canberra: Strategic and Defense Studies Center, Australian National University, 1997.

Bailey, Thomas A. *A Diplomatic History of the American People*. Englewood Cliffs, NJ: Prentice-Hall, 1980.

Barrett, David D. *Dixie Mission: The United States Army Observer Group in Yenan, 1944*. Berkeley: Center for Chinese Studies, University of California Press, 1970.

Bartke, Wolfgang. *Biographical Dictionary and Analysis of China's Party Leadership, 1922–1988*. New York: K. G. Saur, 1990.

———. *Who's Who in the People's Republic of China*. Armonk, NY: M. E. Sharpe, 1981.

Baxter, James P. *The Introduction of the Ironclad Warship*. Cambridge: Harvard University Press, 1933.

Beal, John Robinson. *Marshall in China*. New York: Doubleday, 1970.

Beckmann, George M. *The Modernization of China and Japan*. New York: Harper and Row, 1962.

Beeching, Jack. *The Chinese Opium Wars*. New York: Harcourt Brace Jovanovich, 1976.

Bertram, James. *First Act in China: The Story of the Sian Mutiny*. New York: Viking Press, 1938; reprinted 1973.

Bond, Charles R. *A Flying Tiger's Diary*. College Station: Texas A&M University Press, 1984.

Boorman, Howard L., ed. *Biographical Dictionary of Republican China*. 4 vols. New York: Columbia University Press, 1970.

Borg, Dorothy. *The United States and the Manchurian Crisis of 1933–1938*. Cambridge: Harvard University Press, 1964.

Boyle, John H. *China and Japan at War: The Politics of Collaboration*. Stanford, CA: Stanford University Press, 1972.

Braun, Otto. *A Comintern Agent in China, 1932–1939*. Trans. Jean Moore. London: C. Hurst, 1982.

Brook, Timothy. *Quelling the People: The Military Suppression of the Beijing Democracy Movement*. New York: Oxford University Press, 1992.

Brown, Anthony Cave. *Wild Bill Donovan: The Last Hero*. New York: Times Books, 1982.

Buhite, Russell D. *Patrick J. Hurley and American Foreign Policy*. Ithaca, NY: Cornell University Press, 1973.

Bunge, Frederica M., ed. *China, a Country Study*. Washington, DC: Government Printing Office, 1981.

Byrd, Martha. *Chennault: Giving Wings to the Tiger*. Tuscaloosa: University of Alabama Press, 1987.

Caldwell, John. *China's Conventional Military Capabilities, 1994–2004: An Assessment*. Washington, DC: Center for Strategic and International Studies, 1994.

Carr, Caleb. *The Devil Soldier: The Story of Frederick Townsend Ward*. New York: Random House, 1992.

Cass, Bevan G. *History of the Sixth Marine Division*. Washington, DC: Infantry Journal Press, 1948.

Center for Military History. *China Defensive*. Washington, DC: U.S. Army Military History Institute, 1996.

———. *China Offensive*. Washington, DC: U.S. Army Military History Institute, 1996.

Chace, James. *Acheson: The Secretary of State Who Created the American World*. New York: Simon and Schuster, 1995.

Chang, Gordon H. "To the Nuclear Brink: Eisenhower, Dulles, and the Quemoy–Matsu Crisis." *International Strategy* 12, no. 4 (Spring 1988): 96–123.

Chang Hsin-pao. *Commissioner Lin and the Opium War*. Cambridge: Cambridge University Press, 1964.

Chang, Iris. *The Rape of Nanjing: The Forgotten Holocaust of World War II*. New York: Basic Books, 1997.

Chang Kuo-t'ao. *The Rise of the Chinese Communist Party, Autobiography of Chang Kuo-t'ao*. 2 vols. Lawrence: University of Kansas Press, 1971–1972.

Chang, Parris. *Radicals and Radical Ideology in China's Cultural Revolution*. New York: Columbia University Press, 1973.

Chen Chao. *Liao-Shen Zhanyi* [The Liaoning–Shenyang Campaign]. Shanghai: Renmin Chubanshe, 1959.

Chen Jian. *China's Road to the Korean War: The Making of the Sino–American Confrontation.* New York: Columbia University Press, 1994.
Chen, King C. *China's War with Vietnam, 1979: Issues, Decisions and Implications.* Stanford, CA: Hoover Institution Press, 1987.
Chen Liheng. *Junshi Yuce Xue* [The Science of Military Estimates]. Beijing: Junshi Kexue Chubanshe, 1993.
Chen Tinglong. *The Army of the Republic of China.* Taipei: Army Press, 1993.
Chennault, Anna. *Chennault and the Flying Tigers.* New York: P. S. Eriksson, 1963.
Chennault, Claire L. *Way of a Fighter: The Memoirs of Claire Lee Chennault.* New York: G. P. Putnam's Sons, 1949.
Chesneau, Jean, Francoise Le Barbier, and Marie-Claire Bergere. *China from the 1911 Revolution to Liberation.* New York: Pantheon Books, 1977.
Chi Haotian et al. *Hongsi Zhilu: Zhu De yu Bainian Zhongguo* [The Red Road, Zhu De and 100 Years of China]. Beijing: Guangming Ribao Chubanshe, 1997.
Ch'i Hsi-sheng. *Nationalist China at War: Military and Political Collapse, 1937–1945.* Ann Arbor: University of Michigan Press, 1982.
———. *Warlord Politics in China, 1916–1928.* Stanford, CA: Hoover Institution Press, 1977.
Chiang Kai-shek. *Soviet Russia in China: A Summing-up at Seventy.* New York: Farrar, Straus, and Cuhady, 1958.
Chiang Siang-tseh. *The Nian Rebellion.* Seattle: University of Washington Press, 1954.
Chief of Military History. *Civil War in China, 1945–1950.* Taipei, Taiwan: ROC Army, 1950.
China's National Defense. Beijing: Information Office of the State Council of the People's Republic of China, July 1998.
Chinese Communist Who's Who. Taipei: Institute of International Relations, 1967.
Chinoy, Mike. *China Live: Two Decades in the Heart of the Dragon.* Atlanta: Turner, 1997.
Clubb, Edmund O. "Chiang Kai-shek's Waterloo: The Battle for Hwai-Hai." *Pacific Historical Review* 25, no. 4 (November 1956).
———. *Twentieth Century China.* New York: Columbia University Press, 1964.
Cohen, Warren I., ed. *New Frontiers in American–East Asian Relations.* New York: Columbia University Press, 1983.
Compton, Piers. *The Last Days of General Gordon.* London: Robert Hale, 1974.
Cordier, Henri. *L'Expedition de Chine de 1857–1858: Histoire Diplomatique.* Paris: Ancienne Libraire Germier, 1905; New York: AMS Press, 1974.
Cornelius, Wanda, and Thayne Short. *Ding Hao: America's Air War in China, 1937–1945.* Gretna, LA: Pelican, 1980.
Corr, Gerard R. *The Chinese Red Army: Campaigns and Politics since 1949.* Reading, PA: Osprey Press, 1974.
Couhat, Jean L., ed. *Combat Fleets of the World, 1988/1989.* Annapolis, MD: Naval Institute Press, 1987.
Craine, Eric R. *Burma Roadsters: The 464th CA Anti-Aircraft Battalion in China–Burma–India.* Tucson: Western Research, 1992.
Crowley, James. *Japan's Quest for Autonomy: National Security and Foreign Policy, 1930–1938.* Princeton: Princeton University Press, 1966.
Crozier, Brian, with Eric Chou. *The Man Who Lost China: The First Full Biography of Chiang Kai-shek.* New York: Charles Scribner's Sons, 1976.

Davidson-Houston, James V. *Russia and China: From the Huns to Mao Tse-tung*. London: R. Hale, 1960.

Delios, Rosita. *Modern Chinese Strategy: Present Developments, Future Directions*. New York: St. Martin's Press, 1980.

Detweiler, Donald S., and Charles B. Burdick, eds. *War in Asia and the Pacific, 1937–1949*. Vol. 15: *Civil War in China, 1945–50*. Garland, TX: Garland Press, 1980.

Ding Bingsheng et al., eds. *Di San Yezhanjun Zheng Zhanyi* [The Campaign on the March of the Third Field Army]. Beijing: Junshi Kexueyuan Chubanshe, 1996.

Dirlik, Arif. *The Origins of the Chinese Communist Party*. London: Oxford University Press, 1989.

Dittmer, Lowell. *Liu Shao-ch'i and the Chinese Cultural Revolution: The Politics of Mass Criticism*. Berkeley: University of California Press, 1974.

Domes, Jürgen. *Peng Te-huai: The Man and the Image*. Stanford, CA: Stanford University Press, 1985.

Dupuy, Trevor N. *The Military History of the Chinese Civil War*. New York: Franklin Watts, 1969.

Eastman, Lloyd E. *Seeds of Destruction: Nationalist China in War and Revolution, 1937–1949*. Stanford, CA: Stanford University Press, 1984.

Eberhard, Wolfram. *A History of China*. Berkeley: University of California Press, 1977.

Eiler, Keith E., ed. *Wedemeyer on War and Peace*. Stanford, CA: Hoover Institution Press, 1987.

Esherick, Joseph. *The Origins of the Boxer Uprising*. Berkeley: University of California Press, 1987.

Fairbank, John K. *Chinabound: A Fifty-Year Memoir*. New York: Harper Colophon, 1982.

———. *Trade and Diplomacy on the China Coast*. Cambridge: Harvard University Press, 1953.

———. *The United States and China*. Cambridge: Harvard University Press, 1983.

———, ed. *The Chinese World Order: Traditional Chinese Foreign Relations*. Cambridge: Harvard University Press, 1968.

Fang Tian et al., *Si Ye Zuihou Yizhan* [The Last Campaign of the Fourth Field Army]. Beijing: Guofang Daxue Chubanshe, 1995.

Fathers, Michael, and Andrew Higgins. *Tiananmen: The Rape of Peking*. Toronto: Doubleday Canada, 1989.

Fay, Peter Ward. *The Opium War, 1840–1842: Barbarians in the Celestial Empire in the Early Part of the Nineteenth Century and the War by Which They Forced Her Gates Ajar*. Chapel Hill: University of North Carolina Press, 1975.

Fehrenbach, T. R. *This Kind of War*. New York: Macmillan, 1963.

Feis, Herbert. *The China Tangle: The American Effort in China from Pearl Harbor to the Marshall Mission*. Princeton: Princeton University Press, 1953.

Feuerwerker, Albert. *China's Early Industrialization*. Cambridge: Harvard University Press, 1958.

Fischer, Edward. *The Chancy War: Winning in China, Burma and India in World War Two*. New York: Orion Books, 1991.

Folta, Paul H. *The Defense Industry of the People's Republic of China: Command Structure, Industries, Production, and Foreign Trade*. Washington, DC: National Defense University, 1986.

Foot, Rosemary. *A Substitute for Victory*. Ithaca, NY: Cornell University Press, 1990.

Foreign Relations of the United States, 1945. Vol. 7: *The Far East: China*. Washington, DC: Government Printing Office, 1969.

Foreign Relations of the United States, 1946. Vol. 10: *The Far East: China*. Washington, DC: Government Printing Office, 1972.

Foreign Relations of the United States, 1947. Vol. 7: *The Far East: China*. Washington, DC: Government Printing Office, 1972.

Frank, Benis, and Henry Shaw, Jr. *History of U.S. Marine Corps Operations in World War II: Victory and Occupation*. Historical Branch, Headquarters, U.S. Marine Corps, 1968.

Franz, Michael. *The Taiping Rebellion: History and Documents*. 3 vols. Seattle: University of Washington Press, 1966.

Furuya, Keiji. *Chiang Kai-shek, His Life and Times*. New York: St. Johns University Press, 1981.

Garavente, Anthony J. *He Long and the Rural Revolution in West-Central China, 1927–1935*. Ann Arbor, MI: University Microfilms International, 1988.

Garver, John W. *Foreign Relations of the People's Republic of China*. Englewood Cliffs, NJ: Prentice-Hall, 1993.

Gau Rui, ed. *Zhongguo Junshi Shilue* [A Brief Military History of China]. 3 vols. Beijing: Junshi Kexueyuan Chubanshe, 1991; cited as ZGJSSL.

George, Alexander L. *The Chinese Communist Army in Action: The Korean War and Its Aftermath*. New York: Columbia University Press, 1967.

Gillin, Donald. G. *Warlord: Yen Hsi-shan in Shanxi Province, 1911–1949*. Princeton; Princeton University Press, 1967.

Gillin, Donald, G., with Charles Etter. "Staying On: Japanese Soldiers and Civilians in China, 1945–1949." *Journal of Asian Studies* 42, no. 3 (May 1983): 497–518.

Giquel, Prosper. *The Foochow Arsenal and Its Results: From the Commencement in 1867 to the End of the Foreign Directorate on 16 February 1874*. Trans. H. Lang. Shanghai: 1874.

———. *A Journal of the Chinese Civil War, 1864*. Trans. Steven Leibo. Honolulu: University of Hawaii Press, 1985.

Glynn, Sean, and Alan Booth. *Modern Britain: An Economic and Social History*. New York: Routledge, 1996.

Godwin, Paul H. B. "Changing Concepts of Doctrine, Strategy and Operations in the Chinese People's Liberation Army, 1979–1987." *The China Quarterly*, no. 112 (December 1987): 572–576.

———. *The Chinese Communist Armed Forces*. Montgomery, AL: Air University Press, 1988.

———. "From Continent to Periphery: PLA Doctrine, Strategy and Capabilities towards 2000." *The China Quarterly*, no. 146 (June 1996): 464–487.

———, ed. *The Chinese Defense Establishment*. Boulder, CO: Westview Press, 1983.

Goldstein, Avery. "Trends in the Study of Political Elites in China." *China Quarterly* 139 (September 1994): 714–730.

Goncharov, Sergei N., John W. Lewis, and Xue Litai. *Uncertain Partners: Stalin, Mao, and the Korean War*. Stanford, CA: Stanford University Press, 1993.

Gordon, Charles G. *General Gordon's Private Diary of His Exploits in China*. London: Sampston Low, 1885; New York: Kraus Reprints, 1971.

Graham, Gerald. *The China Station: War and Diplomacy, 1830–1860*. New York: Oxford University Press, 1978.

Great Britain Foreign Office. *Correspondence Relative to the Earl of Elgin's Special Mission to China and Japan, 1857–1859*. London: Foreign Office, 1859; San Francisco: Chinese Materials Center, 1975.

Gregory, J. S. *Great Britain and the Taipings*. New York: Praeger, 1969.

Griffith, Samuel B. II. *The Chinese People's Liberation Army*. New York: McGraw-Hill, 1967.

Guillermaz, Jaques. *A History of the Chinese Communist Party, 1921–1949*. New York: Random House, 1972.

Gurtov, Melvin, and Byong-Moo Hwang. *China under Threat: The Politics of Strategy and Diplomacy*. Baltimore: Johns Hopkins University Press, 1980.

Han, Theodore, and John Li. *Tiananmen Square Spring 1989: A Chronology of the Chinese Democracy Movement*. Berkeley: Institute of East Asian Studies, University of California, 1992.

Hao Yufan and Zhai Zhihai. "China's Decision to Enter the Korean War: China History Revisited." *China Quarterly* 121 (March 1990): 94–115.

Harding, Harry, and Yuan Ming, eds. *Sino–American Relations, 1945–1955: A Joint Assessment of a Critical Decade*. Wilmington, DE: Scholarly Resources Imprints, 1989.

Harrison, James Pinkney. *The Long March to Power: A History of the Chinese Communist Party, 1921–1972*. New York: Praeger, 1972.

Hart, John N. *The Making of an Old China Hand: A Memoir of Colonel David D. Barrett*. Unpublished Manuscript, Hoover Institution, Stanford University.

Hermes, Walter G. *United States Army in the Korean War: Truce, Tent and Fighting Front*. Washington, DC: Center of Military History, U.S. Army, 1965.

Hinton, Harold C. "Conflict on the Ussuri, a Clash of Nationalisms." *Problems of Communism* 20, no. 1–2 (January–April 1971): 48–49.

Hooker, Mary. *Behind the Scenes in Peking*. London: John Murray, 1910; Hong Kong: Oxford University Press, 1987.

Hooten, E. R. *The Greatest Tumult: The Chinese Civil War, 1936–1949*. New York: Brassey's, 1991.

Hoyt, Edwin P. *The Rise of the Chinese Republic: From the Last Emperor to Deng Xiaoping*. New York: McGraw-Hill, 1989.

Hsiung, James C., and Steven I. Levine, eds. *China's Bitter Victory: The War with Japan, 1937–1945*. New York: M. E. Sharpe, 1992.

Hsu, Immanuel C. Y. *The Rise of Modern China*. 5th ed. New York: Oxford University Press, 1995.

Huang Xiaodong, ed. *Sixiang Zhengzhi Gongzuo Qishi Nian* [Seventy Years of Party Political Work]. Beijing: Guofang Daxue Chubanshe, 1991.

Hummel, Arthur W. *Eminent Chinese of the Ch'ing Period (1644–1912)*. Washington, DC: Government Printing Office, 1943.

Hurd, Douglas. *The Arrow War: An Anglo–Chinese Confusion, 1856–1860*. London: Collins Press, 1967.

Iriye, Akira. *Across the Pacific*. New York: Harcourt and Brace, 1967.

Jane's Information Group. *China in Crisis: The Role of the Military*. Surrey, U.K.: Jane's Information Group, 1989.

Jansen, Marius. *Japan and China: From War to Peace, 1894–1972*. New York: Rand-McNally, 1975.

Jencks, Harlan W. "China's Punitive War on Vietnam, a Military Assessment." *Asian Survey* 19 (August 1979): 806–815.
———. *From Muskets to Missiles*. Boulder, CO: Westview Press, 1982.
Jiang Siyi. *Zhongguo Renmin Jiefangjun Dashi Dian* [Major Events of the Chinese People's Liberation Army]. Tianjin: Tianjin Renmin Chubanshe, 1992. Cited as ZGRMJFJDSD.
Jin Jiang and Qin Zhou, trans. *June Four: A Chronicle of the Chinese Democratic Uprising*. Fayetteville: University of Arkansas Press, 1989.
Jin Zha and Xue Qing. *Gonghe Guo Da Junqu Shiming Zhuanqi* [A Collection of the Major Events of the PRC's Military Regions]. Chengdu: Sichuan Renmin Chubanshe, 1995.
Joffe, Ellis. *The Chinese Army after Mao*. Cambridge: Harvard University Press, 1987.
———. *Party and Army: Professional and Political Control in the Chinese Officer Corps, 1949–1964*. Cambridge: Harvard University Press, 1965.
Johnson, Chalmers. *Peasant Nationalism and Communist Power*. Stanford, CA: Stanford University Press, 1962.
Johnson, Robert E. *Far China Station: The U.S. Navy in Asian Waters, 1800–1898*. Annapolis, MD: Naval Institute Press, 1979.
Johnston, Alastair Iain. *Cultural Realism: Strategic Culture and Grand Strategy in Chinese History*. Cambridge: Harvard University Press, 1995.
Jordan, Donald. *The Northern Expedition: China's National Revolution of 1926–1928*. Honolulu: University of Hawaii Press, 1976.
Kang Mei-Yuan Chao Zhanshi [A History of the War to Resist America and Aid Korea]. Beijing: Junshi Kexueyuan Chubanshe, 1990.
Karnow, Stanley. *Vietnam: A History*. New York: Viking Press, 1983.
Kataoka, Tetsuya. *Resistance and Revolution in China: The Communists and the Second United Front*. Berkeley: University of California Press, 1974.
Kau, Michael Y. M., ed. *The Lin Piao Affair: Power Politics and Military Coup*. White Plains, NY: International Arts and Sciences Press, 1975.
Klein, Donald, and Anne B. Clark. *Biographic Dictionary of Chinese Communism: 1921–1965*. Cambridge: Harvard University Press, 1971.
Kraus, Richard Curt. *Class Conflict in Chinese Socialism*. New York: Columbia University Press, 1981.
Kuo, Warren. *Analytical History of the Chinese Communist Party*. Taipei: Institute of International Relations, 1968.
———. *A Comprehensive Glossary of Chinese Communist Terminology*. Taipei: Institute of International Relations, 1978.
Ladany, Laszlo. *The Communist Party of China and Marxism, 1921–1985*. Stanford, CA: Hoover Institution Press, 1988.
Lary, Diana. *Warlord Soldiers: Chinese Common Soldiers, 1911–1937*. New York: Cambridge University Press, 1985.
Lau, D. C., and Roger T. Ames, trans. *Sun Pin: The Art of Warfare*. New York: Ballantine Books, 1996.
Lau Yee-fui et al. *Glossary of Chinese Political Phrases*. Hong Kong: Union Press, 1977.
League of Nations. *Memoranda Submitted by the Chinese Assessor to the Commission of Enquiry of the League of Nations, April–August 1932*. Shanghai: Commercial Press, 1932.
Leavenworth, Charles S. *The Arrow War with China*. London: S. Row, Marston, 1901.

Lee Hong-yung. *The Politics of the Chinese Cultural Revolution.* Berkeley: University of California Press, 1978.

Lewis, John W., and Xue Litai. *China Builds the Bomb.* Stanford, CA: Stanford University Press, 1988.

———. *China's Strategic Seapower: The Politics of Force Modernization in the Nuclear Age.* Stanford, CA: Stanford University Press, 1994.

———. *Uncertain Partners: Stalin, Mao and the Korean War.* Stanford, CA: Stanford University Press, 1993.

Li Bingyan. *Sanshiliu Ji Xin Bian* [A Modern Version of the 36 Stratagems]. Beijing: Jiefangjun Chubanshe, 1991.

Li Hong Chang. *Memoirs of Li Hong Chang.* Ed. William F. Mannix. Boston: Houghton Mifflin, 1913.

Li Jijun. *Junshi Zhanlue Siwei* [Military Strategic Thought]. Beijing: Junshi Kexue Chubanshe, 1996.

Li, Lincoln. *The Japanese Army in North China, 1937–1941.* London: Oxford University Press, 1975.

Li Lu. *Zhongguo Guoxiang Daizhan Zhengzhan Li Jie Cidian* [Compendium to Understand Trends in Chinese Warfare]. Beijing: Guofangdaxue Chubanshe, 1991.

Li Qingshan. *Xin Junshi Geming yu Gao Jishu Zhanzheng* [The Revolution in Military Affairs and High-Technology Warfare]. Beijing: Junshi Kexue Chubanshe, 1995.

Li Tsung-jen. *The Memoirs of Li Tsung-jen.* Trans. Tong Te-kong. Boulder, CO: Westview Press, 1979.

Liang Chin-tun. *The Sinister Face of the Mukden Incident.* New York: St. Johns University Press, 1969.

Liang Chin-tung. *General Stilwell in China, 1942–1944: The Full Story.* New York: St. Johns University Press, 1972.

Liao Gailong and Fan Yuan, eds. *Who's Who in China: Current Leaders.* Beijing: Foreign Languages Press, 1994.

Lilley, James R., and Chuck Downs, eds. *Crisis in the Taiwan Strait.* Washington, DC: National Defense University Press, 1997.

Lindsay, Michael. *The Unknown War: North China, 1937–1945.* London: Bergstrom and Boyle, 1975.

Liu, Frederick Fu (F. F.). *A Military History of Modern China: 1924–1949.* Princeton: Princeton University Press, 1956.

Liu Qi et al. *Huai-Hai Zhanyi Qin Li Ji: Yuan Guomindang Jiangling de Huiyi* [Diaries of the Huai-Hai Campaign: Original Recollections of the Guomindang Generals] Beijing: Wenshi Chubanshe, 1996.

———. *Liao-Shen Zhanyi Qin Li Ji: Yuan Guomindang Jiangling de Huiyi* [Diaries of the Liao-Shen Campaign: Original Recollections of the Guomindang Generals]. Beijing: Wenshi Chubanshe, 1996.

———, eds. *Ping-Jin Zhanyi Qin Li Ji: Yuan Guomindang Jiangling Huiyi* [Diaries of the Ping-Jin Campaign: Original Recollections of Guomindang Generals]. Beijing: Wenshi Chubanshe, 1996.

Lohbeck, Donald. *Patrick J. Hurley.* Chicago: Henry Regnery, 1956.

Lone, Stewart. *Japan's First Modern War: Army and Society in the Conflict with China, 1894–95.* New York: St. Martin's Press, 1994.

Lovejoy, Charles D. Jr., and Bruce W. Watson, eds. *China's Military Reforms.* Boulder, CO: Westview Press, 1986.

Luk, Michael Y. L. *The Origins of Chinese Bolshevism, 1920–1928*. London: Oxford University Press, 1990.

Ma Yunpeng. *Huizhan Ping-Jin* [The Decisive Ping-Jin Campaign: History of the PLA Fourth Field Army]. Shenyang: Liaoning Renmin Chubanshe, 1998.

MacDonald, Callum A. *Korea: The War before Vietnam*. New York: Free Press, 1986.

MacFarquhar, Roderick. *The Origins of the Cultural Revolution*. Vol. 2: *The Great Leap Forward*. New York: Columbia University Press, 1983.

MacGarrigle, George L. *Central Burma*. Washington, DC: U.S. Army Center of Military History, 1996.

Mao Zedong. "A Single Spark Can Start a Prairie Fire," "Struggle in the Chingkang Mountains," and "On Correcting Mistaken Ideas in the Party." *Selected Works of Mao Tse-Tung*. Vol. 1. Beijing: Foreign Languages Press, 1975.

———. "Problems of Guerilla War against Japan," "On Protracted War," and "For the Mobilization of All the Nation's Forces for Victory in the War of Resistance." *Selected Works of Mao Tse-Tung*. Vol. 2. Beijing: Foreign Languages Press, 1965.

———. "On the International United Front against Fascism." *Selected Works of Mao Tse-Tung*. Vol. 3. Beijing: Foreign Languages Press, 1967.

Marshall, Peter J., ed. *The Cambridge Encyclopedia of the British Empire*. New York: Cambridge University Press, 1996.

Maxwell, Neville. "The Chinese Account of the 1969 Fighting at Chen Bao." *China Quarterly*, no. 56 (Fall 1973): 730–739.

———. *India's China War*. Bombay: JAICO Press, 1971.

McCall, Robert W. "The Oyuwan Soviet Area, 1927–1933." *Journal of Asian Studies* 27, no. 1 (November 1967): 41–59.

Meisner, Maurice. *Mao's China: A History of the People's Republic*. New York: Free Press, 1977.

Mi Zhanchen. *The Life of General Yang Hucheng*. Hong Kong: Joint, 1981.

Mi Zhenyu. *Jubu Zhanzheng yu Jundui Jianshe Yanjiu* [Local War and the Study of Army Building]. (Beijing: Jushi Wenyi Chubanshe, 1988).

Michael, Franz. *The Taiping Rebellion: History and Documents*. 3 vols. Seattle: University of Washington Press, 1966.

Miles, Milton E. *A Different Kind of War: The Little-Known Story of the Combined Guerrilla Forces Created in China by the U.S. Navy and the Chinese during World War II*. Garden City, NY: Doubleday, 1967.

———. "U.S. Naval Group China." *United States Naval Institute Proceedings* 72, July 7, 1946. Milton E. Miles Papers, Hoover Institution Archives, Hoover Institution on War Revolution and Peace, Stanford, CA.

The Military Balance, 1996. Singapore: Asian Defense Review, 1996.

Montross, Lynn. *The Chosin Reservoir Campaign*. Washington, DC: Headquarters, U.S. Marine Corps, 1957.

Morley, James W., ed. *The Chinese Quagmire: Japan's Expansion on the Asian Continent, 1933–1941*. New York: Columbia University Press, 1983.

Morse, H. B. *In the Days of the Taipings: An Historical Retrospect*. Salem: Essex Institute, 1927.

Mossman, Billy C. *United States Army in the Korean War: Ebb and Flow, November 1950–July 1951*. Washington, DC: Center of Military History, U.S. Army, 1966.

Mu Xin. *Chen Geng tongzhi zai Shanghai: Zai Zhongyang Teke de Douzheng Jingli* [The

Fighting Experience of Comrade Chen Geng in Shanghai in the Special Service Section of the Central Committee]. Beijing: Wenhua Lishi Chubanshe, 1980.
Muller, David G., Jr. *China as a Maritime Power.* Boulder, CO: Westview Press, 1983.
Murphy, Rhoads. *The Treaty Ports and China's Modernization: What Went Wrong?* Ann Arbor: University of Michigan Press, 1970.
Nan Li. ''The PLA's Evolving Warfighting Doctrine, Strategy and Tactics, 1985–1995: A Chinese Perspective.'' *The China Quarterly*, no. 146 (June 1996): 443–463.
National Archives. *Record Group 226.* Records of the Office of Strategic Services. Washington, DC.
National Archives. *Record Group 331.* Records of the SCAP G-3. Washington, DC.
National Archives. *Record Group 334.* Records of the Army Military Advisory Group in Nanking: 1942–1949. Washington, DC.
National Archives. *Record Group 353.* Records of the State-War-Navy-Coordinating-Committee (SWNCC). Washington, DC.
Nie Rongzhen. *Inside the Red Star: The Memoirs of Marshal Nie Rongzhen.* Trans. Zhong Renyi. Beijing: New World Press, 1988.
Nutting, Anthony. *The Last Days of General Gordon.* London: Constable, 1966.
———. *Martyr and Misfit.* London: Constable, 1966.
Office of Naval Intelligence. *Chinese Exercise Strait 961: 8–25 March 1996.* Washington, DC: Office of Naval Intelligence, 1996.
Ormont, Arthur. *Chinese Gordon: Hero of Khartoum.* New York: G. P. Putnam's Sons, 1966.
Owen, Joseph R. *Colder than Hell: A Marine Rifle Company at Chosin Reservoir.* Annapolis, MD: Naval Institute Press, 1996.
Pan Shiying et al. *Dangdai Zhongguo Junshi Sixiang Chingyao* [The Essence of Modern Chinese Military Thinking]. Beijing: Jiefangjun Chubanshe, 1992.
Pan Xiangting, ed. *Gao Jishu Tiaojian Xia Meijun Jubu Zhanzheng* [American High-Technology Limited War under Modern Conditions]. Beijing: Jiefangjun Chubanshe, 1994.
Payne, Robert. *Chiang Kai-shek.* New York: Weybright and Talley, 1969.
Peng Dehuai. *Memoirs of a Field Marshal.* Trans. Zheng Longpu. Beijing: Foreign Languages Press, 1984.
Pepper, Suzanne. *Civil War in China: The Political Struggle, 1945–1949.* Berkeley: University of California Press, 1978.
Perry, Elizabeth. *Rebels and Revolutionaries in North China, 1845–1945.* Stanford, CA: Stanford University Press, 1980.
Pillsbury, Michael, ed. *Chinese Views of Future Warfare.* Washington, DC: National Defense University Press, 1997.
Powell, Ralph L. *The Rise of Chinese Military Power: 1895–1912.* Princeton: Princeton University Press, 1955.
Purcell, Victor. *The Boxer Uprising: A Background Study.* New York: Cambridge University Press, 1963.
Pye, Lucian W. *Warlord Politics: Conflict and Coalition in the Modernization of Republican China.* New York: Praeger, 1971.
Qian Li et al. *Ezhi Zhongguo: Shenhua yu Xianshi* [Containing China: Truth and Reality]. Beijing: Zhongguo Yanshi Chubanshe, 1996.
Rawlinson, John L. *China's Struggle for Naval Development, 1839–1895.* Cambridge: Harvard University Press, 1967.

Reports of General MacArthur. *MacArthur in Japan: The Occupation: Military Phase.* Vol. 1: *Supplement.* Prepared by his General Staff. Presidio, 1966.

Robinson, Thomas W., ed. *The Cultural Revolution in China.* Los Angeles: University of California Press, 1971.

Romanus, Charles F., and Riley Sunderland. *United States Army in World War II: The China-Burma-India Theater: Time Runs Out in the CBI.* Washington, DC: Government Printing Office, 1959.

———. *The United States Army in World War Two. The China-Burma-India Theater: Stilwell's Command Problems.* Washington, DC: Government Printing Office, 1956.

———. *The United States Army in World War Two. The China-Burma-India Theater: Stilwell's Mission to China.* Washington, DC: Government Printing Office, 1953.

Rue, John E. *Mao Tse-tung in Opposition, 1927–1935.* Stanford, CA: Stanford University Press, 1966.

Ryan, Mark A. *Chinese Attitudes toward Nuclear Weapons: China and the United States during the Korean War.* Armonk, NY: M. E. Sharpe, 1989.

Saitch, Tony, ed. *New Perspectives on the Chinese Communist Revolution.* Armonk, NY: M. E. Sharpe, 1995.

Salisbury, Harrison E. *The Long March: The Untold Story.* New York: Harper and Row, 1985.

Schnabel, James F. *United States Army in the Korean War: Policy and Direction, the First Year.* Washington, DC: Center for Military History, U.S. Army, 1988.

Schurmann, Franz, and Orville Schell. *The China Reader: Republican China—Nationalism, War, and the Rise of Communism, 1911–1949.* New York: Random House, 1967.

Schwartz, Benjamin. *Chinese Communism and the Rise of Mao.* Cambridge: Harvard University Press, 1951.

Scott, Robert L., Jr. *Flying Tiger: Chennault of China.* Westport, CT: Greenwood Press, 1959.

Seagrave, Sterling. *The Soong Dynasty.* New York: Harper and Row, 1985.

Segal, Gerald. *Defending China.* London: Oxford University Press, 1985.

Selden, Mark. *The Yenan Way in Revolutionary China.* Cambridge: Harvard University Press, 1971.

Shanxi Renmin Zhengzhi Xieshang Hui. *Wo Suo Zhidao de Yan Xishan* [The Yan Xishan I Knew]. Taiyuan: Taiyuan Yinshuachang, 1986.

Sherick, Joseph E. *The Origins of the Boxer Uprising.* Berkeley: University of California Press, 1987.

Sheridan, James E. *Chinese Warlord: The Career of Feng Yu-hsiang.* Stanford, CA: Stanford University Press, 1966.

Shoup, David M. *The Marines in China, 1927–1928: The China Expedition Which Turned Out to Be the China Exhibition.* Hamden, CT: Archon Books, 1987.

Smedley, Agnes. *The Great Road.* New York: Monthly Review Press, 1956.

Snow, Edgar. *Red Star over China.* New York: Grove Press, 1956.

Solomon, Richard H. *Mao's Revolution and the Chinese Political Culture.* Berkeley: University of California Press, 1971.

Spence, Jonathan D. *To Change China: Western Advisers in China, 1620–1960.* New York: Penguin Press, 1969.

———. *The Search for Modern China.* New York: W. W. Norton, 1990.

Spencer, Otha C. *Flying the Hump: Memories of an Air War*. College Station: Texas A&M University Press, 1992.

Sperry, Roland. *China through the Eyes of a Tiger*. New York: Pocket Books, 1990.

Spurr, Russell. *Enter the Dragon: China's Undeclared War against the U.S. in Korea, 1950–51*. New York: Newmarket Press, 1988.

Stilwell, Joseph. *The Stilwell Papers*. Arranged by Theodore White. New York: Da Capo Press, 1991, reprint.

Strategy Research Department of the Academy of Military Science, ed. *Mao Zedong Junshi Zhanlue Lun* [A Discussion of Mao Zedong's Military Strategy]. Beijing: Junshi Kexue Chubanshe, 1993.

Stratton, Ray Olin. *SACO—The Rice Paddy Navy*. Pleasantville, NY: C. S. Palmer, 1950.

Su Zefeng. *Jing-Jin Zhanyi* [The Beijing-Tianjin Campaign]. Shanghai: Renmin Chubanshe, 1959.

Sun Zhen, ed. *PLA Forces*. Hong Kong: CONMILIT Press, 1986.

Swanson, Bruce. *Eighth Voyage of the Dragon—A History of China's Quest for Seapower*. Annapolis, MD: Naval Institute Press, 1982.

Tanaka, Gichi. *Tanaka Memorial: Japanese Imperialism Stripped, the Secret Memorandum of Tanaka, Premier of Japan*. New York: Workers Library, 1927.

Tanaka, Yuki. *Hidden Horrors: Japanese War Crimes in World War II*. Boulder, CO: Westview Press, 1996.

Tang Tsou. *America's Failure in China, 1941–1950*. Chicago: University of Chicago Press, 1963.

Teiwes, Frederick C. *Politics and Purges in China: Rectification and the Decline of Party Norms: 1950–1965*. White Plains, NY: M. E. Sharpe, 1979.

———. *Politics at Mao's Court: Gao Gang and Party Factionalism in the Early 1950's*. Armonk, NY: M. E. Sharpe, 1990.

Teng, Ssu-yu. *The Taiping Rebellion and the Western Powers: A Comprehensive Survey*. Oxford: Clarendon Press, 1971.

Teng, Tony Y. Y. *Prince Kung and the Survival of Ch'ing Rule, 1858–1898*. Ann Arbor, MI: University Microfilms International, 1980.

Tetsuya Kataoka. *Resistance and Revolution in China: The Communists and the Second United Front Army*. Berkeley: University of California Press, 1974.

Thorton, Richard C. *The Comintern and the Chinese Communists, 1928–1931*. Seattle: University of Washington Press, 1969.

Tien, Chen-Ya. *Chinese Military Theory: Ancient and Modern*. New York: Mosaic Press, 1992.

Toland, John. *In Mortal Combat: Korea, 1950–1953*. New York: William Morrow, 1991.

Tong Bai, ed. *Junren Shouce* [Soldier's Handbook]. Beijing: Changzheng Chubanshe, 1987.

Tong Te-kong and Li Tsung-jen. *The Memoirs of Li Tsung-jen*. Boulder, CO: Westview Press, 1979.

Townsend, James R. *Politics in China*. Boston: Little, Brown, 1974.

Tsu, Susan Fu. *A Study of Chang Hsueh-liang's Role in Modern Chinese History*. Ann Arbor, MI: University Microfilms International, 1984.

Tuchman, Barbara W. *Stilwell and the American Experience in China, 1911–45*. New York: Macmillan, 1971.

U.S. Department of State (Far Eastern Series 30). *United States Relations with China:*

With Special Reference to the Period 1944–1949. Washington, DC: Government Printing Office, 1949.

U.S. Navy. *Official History of the United States Naval Observer Group, Chongqing, China.* Milton E. Miles Papers, Hoover Institution Archives, Hoover Institution on War, Revolution, and Peace, Stanford, CA.

Van Slyke, Lyman P. *Enemies and Friends: The United Front in Chinese Communist History.* Stanford, CA: Stanford University Press, 1967.

Von Senger, Harro. *The Book of Stratagems: Tactics for Triumph and Survival.* New York: Penguin Books, 1991.

Waldron, Arthur. ''The Warlord: Twentieth-Century Chinese Understandings of Violence, Militarism, and Imperialism.'' *The American Historical Review* 96, No. 4 (October 1991): 1073–1100.

Wang Xuepeng. *Zhongguo Dangdai Jiangjun Fengcai* [Selections on China's Contemporary Generals]. Beijing: Zhongguo Dangshi Chubanshe, 1993.

War Department, Adjutant General's Office. *Reports on Military Operations in South Africa and China.* Washington, DC: Government Printing Office, 1901.

Wedemeyer, Albert C. *Wedemeyer Reports!* New York: Henry Holt, 1958.

Wee, William. *Counter-Revolution in China: The Nationalists in Jiangxi during the Soviet Period.* Ann Arbor: University of Michigan Press, 1985.

White, Lynn. *Policies of Chaos: The Organizational Causes of Violence in China's Cultural Revolution.* Princeton: Princeton University Press, 1989.

Whiting, Allen S. *China Crosses the Yalu: The Decision to Enter the Korean War.* Stanford, CA: Stanford University Press, 1960.

Whitson, William W. *Military Campaigns in China, 1924–1950.* Taipei: Military History Office, War History Bureau, 1966.

Whitson, William W., with Chen-Hsia Huang. *The Chinese High Command: A History of Communist Military Politics, 1927–71.* New York: Praeger, 1973.

———. *Civil War in China 1946–1950.* 2 vols. Washington, DC.: Office of Military History, U.S. Army, 1967.

Who's Who in China: Current Leaders. Beijing: Foreign Languages Press, 1991, 1995.

Wilbur, C. Martin. *Sun Yat-sen, Frustrated Patriot.* New York: Columbia University Press, 1976.

Wilbur, C. Martin, and Julie Lien-ying How. *Documents on Communism, Nationalism, and Soviet Advisers in China, 1918–1927.* New York: Columbia University Press, 1956.

Wilson, Andrew. *The ''Ever-Victorious Army'': A History of the Chinese Campaign under Lt. Col. C. G. Gordon and the Suppression of the Tai-Ping Rebellion.* London: William Blackwood, 1868; reprinted Novato, CA: Presidio Press, 1991.

Wilson, Dick. *The Long March: The Epic of Communism's Survival, 1935.* New York: Viking Press, 1971.

Wortzel, Larry M. *The ASEAN Regional Forum: Asian Security without an American Umbrella.* Carlisle, PA: Strategic Studies Institute, 1996.

———. ''China Seeks Traditional Great Power Status.'' *Orbis* 2, no. 38 (Spring 1994): 157–176.

———. *China's Military Potential.* Carlisle, PA: Strategic Studies Institute, 1998.

———, ed. *China's Military Modernization: International Implications.* Westport, CT: Greenwood Press, 1988.

Wright, Mary C., ed. *China in Revolution: The First Phase*. New Haven, CT: Yale University Press, 1968.

Wu Tien-wei. *The Sian Incident: A Pivotal Point in Modern Chinese History*. Ann Arbor: University of Michigan Press, 1976.

Wu Zhoulong et al. *Sunzi Jiaoshi* [Explaining Sun-Tzu]. Beijing: Junshi Kexue Chubanshe, 1996.

Xia Dongyuan. *Yang-Wu Yondong Shi* [A History of the Westernization Movement]. Shanghai: Huadong Shifan Daxue, 1992.

Xing Huo Bianjibu, ed. *Jiefangjun Jianglingzhuan* [A Collection on China's Generals]. Beijing: Jiefangjun Chubanshe, 1986. Cited as JFJJLZ.

Yao Ming-le. *The Conspiracy and Death of Lin Biao*. New York: Alfred A. Knopf, 1983.

Yao Yanjin, ed. *Deng Xiaoping Xin Shiqi Junshi Lilun Yanjiu* [A Study of Deng Xiaoping's Military Theory for the New Age]. Beijing: Jushi Kexue Chubanshe, 1994.

Yeaton, Ivan D. *Memoirs of Colonel Ivan D. Yeaton, 1919–1953*. Stanford, CA: Hoover Institution on War, Revolution, and Peace, 1976.

Yoshihashi, Takehikko. *Conspiracy at Mukden: The Rise of the Japanese Military*. New Haven, CT: Yale University Press, 1963.

Young, Ernest. *The Presidency of Yuan Shi-k'ai*. Ann Arbor: University of Michigan Press, 1977.

Young, John R. *The Dragon's Teeth: Inside China's Armed Forces*. New York: Orion Books, 1987.

Yu, Maochun. *OSS in China: Prelude to the Cold War*. New Haven, CT: Yale University Press, 1996.

Yue Shuiyu. *Sunzi Bingfa yu Gao Jishu Zhanzheng* [Sun-Tzu's Military Thought and High-Technology Warfare]. Beijing: Guofang Daxue Chubanshe, 1998.

Yung, Christopher D. *People's War at Sea: Chinese Naval Power in the Twenty-First Century*. Alexandria, VA: Center for Naval Analyses, 1996.

Zhang Shuguang. *Deterrence and Strategic Culture: Chinese–American Confrontations, 1949–1958*. Ithaca, NY: Cornell University Press, 1992.

———. *Mao's Military Romanticism: China and the Korean War, 1950–1953*. Lawrence: University Press of Kansas, 1995.

Zhong Kan. *Kang Sheng Ping Zhuan* [A Critical Biography of Kang Sheng]. Beijing: Hong Qi Chubanshe, 1982.

Zhong Yufei. *Liao-Shen Zhanyi* [The Liaoning–Shenyang Campaign]. Beijing: Qingnian Chubanshe, 1964.

Zhongguo Dabaike Quanshu: Junshi [A Complete Encyclopedia of China]. 2 vols. Beijing: Zhongguo Dabaike Quanshu Chubanshe, 1989. Cited as ZGDBKQS.

Zhonggong Dangshi Jianming Cidian [A Concise Dictionary of China's Party History]. Beijing: Jiefangjun Chubanshe, 1986. Cited as ZGDSJMCD.

Zhongguo Renmin Geming Zhanzheng Ditu Xuan, 1927–1945 [A Collection of Maps of the Chinese People's Revolutionary War, 1927–1945]. Beijing: Ditu Chubanshe, 1981.

Zhongguo Renmin Jiefangjun [The Chinese People's Liberation Army]. 3 vols. Beijing: Dangdai Zhongguo Chubanshe, 1994. Cited as ZGRMJFJ.

Zhongguo Renmin Jiefangjun Zhanshi [A Combat History of the Chinese People's Liberation Army]. 3 vols. Beijing: Junshi Kexueyuan Chubanshe, 1987. Cited as ZGRMJFJZS.

Zhongguo Renmin Jiefangjun Zhanshi Jianbian [A Concise Combat History of the Chinese People's Liberation Army]. Beijing: Jiefangjun Chubanshe, 1992. Cited as ZGRMJFJZSJB.

Zhu Mingquan. *Meiguo Guojia Anquan Zhengce* [U.S. National Security Policy]. Tianjin: Renmin Chubanshe, 1996.

Zhu Youwen. *Gao Jishu Tiaojian Xia de Xinxi Zhan* [Information Warfare under High-Technology Conditions]. Beijing: Junshi Kexue Chubanshe, 1994.

Index

Page numbers in **bold** indicate main entries.

ABOUT THE AUTHOR AND CONTRIBUTORS

VANCE H. MORRISON, USN (Ret.) served as a surface warfare officer and intelligence subspecialist, and had combat tours in Vietnam and Libya. He commanded the USS *Francis Hammond* and the USS *Richmond K. Turner*. A Chinese linguist, Captain Morrison had a tour with the Naval Security Group in Japan and was U.S. Naval Attaché to the PRC during the Tiananmen Incident in 1989. Since retirement from the Navy, he has been a senior member of the technical staff of General Research Corporation International.

SUSAN M. PUSKA has served in a variety of assignments with the U.S. Army in Asia, Europe, and the United States during the last twenty-two years. She served as an Assistant Army Attaché in Beijing during 1992–1994. Most recently, since 1996 she has worked as the China Desk Officer, Office of the Deputy Under Secretary of the Army, International Security Affairs. She holds an advanced degree in Asian Studies from the University of Michigan and is a graduate of Johns Hopkins University–Nanjing University Center for American and Chinese Studies in Nanjing, China.

DAVID E. REUTHER, over a twenty-seven-year career with the U.S. Department of State, was deeply involved in Asian and Middle Eastern issues. As a Foreign Service officer, he served twelve years in Beijing, Bangkok, and Taipei. Recently retired, he now works on contract at the Department of Defense, is a private researcher and lecturer, and is pursuing an advanced degree at the George Washington University.

LARRY M. WORTZEL is the author of *Class in China: Stratification in a Classless Society* (Greenwood Press, 1987) and the editor of *China's Military Modernization: International Implications* (Greenwood Press, 1988). He has also published several articles and book reviews on China in scholarly journals. He is a colonel in the United States Army and is a military intelligence officer and foreign area officer specializing in China and East Asia. He is presently the director of the Strategic Studies Institute, U.S. Army War College. From 1995 to 1997 he was U.S. Army Attaché at the American Embassy in Beijing, China. From 1988 to 1990, Wortzel served as the Assistant Army Attaché in Beijing, a period that spanned the Tiananmen Square Massacre. He has also served as an analyst of military affairs in China for the U.S. Pacific Command and as a strategist for Asia in the Department of the Army Headquarters. He has a Ph.D. in Political Science from the University of Hawaii—Manoa.

www.ingramcontent.com/pod-product-compliance
Ingram Content Group UK Ltd.
Pitfield, Milton Keynes, MK11 3LW, UK
UKHW020234290926
15470UKWH00002B/536

9 780313 293375